W9-AEV-321

EMBRYOS, GENES, AND EVOLUTION

EMBRYOS, GENES, AND EVOLUTION

The Developmental - Genetic Basis of Evolutionary Change

RUDOLF A. RAFF

•••••• AND ••••••

THOMAS C. KAUFMAN

•••

Illustrated by Elizabeth C. Raff

INDIANA UNIVERSITY PRESS
Bloomington and Indianapolis

©1983 by Rudolf A. Raff and Thomas C. Kaufman

Introduction and Preface ©1991 by Indiana University Press

Originally published 1983 by Macmillan Publishing Company

Library of Congress Cataloging-in-Publication Data

Raff, Rudolf A.
    Embryos, genes, and evolution : the developmental-genetic
basis of evolutionary change / Rudolf A. Raff and Thomas C.
Kaufman : illustrated by Elizabeth C. Raff.
        p.   cm.
Reprint. Originally published: New York : Macmillan, © 1983.
Includes bibliographical references (p.) and index.
    ISBN 0-253-34790-4 (cloth : alk. paper). — ISBN
0-253-20642-1 (paper : alk. paper)
    1. Evolution.   2. Embryology.   3. Developmental genetics.
I. Kaufman, Thomas C.   II. Title.
QH371.R27   1991
575—dc20   90-5351
                                                       CIP

ACKNOWLEDGMENTS FOR QUOTATIONS

We thank the following organizations for permission to
quote parts of the copyrighted material listed here:

American Association for the Advancement of Science, The
gene and the ontogenetic process, F. R. Lillie, *Science*
66:361–368, 1927; American Scientist, Evolution as viewed
by one geneticist, Richard Goldschmidt, *American Scientist*
40:89–98, 1952; Harcourt Brace Jovanovich, Inc., *Complete
Poems*, by Carl Sandburg, 1950; Methuen and Company, Ltd.,
Developmental Genetics and Lethal Factors by E. Hadorn, 1961;
Penguin Books, Ltd., *On the Nature of the Universe* by
Lucretius, translated by R. E. Latham, 1975; G. P. Putnam's
Sons, *The Once and Future King* by T. H. White, 1958; Yale
University Press, *The Material Basis of Evolution* by Richard
Goldschmidt, 1940; and Paul Simon, "Patterns" by Paul
Simon ©1965, 1967.

To Richard Goldschmidt
1878–1958

Contents

Five
Interactions Within Embryos

Six
Timing of Developmental Events: Evolution Through Heterochrony

Seven
Genetic Control of Development

Eight
Homoeosis in Ontogeny and Phylogeny

Nine

Pattern Formation

Ten

Adaptations for Gene Expression in Development

Eleven

The Eucaryotic Genome and the C-Value Paradox

Twelve

Regulatory Hierarchies and Evolution: A Synthesis

Introduction to the
Indiana Edition

When we wrote *Embryos, Genes, and Evolution* our goal was to integrate developmental biology and evolutionary biology in order to provide a basis for understanding how novel morphologies arise in the course of the evolutionary history of organisms. We perhaps at the time underestimated the great difficulties in bridging the immense gap in outlook between these two great fields. That difficulty was colorfully expressed by Peter Lawrence, who, in his 1983 review of the book for *Nature*, suggested that the effort was "like trying to make a smooth paste of diamonds and marshmallows." Nevertheless, if we are to understand evolution, the novel concoction has to be made. The writing of *Embryos, Genes, and Evolution* allowed us to formulate approaches to a synthesis and provided us with a means of assessing the range of developmental processes that might be mechanistically relevant to evolutionary modifications of ontogeny. We were able to go beyond the major focus of the field, which up until that time was on heterochrony.

Although we intended *Embryos, Genes, and Evolution* to serve as a new departure in studying the developmental basis of morphological evolution, the field still lacked an experimental program. We had to borrow most of the data we discussed from studies not designed to be direct studies of development/evolution. Nevertheless, we were able to formulate many of the questions that should be asked and we pointed to potential experimental systems. The major change since the book was first published has been the rapid growth of an explicit experimental program, which has refined appropriate experimental questions, found suitable experimental systems, and devised approaches and methods for such studies. The recent studies have illuminated the major issues raised by *Embryos, Genes, and Evolution* and have provided the foundation for a new discipline in biology. This new discipline can now add to our understanding of how development has evolved, and evolutionary insights illuminate studies of developmental processes. In this introductory essay to Indiana's paperback edition, we discuss some of the interesting advances made since the book was first published. Because the literature has grown rapidly, we also provide a reference list, which, although by no means complete, should enable the reader to enter the newer literature.

Several of the issues that we discussed in *Embryos, Genes, and Evolution* have become subjects of experimental investigation. Thus, heterochrony, a term coined by the pioneer of this field, Ernst Haeckel, refers to evolutionary changes in relative timing of developmental events. Clearly heterochronic patterns of change are common in evolving lineages. The classic categories of heterochrony involve changes in relative timing between maturation of the somatic tissues and maturation of the gonads. These "deBeerian" heterochronies (named for deBeer, who first categorized kinds of heterochronies) include such famous kinds of heterochrony as neoteny, in which the body grows to full size but retains larval features into sexual maturity.

Heterochrony has been generally viewed as the major "mechanism" of evolutionary change in development (deBeer, 1958; Gould, 1977). This view has led to important predictions, and to many research papers that have sought to explain evolutionary trends as heterochronic. It has also led to attempts to find developmental mechanisms underlying heterochrony, and even genes whose mutations produce heterochronic effects (Ambros and Horvitz, 1984; McKinney, 1988; Raff and Wray, 1989). Nevertheless, work since 1983 has very strikingly demonstrated that heterochrony is generally a pattern of change, which can result from many kinds of underlying changes. These underlying changes are, in fact, generally not heterochronic in nature. The heterochronic results stem from the temporal structure of ontogeny (Raff, 1990; Raff and Wray, 1989). This perspective relieves us from having to consider genes that control timing *per se* to be unusually subject to mutational change.

Because of the development of classic ideas on heterochrony, most studies have involved late stages of development and focus on global, whole soma versus germ line changes. Yet, it is becoming increasingly clear that all stages of development can undergo heterochronic modifications, and that these are often very local in effect. The expanded sphere of study of heterochrony includes the genetics of timing decisions (Ambros and Horvitz, 1984), molecular heterochronies (Parks et al., 1988; Regier and Vlahos, 1988; Wray and McClay, 1989), local heterochronies in remodeling of particular anatomical features (Alberch and Alberch, 1981; Hall, 1984; Hanken, 1989; Larson, 1983; Wake and Larson, 1987), and a consideration of plants as well as animals (Lord and Hill, 1987; Poethig, 1988).

By now there is no question but that changes in body structure that we see in the fossil record had their origins in underlying evolutionary modifications of development. The connection between developmental processes and evolutionary change is more profound than simply an accumulation of changes as successive ontogenies track the demands of natural selection. Modes of development themselves influence aspects of life history, choice of adaptive zone, competition, and susceptibility to extinction (Gould, 1977; Hansen, 1982; Jablonski, 1986; Mileikovski, 1971; Strathmann, 1985). Evolution of developmental processes mechanistically links evolving genotype with novel phenotype, and, very possibly, existing developmental processes may constrain and channel evolution (Alberch, 1982; Maynard Smith et al., 1985).

Developmental constraint has emerged as a second major theoretical idea of a principle governing the evolution of development (Maynard Smith et al., 1985). This concept is now open to experimental study. The currently dominant view is that Darwinian natural selection provides the primary force for evolutionary change. The evolving lineage is seen as a responder to external selection, with new genotypes being selected to produce an advantageous phenotype. The responding system is often regarded as being free to vary in any direction, with phenotypes not limited except by selection. However, it is quite clear that variation is bounded by history, and thus no one really completely believes the scenario of variation unconstrained except by selection. There is in fact a long-standing dialectic in evolutionary thought over the relative roles of internal versus external forces driving evolution.

In the broadest sense developmental constraint is any aspect of the ontogeny of an organism that limits or channels the potential evolution of the organism's lineage. At first sight the concept is plausible and interpretations of numerous examples suggest a vast pervasiveness of constraints. But, by its common invocation the idea of development constraint may suffer from being able to explain too much without really explaining anything. Very few of the examples have been examined mechanistically. In fact, it may be hard to do so. Most cited evidence for the operation of developmental constraints comes from the existence of features or processes in development that are conserved over very long evolutionary spans. The logic is that if a feature has not changed it must be constrained from changing. The problem is in knowing if internal factors really have played a role.

Gould (1989a) has pointed out that constraint has both positive and negative connotations. The concept after all implies both that internal factors can limit or even close off possible directions of evolution, and that internal factors can favor or even determine particular directions in which lineages can respond to selective pressures. The most interesting and important results of evolution are not simply modifications of an ancestral body pattern, but the appearances of novel features. If novelties owe any of their attributes to the actions of internal factors, then developmental constraints assume a considerable significance as positive forces in evolution.

It is evident from the fossil record, as well as from existing diversity, that not all possible morphologies exist for any given body plan (Alberch, 1982). This generalization is well illustrated by the apparent limitations to evolutionary modification of digit patterns in the tetrapod limb (Holder, 1983) and in other aspects of limb evolution (Oster et al., 1988). There are a lot of possible ways to arrange the fingers on a vertebrate hand; yet among what has actually been realized by evolution there seem to be distinctly favored and non-favored or forbidden patterns.

When people say "ontogeny," generally they are implying both pattern and process. The pattern is the particular developmental sequence observed for a species under its normal conditions of development. But there are many processes underlying any ontogeny. It is becoming clear that at least some genetic controls in development operate in a hierarchical fashion; the most convincing cases, from *Drosophila*, are the cascade of steps involved in segmentation and determination of segment identity (Akam, 1987; Ingham, 1988) and those controlling sex determination (Hodgkin 1987, 1989).

The case of establishment of the anterior-posterior axis of the *Drosophila* embryo offers the mechanistically best defined example of a cascade of developmental events resulting from a change in gene action at the very start of development. The maternally active gene *bicoid* produces an mRNA that is stored in the anterior portion of the *Drosophila* egg. The *bicoid* mRNA is held in position by the products of two other maternally active genes, *swallow* and *exuperentia,* which apparently interact with the cytoskeleton of the egg (Frohnhöfer and Nüsslein-Volhard, 1987; Macdonald and Struhl, 1988). Mutations in *bicoid* affect the proper establishment of the head of the larva, with loss of *bicoid* function producing headless larvae (Nüsslein-Volhard et al., 1987). The *bicoid* protein contains a homeobox domain and functions as a transcriptional activator, which has the specific function of turning on the zygotically active gene *hunchback* (Driever and Nüsslein-Volhard, 1989; Struhl et al., 1989). This gene in turn is part of a cascade of genetic events which ultimately specify the position, polarity, and identity of segments in this embryo. *Bicoid* is one of a small number of *Drosophila* genes, about 40, that are maternal in action and specify anterior-posterior, dorsal-ventral, and terminal positional information (Anderson, 1989; Nüsslein-Volhard et al., 1987). The limited number of these genes is significant: it takes very large numbers of genes to execute the differentiation of individual structure in development. For example, several hundred structural genes are required to assemble the structure of a sperm cell, and about 1,600 genes in *Drosophila* affect spermatogenesis.

The small number of maternal acting genes required for pattern formation is consistent with their setting initial conditions. Given that subsequent patterning builds on the interpretation of the graded distribution of maternal gene products (Driever and Nüsslein-Volhard, 1988a, 1988b), one might expect these genes to be candidates for constraint. Because the regulation of axial determination and pattern formation is not presently known in other insects, it is not clear how constrained these genetic systems are. Even if the basic machinery is retained widely in insects, significant variation is present in distribution patterns of maternal gene products and possibly even in the zygotic genes which interpret the maternal information. The existence of two quite different methods of laying out segmental patterns, short versus long germ band insects, is a strong indication that even very early acting global pattern specifying systems can evolve.

Given the presence of the *bicoid* gradient, one might wonder how graded information can be interpreted in a discontinuous manner by other genes downstream in the ontogenetic pathway. Recent work has shown that the concentration of the *bicoid* protein is instructive in its relay by multiple protein binding sites of genes such as *hunchback* that serve as receptors in the control regions of loci that interpret the *bicoid* gradient (Driever et al., 1989; Struhl et al., 1989). Thus, the spatial pattern of expression of downstream genes is regulated by relative concentration of the regulatory protein or the number of receptor sites. It is just such a set of protein to binding site interactions which characterizes the entire cascade of events leading from global polarity to large domains and finally to individual segments (Levine and Harding, 1989). The molecular mechanism is simple in principle and elegant.

The final step in this developmental pathway lies in the functions of the homoeotic loci, which specify the individual identities of each segment. Homoeotic genes are those which specify the identity of serially repeated parts (Stern, 1990). These loci were first discovered as mutations in *Drosophila* that transform one segment into the identity of a different segment, such as metathorax to mesothorax. Homoeosis is best studied in insects, but is widespread, and even occurs in plant mutations that transform the identities of flower parts (Meyerowitz et al., 1989). We discussed the genetic structure of the homoeotic genes of *Drosophila* in *Embryos, Genes, and Evolution*. Since then, the molecular basis of homoeosis has been defined. The homoeotic genes produce DNA-binding proteins (Desplan et al., 1985; Muller et al., 1988). These are transcriptional regulators for other genes in the cascade (Winslow et al., 1989). One of the characteristics of these proteins of evolutionary importance is that the amino acid sequence necessary for binding to DNA is contained within a highly conserved domain referred to as the homeobox. This protein motif has been found in all metazoa thus far examined, whether segmented or not, indicating a very early origin. As noted in *Embryos, Genes, and Evolution*, these genes occur in *Drosophila* in two major clusters. The Antennapedia complex (ANT-C) controls segmental identity in the anterior of the embryo and adult (Kaufman et al., 1989). The bithorax cluster (BX-C) performs the same function in the posterior (Peifer et al., 1987; Sánchez Herrero et al., 1985). These two complexes exist as a single unit in a more primitive insect, the red flour beetle *Tribolium* (Beeman et al., 1989).

That two insects have similar clusters of homoeotic genes is perhaps not terribly surprising. But it is astonishing that similar clusters also exist in the mouse and in humans (Acampora et al., 1989; Schughart et al., 1989). These gene clusters, based on the similarities of their homeoboxes, have the same linear array on the chromosome in flies and mice and are expressed in a similar anterior to posterior order along the axes of both animals (Duboule and Dolle, 1989; Graham et al., 1989). Not only has the homeobox protein motif been maintained over a very long evolutionary time frame, but the organization of these regulatory gene complexes must also have had an extremely early origin. Current studies of the fossil history of the radiation of metazoan phyla suggest that the spectacular radiation of animal phyla occurred in the latest Precambrian and was complete by the mid Cambrian, about 540 million years ago (Gould, 1989b; McMenamin and McMenamin, 1990; Whittington, 1985). The maintenance of the higher order organization of homeobox-containing genes over such vast phylogenetic distances would argue strongly that the clusters have some important functional significance, with coupled loci constrained from separation into distinct genes.

Our ideas of developmental constraint will be heavily influenced by our concepts of process. The common notion is that early development is particularly refractory to evolutionary change because subsequent development is hierarchically dependent on early events (Riedl, 1978). This view has been most formally stated by Arthur (1988), but it is a widely held view and if true would constitute a major kind of developmental constraint. It curiously does not appear to be true (at least at all stages of development).

Closely related species with radically different modes of development offer powerful systems for studies of these evolutionary changes because cellular homologies are readily identified, developmental processes can be readily compared, and molecular probes generally cross-react well. For instance, of the Australian sea urchins of the genus *Heliocidaris*, *H. tuberculata* develops via a pluteus larva in the mode of development characteristic of sea urchins. The other species, *H. erythrogramma*, develops directly via a highly modified larva. Divergence of nuclear single-copy and mitochondrial DNAs indicates that these species split no more than about ten million years ago and that direct development has arisen within that time.

Early development of *H. erythrogramma* has been radically remodeled (Henry and Raff, 1990; Wray and Raff, 1989, 1990). Modifications are obvious in cleavage pattern and partitioning of maternal materials. Cell lineage tracing and experimental removal of individual blastomeres indicate these effects to be profound, with the result that there is a maternally specified dorsoventral axis and asymmetry in cell fates. Changes in cleavage have significantly altered the geometry of the embryo. There are also heterochronies in cell lineage differentiation, and cell lineage tracing studies show extensive changes in cell lineage determination. The fate map of the embryo is modified to reflect changes in the relative size of larval structures in the direct developer. Generally, the same cell lineages are involved. However, there are also shifts in cell fate. An example is the case of the ciliated band. In typical development the ciliated band arises primarily from animal tiers of cells: in *H. erythrogramma* the ciliated band arises from vegetal cells. The most radical change is in establishment of founder cells. The patterns of cell fate splits have been extensively modified. Thus, the same cell types are produced, but the founder cells are set up through new pathways. These changes are radical. However, experimental studies of the regulative ability of typical sea urchin early development suggests that early development in these forms is highly flexible and that evolutionary modifications should be more readily tolerated than previously suspected (Henry et al., 1989; Raff et al., 1990). Evolutionary changes in sea urchin development have occurred in a number of processes: specification of axes has become maternal, cell lineage decisions are transformed to new patterns, and gastrulation is driven by a mechanism different from that in typical sea urchin development. Analogous and equally dramatic changes have been observed in early development between closely related species of ascidians (Jeffery and Swalla, 1990) and frogs (Elinson, 1987).

Although developmental constraints can be argued to bias the direction of evolution of ontogeny, constraints can also be evaded. In a striking study Alberch (1987) showed that dramatic remodeling of larval structures to serve very different adult functions in amphibian metamorphosis has required the evolution of "developmental compartmentalization." In the example he studied, the epibranchial cartilages, important in feeding in salamanders, this remodeling has allowed compartmentalization of cells such that the adult structure arises from a distinct group of imaginal cells, while the cells of the larval cartilage die and are resorbed. Compartmentalization can thus uncouple larval from adult ontogenies.

In *Embryos, Genes, and Evolution* we suggested that inductive events offer an immense potential for evolutionary changes, and we offered experimental examples available at that time. One of those examples was the classic experiment of Hampé, in which by insertion of a mica sliver into the hind limb bud of the chick to separate the presumptive territories of the tibia and fibula he was able to obtain development of a full-length fibula reminiscent of that of the fossil ancestral bird *Archaeopteryx*. This experiment has been reinvestigated and reinterpreted by Gerd Müller (1989). Müller has found that the original interpretations of the experiment are not tenable. In operated chicks, the fibula has not become longer as in the ancestral limb, but the tibia has become shorter. Proportions have changed, but not to produce an *Archaeopteryx* limb. The consequence is a significant change in how we must interpret the experiment. Because the skeletal pattern of bone growth of the affected part of the limb has become more reptilian, several muscles assume a reptilian pattern also. These changes, Müller points out, do not mean that quiescent ancestral genes have been reactivated; rather, muscle patterns respond to a new epigenetic situation, one similar to that found in reptilian limbs.

Another difficult problem in vertebrate morphological evolution also has become explicable in terms of cellular behavior during development. Turtles differ from all other vertebrates in that their limb girdles lie inside their rib cages rather than outside. Burke (1989) has studied the cellular basis of turtle carapace development, and defined inductive processes which underlie growth of this structure. She also has suggested a model for changes in migration of cell precursors of limb girdle and carapace structures that allow the radical reorganization of these structures relative to each other.

These and other studies demonstrate the sophistication now possible in evolutionary investigations of inductive interactions and other aspects of complex cellular events underlying morphogenesis. The molecular bases for cellular interactions in development are also becoming better understood. Thus, in establishment of the polarity of the amphibian body plan, peptide growth factors are involved in communication between cells of the embryo (Ruiz i Altaba and Melton, 1990). Growth factors and homeobox genes apparently interact in establishment of cell fates and pattern. Particular signal and signal receptor motifs are widely distributed among distant animal groups and play a wide variety of roles in cell-cell signaling and cell fate determination. Thus, the epidermal growth factor domain is present in mammalian growth factor and in other mammalian proteins (Doolittle, 1985). This motif is also present in proteins encoded by insect neurogenic-determining loci (Kopczynski et al., 1988; Vässin et al., 1987; Wharton et al., 1985) and in genes that determine cell fates in nematode development (Greenwald, 1985; Yochem et al., 1988; Yochem and Greenwald, 1988). These commonalities reveal an important insight into the evolution of the regulation of development. Regulatory molecules and receptors are retained over great evolutionary distances, but have been tied into diverse intracellular response systems.

A last point worth noting is the role of phylogenetics in understanding the evolution of development. We discussed phylogenetic patterns and molecular studies of phylogeny in *Embryos, Genes, and Evolution.* We were correct in tying phylogeny into the problem, but we really addressed only part of the significance of phylogenetic data to analysis of specific cases of the evolution of development. Our emphasis in 1983 was on the fact that evolutionary history reveals patterns and timing of evolutionary events. These provide important constraints on what we envision in terms of mechanisms of change and in the time frames of their operation. Molecular data provided us with the important insight that morphological, and thus developmental, evolution is largely independent of evolutionary change in the major part of the genome. These insights remain valid. However, if one is studying evolution of features of any specific organisms, it becomes necessary to know genealogy to be able to assign the polarity of change. In broader studies of evolution of development (e.g. in considering such problems as the evolution of the diverse larval forms of invertebrate phyla) it becomes vital to have methods for inferring relationships not involving developmental patterns. For example, is segmentation in the annelids homologous to segmentation in the supposedly closely related arthropods? The *engrailed* gene, which is involved in segmentation in arthropods, is expressed differently in annelids (Patel et al., 1989). Since many of our inferences about such relationships have traditionally been based on embryology, a serious problem is posed: whence the new characters? The only choice seems to lie in the genome itself, with determinations of gene organization and sequence providing the data.

Molecular phylogeny has now become a large-scale industry. Although molecular phylogenetics is still suffering from teething troubles, molecular data are already beginning to revolutionize our views of eukaryotic relationships and animal phylogeny (Fernholm et al., 1989; Field et al., 1988; Olsen, 1987; Sogin et al., 1989). Attempts to calibrate rates of molecular evolution by the fossil record are becoming more sophisticated in estimating the uncertainties involved (Marshall, 1990). A particularly striking example of the power of molecular phylogenetics is provided by the unsuspected and rapid morphological evolution in false-truffles revealed by mitochondrial DNA comparisons (Bruns et al., 1989). These highly derived forms turn out to be closely related to a genus of very typical mushrooms. Without the molecular data, the rapid morphological divergence could not have been recognized. The demonstration of such examples provides good cases which expand our ability to ask penetrating questions about the evolution of developmental processes. Experimental studies may now yield answers to a relationship between development and evolution that has been visible, albeit elusive, for over a century.

Rudolf A. Raff and Thomas C. Kaufman, Bloomington, 1990

References

Acampora, D., M. D'Esposito, A. Faiella, M. Pannese, E. Migliaccio, F. Morelli, A. Stornaiuolo, V. Nigro, M. Simeone, and E. Boncinelli, 1989. The human HOX gene family. *Nucleic Acid Res.* **17**:10385-10402.

Akam, M., 1987. The molecular basis for metameric pattern in the *Drosophila* embryo. *Development* **101**:1-22.

Alberch, P., 1982. Developmental constraints in evolutionary processes. In *Evolution and Development* (J. T. Bonner, ed.), Springer-Verlag, Berlin, pp. 313-330.

Alberch, P., 1987. Evolution of a developmental process: Irreversibility and redundancy in amphibian metamorphosis. In *Development as an Evolutionary Process* (R. A. Raff and E. C. Raff, eds.), A. R. Liss, New York, pp. 23-46.

Alberch, P., and J. Alberch, 1981. Heterochronic mechanisms of morphological diversification and evolutionary change in the neotropical salamander *Bolitoglossa occidentalis* (Amphibia: Plethodontidae). *J. Morphol.* **167**:249-264.

Ambros, V., and H. R. Horvitz, 1984. Heterochronic mutants of the nematode *Caenorhabditis elegans*. *Science* **226**:409-416.

Anderson, K. U., 1989. *Drosophila:* The maternal contribution. In *Genes and Embryos* (D. M. Glover and B. D. Hanes, eds.), IRL Press, Oxford, pp. 1-38.

Arthur, W., 1988. *A Theory of Development.* Wiley, Chichester.

Beeman, R. W., J. J. Stuart, M. S. Haas, and R. E. Denell, 1989. Genetic analysis of the homeotic gene complex (HOM-C) in the beetle *Tribolium casteneum*. *Dev. Biol.* **133**:196-209.

Bruns, T. D., R. Fogel, T. J. White, and J. D. Palmer, 1989. Accelerated evolution of a false-truffle from a mushroom ancestor. *Nature* **339**:140-142.

Burke, A. C., 1989. Development of the turtle carapace: Implications for the evolution of a novel bauplan. *J. Morph.* **199**:363-378.

deBeer, G., 1958. *Embryos and Ancestors.* Oxford University Press, Oxford.

Desplan, C., J. Theis, and P. H. O'Farrell, 1985. The *Drosophila* developmental gene, *engrailed*, encodes a sequence specific DNA binding activity. *Nature* **318**:630-635.

Doolittle, R. F., 1985. The genealogy of some recently evolved vertebrate proteins. *Trends in Biochemical Sciences* **10**:233-237.

Driever, W., and C. Nüsslein-Volhard, 1988a. A gradient of *bicoid* protein in *Drosophila* embryos. *Cell* **54**:83-93.

Driever, W., and C. Nüsslein-Volhard, 1988b. The *bicoid* protein determines position in the *Drosophila* embryo in a concentration-dependent manner. *Cell* **54**:95-104.

Driever, W., and C. Nüsslein-Volhard, 1989. The *bicoid* protein is a positive regulator of *hunchback* transcription in the early *Drosophila* embryo. *Nature* **337**:138-143.

Driever, W., G. Thoma, and C. Nüsslein-Volhard, 1989. Determination of spatial domains of zygotic gene expression in the *Drosophila* embryo by the affinity of binding sites for the *bicoid* morphogen. *Nature* **340**: 363-367.

Duboule, D., and P. Dolle, 1989. The structural and functional organization of the murine HOX gene family resembles that of the *Drosophila* homeotic genes. *EMBO* **8**:1497-1505.

Elinson, R. P., 1987. Changes in developmental patterns: Embryos of amphibians with large eggs. In *Development as an Evolutionary Process* (R. A. Raff and E. C. Raff, eds.), A. R. Liss, New York, pp. 1-21.

Fernholm, B., K. Bremer, and H. Jönvall (eds.), 1989. *The Hierarchy of Life: Molecules and Morphology in Phylogenetic Analysis*. Excerpta Medica, Amsterdam.

Field, K. G., G. J. Olsen, D. J. Lane, S. J. Giovannoni, M. T. Ghiselin, E. C. Raff, N. R. Pace, and R. A. Raff, 1988. Molecular phylogeny of the animal kingdom. *Science* **239**:748-753.

Frohnhöfer, H. G., and C. Nüsslein-Volhard, 1987. Maternal genes required for the anterior localization of *bicoid* activity in the embryo of *Drosophila*. *Genes and Development* **1**:880-890.

Gould, S. J., 1977. *Ontogeny and Phylogeny*. Harvard University Press, Cambridge.

Gould, S. J., 1989a. A developmental constraint in *Cerion*, with comments on the definition and interpretation of constraint in evolution. *Evolution* **43**:516-539.

Gould, S. J., 1989b. *Wonderful Life.* W. W. Norton, New York.

Graham, A., N. Papalopulu, and R. Krumlauf, 1989. The murine and *Drosophila* homeobox gene complexes have common features of organization and expression. *Cell* **57**:367-378.

Greenwald, I., 1985. *lin-12*, a nematode homoeotic gene, is homologous to a set of mammalian proteins that includes epidermal growth factor. *Cell* **43**:583-590.

Hall, B. K., 1984. Developmental processes underlying heterochrony as an evolutionary process. *Can. J. Zool.* **62**:1-7.

Hampé, A., 1960. Le compétition entre les éléments osseux du zeugopode de poulet. *J. Embryol. Exp. Morphol.* **8**:241-245.

Hankin, J., 1989. Development and evolution in amphibians. *American Scientist* **77**:336-343.

Hansen, T. A., 1982. Modes of larval development in early Tertiary neogastropods. *Paleobiol.* **8**:367-377.

Henry, J. J., and R. A. Raff, 1990. Evolutionary change in the process of dorsoventral axis determination in the direct developing sea urchin, *Heliocidaris erythrogramma*. *Dev. Biol.* **141**:55–69.

Henry, J. J., S. Amemiya, G. A. Wray, and R. A. Raff, 1989. Early inductive interactions are involved in restricting cell fates of mesomeres in sea urchin embryos. *Dev. Biol.* **136**:140-153.

Hodgkin, J., 1987. Sex determination and dosage compensation in *Caenorhabditis elegans*. *Ann. Rev. Genet.* **21**:133-154.

Hodgkin, J., 1989. *Drosophila* sex determination: A cascade of regulated splicing. *Cell* **56**:905-906.

Holder, N., 1983. Developmental constraints and the evolution of vertebrate digit patterns. *J. Theoret. Biol.* **104**:451-471.

Ingham, P. W., 1988. The molecular genetics of embryonic pattern formation in *Drosophila*. *Nature* **335**:25-34.

Jablonski, D., 1986. Larval ecology and macroevolution in marine invertebrates. *Bull. Marine Sci.* **39**:565-587.

Jeffery, W. R., and B. J. Swalla, 1990. Anural development in ascidians: Evolutionary modification and elimination of the tadpole larva. *Seminars in Devel. Biol.* **1**, (4):253–261.

Kaufman, T. C., M. A. Seeger, and G. Olsen, 1989. Molecular and genetic organization of the Antennapedia gene complex of *Drosophila melanogaster*. In *Advances in Genetics*, Vol. 27 (T. Wright, ed.), Academic Press, San Diego, pp. 309-362.

Kopczynski, C. C., A. K. Alton, K. Fechtel, P. J. Kooh, and M. A. T. Muskavitch, 1988. *Delta*, a *Drosophila* neurogenic gene, is transcriptionally complex and encodes a protein related to blood coagulation factors and epidermal growth factor of vertebrates. *Genes and Devel.* **2**:1723-1735.

Larson, A., 1983. A molecular phylogenetic perspective on the evolution of a lowland tropical salamander fauna. II. Patterns of morphological evolution. *Evolution* **37**:1141-1153.

Lawrence, P., 1983. Developing theories of evolution. *Nature* **304**:378-379.

Levine, M. S., and K. W. Harding, 1989. *Drosophila:* The zygotic contribution. In *Genes and Embryos* (D. M. Glover and B. D. Hanes, eds.), IRL Press, Oxford, pp. 39-94.

Lord, E. M., and J. P. Hill, 1987. Evidence for heterochrony in the evolution of plant form. In *Development as an Evolutionary Process* (R. A. Raff and E. C. Raff, eds.), A. R. Liss, New York, pp. 47-70.

Macdonald, P. M., and G. Struhl, 1988. Cis-acting sequences responsible for anterior localization of *bicoid* mRNA in *Drosophila* embryos. *Nature* **336**:595-598.

Marshall, C. R., 1990. The fossil record and estimating divergence times between lineages: Maximum divergence times and the importance of reliable phylogenies. *J. Mol. Evol.* **30**:400-408.

Maynard Smith, J., R. Burian, S. Kauffman, P. Alberch, J. Campbell, B. Goodwin, R. Lande, D. Raup, and L. Wolpert, 1985. Developmental constraints and evolution. *Q. Rev. Biol.* **60**:265-287.

McKinney, M. L. (ed.), 1988. *Heterochrony in Evolution. A Multidisciplinary Approach.* Plenum Press, New York.

McMenamin, M. A. S., and D. L. S. McMenamin, 1990. *The Emergence of Animals: The Cambrian Breakthrough.* Columbia University Press, New York.

Meyerowitz, E. M., D. R. Smyth, and J. L. Bowman, 1989. Abnormal flowers and pattern formation in floral development. *Development* **106**:209-217.

Mileikovski, S. A., 1971. Types of larval development in marine bottom invertebrates, their distribution and ecological significance: A reevaluation. *Marine Biol.* **10**:193-213.

Müller, G. B., 1989. Ancestral patterns in bird limb development: A new look at Hampé's experiment. *J. Evol. Biol.* **2**:31-47.

Muller, M., M. Affolter, W. Leupin, G. Otting, K. Wuthrich, and W. J. Gehring, 1988. Isolation and sequence-specific DNA binding of the *Antennapedia* homeodomain. *EMBO* **7**:4299-4304.

Nüsslein-Volhard, C., H. G. Frohnhöfer, and R. Lehman, 1987. Determination of anteroposterior polarity in *Drosophila*. *Science* **238**: 1675-1681.

Olsen, G. J., 1987. Earliest phylogenetic branchings: Comparing rRNA-based evolutionary trees inferred with various techniques. *Cold Spring Harbor Symp. Quant. Biol.* **52**:825-837.

Oster, G. F., N. Shubin, J. D. Murray, and P. Alberch, 1988. Evolution and morphogenetic rules: The shape of the vertebrate limb in ontogeny and phylogeny. *Evolution* **42**:862-884.

Parks, A. L., B. A. Parr, J.-E. Chin, D. S. Leaf, and R. A. Raff, 1988. Molecular analysis of heterochronic changes in the evolution of direct developing sea urchins. *J. Evol. Biol.* **1**:27-44.

Patel, N. H., E. Martin-Blanco, K. G. Coleman, S. J. Poole, M. C. Ellis, T. B. Kornberg, and C. S. Goodman, 1989. Expression of *engrailed* protein in arthropods, annelids, and chordates. *Cell* **58**:955-968.

Peifer, M., F. Karch, and W. Bender, 1987. The bithorax complex: Control of segmental identity. *Genes Dev.* **1**:891-898.

Poethig, R. S., 1988. Heterochronic mutations affecting shoot development in maize. *Genetics* **119**:959-973.

Raff, R. A. (ed.), 1990. Heterochronic changes in development. *Seminars in Devel. Biol.* **1**, (4).

Raff, R. A., and G. A. Wray, 1989. Heterochrony: Developmental mechanisms and evolutionary results. *J. Evol. Biol.* **2**:409-434.

Raff, R. A., G. A. Wray, and J. J. Henry, 1991. Implications of radical evolutionary changes in early development for concepts of developmental constraints. In *New Perspectives on Evolution* (L. Warren and M. Meselson, eds.), A. R. Liss, New York, in press.

Regier, J. C., and N. S. Vlahos, 1988. Heterochrony and the introduction of novel modes of morphogenesis during the evolution of moth choriogenesis. *J. Mol. Evol.* **28**:19-31.

Riedl, R., 1978. *Order in Living Organisms.* John Wiley, Chichester.

Ruiz i Altaba, and D. A. Melton, 1990. Axial patterning and the establishment of polarity in the frog embryo. *Trends in Genetics* **6**:57-64.

Sánchez-Herrero, E., I. Vernós, R. Marco, and G. Morata, 1985. Genetic organization of *Drosophila* bithorax complex. *Nature* **313**:108-113.

Schughart, K., C. Kappen, and F. H. Ruddle, 1989. Duplication of large genomic regions during the evolution of vertebrate homeobox genes. *Proc. Natl. Acad. Sci.* **86**:7067-7071.

Sogin, M. L., J. H. Gunderson, H. J. Elwood, R. A. Alonso and D. A. Peattie, 1989. Phylogenetic meaning of the kingdom concept: An unusual ribosomal RNA from *Giardia lamblia*. *Science* **243**:75-77.

Stern, C. D. (ed.), 1990. The evolution of segmental patterns. *Seminars in Devel. Biol.* **1**:75-145.

Strathmann, R. R., (1985). Feeding and nonfeeding larval development and life-history evolution in marine invertebrates. *Ann. Rev. Ecol. Syst.* **16**:339-361.

Struhl, G., K. Struhl, and P. M. Macdonald, 1989. The gradient morphogen *bicoid* is a concentration-dependent transcriptional activator. *Cell* **57**:1259-1273.

Vässin, H., K. A. Bremer, E. Knust, and J. A. Campos-Ortega, 1987. The
neurogenic gene Delta of *Drosophila melanogaster* is expressed in
neurogenic territories and encodes a putative transmembrane protein with EGF-like repeats. *EMBO J.* **6**:3431-3440.

Wake, D. B., and A. Larson, 1987. Multidimensional analysis of an evolving lineage. *Science* **238**:42-48.

Wharton, K. A., K. M. Johansen, T. Xu, and S. Artavanis-Tsakonas,
1985. Nucleotide sequence from the neurogenic locus *Notch* implies a gene product that shares homology with proteins containing EGF-like repeats. *Cell* **43**:567-581.

Whittington, H. B., 1985. *The Burgess Shale.* Yale University Press, New
Haven.

Winslow, G. M., S. Hayashi, M. Krasnow, D. S. Hogness, and M. P.
Scott, 1989. Transcriptional activation by the *Antennapedia* and
fushi tarazu proteins in cultured *Drosophila* cells. *Cell* **57**:1017-1030.

Wray, G. A., and D. R. McClay, 1989. Molecular heterochronies and het-
erotopies in early echinoid development. *Evolution* **43**:803-813.

Wray, G. A., and R. A. Raff, 1989. Evolutionary modification of cell lineage and fate in the direct developing sea urchin *Heliocidaris
erythrogramma. Dev. Biol.* **132**: 458-470.

Wray, G. A. and R. A. Raff, 1990. Novel origins of lineage founder cells in
the direct-developing sea urchin *Heliocidaris erythrogramma. Dev.
Biol.* **141**:41–54.

Yochem, J., and I. Greenwald, 1988. *lin-12* and *glp-1* encode very similar
potential transmembrane proteins. *The Worm Breeder's Gazette* **10**:44.

Yochem, J., K. Weston, and I. Greenwald, 1988. The *Caenorhabditis elegans
lin-12* gene encodes a transmembrane protein with overall similarity to *Drosophila Notch. Nature* **335**:547-550.

Preface

This book has its immediate origins in a course we have taught for the past several years at Indiana University on the developmental-genetic mechanisms that generate evolutionary changes in morphology. The underlying theme that evolution cannot be understood without an understanding of the developmental processes that produce form in ontogeny is an old one; indeed, in the late nineteenth century it was a dominant part of evolutionary theory. However, throughout most of the twentieth century the obvious connection between phylogenetic transformations in shape of organisms and the underlying modifications of the genetic systems controlling ontogeny has been largely ignored except by a few outsiders to the neo–Darwinian synthesis, which was built of other elements. That synthesis was incomplete.

Our own fascination with this subject goes back to our undergraduate days, in which we were first exposed to the incredible diversity of body plans of marine invertebrate phyla and to the elegant functional anatomy of vertebrates. Equally important, both of us, in our research, have been engaged in somewhat different ways in trying to determine how genes direct the processes that make up embryonic development. Thus, there is a distinct mental set to our approach to evolution as there is to our approach to development, and this colors our choice of the topics considered in this book. The essential position is that there is a genetic program that governs ontogeny, and that the momentous decisions in development are made by a relatively small number of genes that function as switches between alternate states or pathways. The significance of this view, if correct, is that evolutionary changes in morphology occur mechanistically, as a result of modifications of these genetic switch systems. If our prediction that there are a relatively small number of such gene switches is correct, then the potential exists for geologically rapid and dramatic evolutionary changes. Such macroevolutionary events are apparently associated with the origins of new groups of organisms.

The text is divided into four informal sections. The first group of chapters deals with the history of the problem, with rates of evolution, and with the noncongruence of morphological and molecular evolution. The second set of chapters deals with the evolutionary role of developmental processes and considers the organization of eggs and early embryos, interactions between regions of embryos, and timing. Change in relative timing of developmental processes provides one of the best-documented mechanisms for achieving evolutionary changes in shape. Indeed, most previous treatments of the role of developmental processes in evolution, notably those of de Beer and Gould, have focused

XXV

on timing. Other modes of disassociating processes with respect to one another have been discussed less frequently, but may be of equal importance. The third part of the book considers the genetics of development; and here we demonstrate that genes indeed control ontogeny in very specific ways and that there is a genetically determined developmental program. Finally, although ontogeny can be analyzed by classic genetic methodology, we are not limited to this approach in the analysis of gene expression. Advances in techniques for the cloning of genes and for high-resolution studies of DNA and RNA have made it possible to examine directly genes and gene expression during development. The results of such studies are discussed in the remaining chapters. In the final chapter we attempt to arrive at a synthetic integration of the developmental-genetic basis for morphological evolution.

One other point of organization should be mentioned. To avoid annoying interruptions of the flow of the text by intrusive citations or footnotes, we have worked most references into the text by author name without dates. This procedure provides sufficient information to locate the cited work in the bibliography at the end of the book.

As in any project of this kind, the advice and support of many people have been vital. We are grateful to our many colleagues who patiently answered our numerous questions and provided us with information, references, reprints, preprints, sketches, and photographs. We also thank the students who have studied this problem with us in our course for their perceptive questions and for insights they have provided. We owe particular thanks to our colleague Elizabeth C. Raff, who so superbly illustrated this book and who so mercilessly red-penciled the awkward writing of our first drafts.

Because so many of the topics we have had to consider fall well beyond the range of our own expertise, it has been especially important to us that the chapters be critically read by experts. These readers generously put forth considerable time, effort, and thought in their reviews and provided us with priceless criticism and suggestions as well as very much appreciated encouragement. To John Tyler Bonner, Peter Bryant, Hampton Carson, Robert Edgar, Gary Freeman, Stephen J. Gould, Donna Haraway, Vernon Ingram, Burke Judd, Raymond Keller, William Klein, Jane Maienschein, Elizabeth Raff, Steven Stanley, Alan Templeton, Robert Tompkins, David Wake, and J. R. Whittaker we are profoundly grateful. Naturally, like all good academics, we have not followed all of the advice proffered—and no doubt we have committed errors for which only we can accept responsibility.

We have been fortunate in having our manuscript skillfully typed by Ann Martin and in having the help of Monica Bonner, who with extraordinary patience, good spirits, and intelligence kept us from being submerged by organizational problems. We also owe much to the staffs of the libraries of Indiana University and the Marine Biological Laboratory at Woods Hole, Massachusetts for their help in locating materials and for their forbearance with holders of long-overdue books.

Rudolf A. Raff and Thomas C. Kaufman

EMBRYOS, GENES, AND EVOLUTION

One

Embryos and Ancestors

"Perhaps I ought to explain," added the badger, lowering his papers nervously and looking at the Wart over the top of them, "*that all embryos look very much the same.* They are what you are before you are born—and, whether you are going to be a tadpole or a peacock or a cameleopard or a man, when you are an embryo you just look like a peculiarly repulsive and helpless human being. I continue as follows:

The embryos stood in front of God, with their feeble hands clasped politely over their stomachs and their heavy heads hanging down respectfully, and God addressed them. He said: 'Now, you embryos, here you are, all looking exactly the same, and We are going to give you the choice of what you want to be. When you grow up you will get bigger anyway, but We are pleased to grant you another gift as well. You may alter any parts of yourselves into anything you think would be useful to you in later life.'"

T. H. White, The Once and Future King

The Problem of Morphology

Organisms have morphologies, and they exhibit behaviors and physiological adaptations. In our perspective of long eons of geological time these characteristics appear to be almost protean: the lobe-fin of the crossopterygian becomes the limb of the amphibian, the wing of a bird, the human arm and hand. This is the visible record of evolution. By what mechanisms are evolutionary changes in morphology achieved?

In fact, we already know the answer to this question, at least in a formalistic sense. Garstang provided it in 1922 when he pointed out that an evolutionary sequence or phylogeny is not simply a succession of adult forms. Each generation of adults has been produced by a series of developmental processes, ontogeny, from an apparently structureless egg to the complex morphology of the adult. Thus, in order for an evolutionary change to be expressed as an altered body structure, a new morphology, a change, must occur in ontogeny.

One might expect that the role of developmental processes in evolution should be a major component of current studies of evolution; however, this is not so. Embryological development, which was so vital a part of evolutionary theory in the late nineteenth century, has been considered largely irrelevant in the twentieth. Later in this chapter we shall discuss the reasons for this strange divorce. Of course, apprecia-

1

tion of the significance of the relationship between development and evolution never totally died out. Garstang, J. S. Huxley, de Beer, and Goldschmidt all explicitly dealt with this relationship in a serious way during the period from the 1920s to the 1950s. And the recent publication of Gould's book *Ontogeny and Phylogeny* indicates that current interest in the subject may be not only alive, but lively as well.

Our own fascination was stimulated several years ago by reading de Beer's book *Embryos and Ancestors*, which so cogently argues that changes in timing of developmental events can have dramatic evolutionary results. Unfortunately, beyond a brief general discussion of genes controlling rates of developmental processes, de Beer paid little attention to the role of genetic regulation in development or evolution. There was simply not enough known about developmental genetics when de Beer wrote the first edition, published in 1930, for him to have dealt with it in a very profound way. By 1958, when the third and last edition of *Embryos and Ancestors* was published, much more was known, but de Beer cited little of the advances in developmental genetics made subsequent to the 1930s. His main emphasis lay in another direction.

In fact, the developmental-genetic basis for evolutionary change has simply never been explored in detail. This is what we intend to accomplish in this book. Our premise is that developmental processes are under genetic control and that evolution should be envisaged as resulting from changes in the genes regulating ontogeny.

It is interesting that this was the view advanced by Goldschmidt in 1940 in his *Material Basis of Evolution*, though not enough was known about genes and their functions in development at that time for the synthesis to be successful. Goldschmidt's ideas have been ignored for the past 35 years because of his idiosyncratic (and erroneous) view of the nature of genes, but his definition of evolution provides a perfectly clear statement of the theme of this book:

> *Evolution means the transition of one rather stable organic system into a different but still stable one. The genetic basis of the process, the change from one stable genetic constitution to another, is one side of the problem. No evolution is possible without a primary change within the germ plasm; i.e., predominantly within the chromosomes, to a new stable architecture. But there is also another side to the problem. The germ plasm controls the type of the species by controlling the developmental process of the individual...the specificity of the germ plasm is its ability to run the system of reactions which make up the individual development, according to a regular schedule which repeats itself,* ceteris paribus, *with the purposiveness and orderliness of an automaton. Evolution, therefore, means the production of a changed process of development, controlled by the changed germ plasm....*

The term germ plasm used by Goldschmidt refers to the genetic material—in modern terms, the DNA of the genome.

What sorts of genes govern ontogeny, and by what modes do they evolve?

At present the best-understood genes are those that code for specialized RNAs or proteins vital to the overall structure and function of cells, such as ribosomal RNAs; various enzymes; structural proteins, such as tubulin or collagen; or carrier proteins, such as hemoglobin.

Estimations of the relevance of such structural genes to the control of development and morphogenesis span a considerable range. Our view is that structural genes have a very limited regulatory function in development, but the opposite has been suggested. An example of a developmental hypothesis granting an extensive role to structural genes and their products is that advanced by Monod in *Chance and Necessity*. Monod proposed that the generation of structural complexity results from what he called the molecular epigenesis of proteins. By this term he was referring to a well-known feature of proteins, that the amino acid sequence of a protein determines the three-dimensional conformation the protein assumes within the environment of the cell. Further, proteins interact with other proteins in specific ways to yield supramolecular structures. To quote Monod, "Order, structural differentiation, acquisition of functions—all these appear out of a random mixture of molecules individually devoid of any activity, any intrinsic functional capacity other than that of recognizing the partners with which they will build the structure." He goes on to suggest that this process both underlies and serves as a paradigm for a series of autonomous epigenetic events culminating in the development of the entire organism. An extreme interpretation of this idea may conjure up the epigenetic fantasy that, given a mixture of the proper macromolecules, it would be possible to obtain the assembly of a mouse from solution.

Monod's proposal, even stopping short of the extreme, is untenable as a model for development. Nor does evolution of structural genes account for morphological evolution. The work of A. C. Wilson and his collaborators indicates that, at least with respect to living groups of organisms such as frogs and mammals, evolution of structural genes for proteins is largely irrelevant to morphological evolution. Humans and chimpanzees have rapidly diverged morphologically, but they are 99% similar in their protein sequences. On the other hand, frogs, an old group, have exhibited rather slow morphological rates of evolution, but their protein sequences have evolved at rates comparable to those of other organisms. This realization has led King and Wilson to propose that changes in regulatory genes rather than structural genes provide the basis for morphological evolution.

Because there is a hierarchy of interacting controls governing gene expression and ontogeny, regulatory genes fall into a number of categories and are more difficult to define as a group than structural genes. Essentially, structural genes supply the materials for development and regulatory genes both provide and interpret the blueprint. Structural genes are relatively easy to understand because the products they code for can be readily isolated and studied, and their functions defined. Not surprisingly, regulatory genes have proven a great deal more elusive than structural genes. Some regulatory genes, or elements, produce no products; others do, but their products exist only in minute amounts. The best-known example is the lac-repressor protein of the bacterium *E. coli*. This product of a regulatory gene controls the expression of the genes involved in lactose metabolism. Only 10 molecules of repressor are present per cell.

Regulatory genes function throughout the developmental process and govern ontogeny in three major ways: first, by controlling the timing of events; second, by making binary choices, and thus decisions about the fates of groups of cells or regions within the embryo; and third, by integrating the expression of structural genes to produce stable, differentiated tissues. All three modes of regulation have a considerable bearing on evolution.

That changes in timing of developmental events serve as a major and flexible mechanism for achieving significant morphological evolution was argued by de Beer in his seminal work *Embryos and Ancestors* and more recently by Gould in *Ontogeny and Phylogeny*. These authors were less concerned with mechanisms for the genetic control of developmental processes than with defining the kinds of modifications possible in relative timing of events in ontogeny and in demonstrating their evolutionary sequels. A variety of evolutionary events have been proposed as representing consequences of timing changes. The most widely cited are cases of neoteny in which new kinds of adult body plans result from larval stages becoming sexually mature and losing the ancestral adult stage. We explore timing as a mode of regulatory evolution in Chapter 6.

However, genetic regulation of ontogeny is not limited to control of timing. Recent work, especially with the fruit fly, *Drosophila melanogaster*, which has become something of a eucaryotic *E.coli* to investigators of gene organization and function in this decade, has revealed that a hierarchy of regulatory genes control the organization of the developing embryo. These genes act as switches that determine which of two alternate choices a cell or group of cells will make in development. Once the decision is made, the cells are restricted in which subsequent choices they can make as their developmental fates become further specified. Regulatory genes of this type are accessible to study because of the spectacular effects manifested by mutations that abolish or modify their functions as binary switches. In *Drosophila* these mutations produce transformations in which a modified pattern of morphogenesis will substitute one structure for another, such as legs in place of antennae or an extra set of wings instead of halteres. Modification or creation of new sets of this class of regulatory gene provide an impressive potential for production of radical evolutionary modifications or morphological novelties. That such a mode of evolution has occurred, and indeed has been critical in the evolution of insects and other organisms, is clear, and we have much more to say about it later in Chapters 8 and 9.

Like changes in timing or organizational integration, changes in regulatory genes controlling tissue differentiation have great evolutionary potential. Whereas changes in the first two classes of regulators yield changes in morphology, changes in the third class produce novel tissues. An example that we discuss in greater detail in Chapter 12 is the mammary gland, the evolution of which has involved the origin of a new tissue, novel proteins, new regulatory genes, and a set of behavioral patterns crucial in the evolution of mammalian reproduction and infant care.

The three modes of regulation of development we have so briefly outlined do not stand apart from each other. All have been involved in varying degrees in the morphological evolution of particular lineages of organisms.

Perhaps the major difficulty we face in our project of attempting to understand morphological evolution in terms of developmental-genetic mechanisms is that the generation of morphology is extremely poorly understood at the molecular level. This is true not only from the perspective of mechanisms of morphogenesis per se—cell motility, cell-cell interactions, and pattern formation—but also because of a conceptual difference in the ways in which we score the information content of morphological structure as opposed to genetic information. As an illustration of this difference consider some approaches taken to morphogenesis from a nonmolecular genetic point of view by D'Arcy Thompson, who pioneered the application of mathematics to problems of shape in his book *On Growth and Form* first published in 1917 (Fig. 1–1). His aim was simple: "We want to see how, in some cases at least, the forms of living things, and the parts of living things, can be explained by physical considerations, and to realize that in general no organic forms exist save such as are in conformity with physical and mathematical laws." Thompson made his point in a book that has enlightened generations of biologists, who have learned from it the mathematical rules that underly the shapes of interfaces between cells and the morphologies of radiolaria or spiral shells and rams' horns; why the vertebrate skeleton and bridges obey the same laws of engineering; and how by using transformations of Cartesian grids, it is possible to represent the evolutionary changes of shapes of complex objects, such as skulls, fishes, and isopods. Thompson removed the sense of ineffable mystery from biological form, and indeed, elegantly showed that complex biological entities obey analyzable physical and mathematical rules. However, he paid little attention to genetic or molecular events, perhaps wisely, because these are still not completely understood, and instead focused on physical forces acting on the organism as immediate causes of morphology.

Thompson was much less successful in dealing with changes in shape during growth. Mathematical analysis of relative growth of the parts of an organism during its development (allometry) was devised by Huxley in the early 1930s. Basically, such growth relations can be expressed by the simple expression $y = bx^\alpha$, where x and y represent dimensions of two structures being compared. Allometry has been of considerable value in understanding evolutionary changes, but again, a genetic or molecular appreciation of the alterations in bodily proportions that accompany growth remains elusive and is assuredly more complex than the simple allometric equation implies.

Similarly, computer simulations of mollusk shells, produced by Raup and Michelson, show that the generation of objects with quite sophisticated morphologies may require only a small number of parameters (Fig. 1–2). For example, snail shells are tapered tubes wound in a spiral about a fixed axis. Only four parameters are required to generate com-

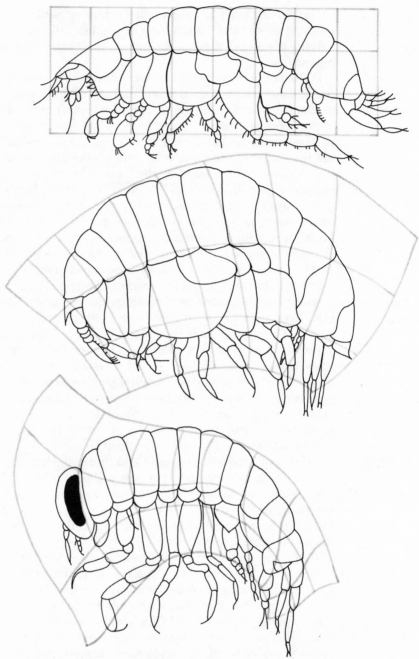

FIGURE 1–1. *Transformations in the shapes of some isopods. The species at the top of the figure is projected onto rectangular coordinates. Deformations in the corresponding grids for the two other species illustrate evolutionary changes in proportions.* [Redrawn from D. W. Thompson, *On Growth and Form*, Cambridge University Press, Cambridge, 1961, p. 295.]

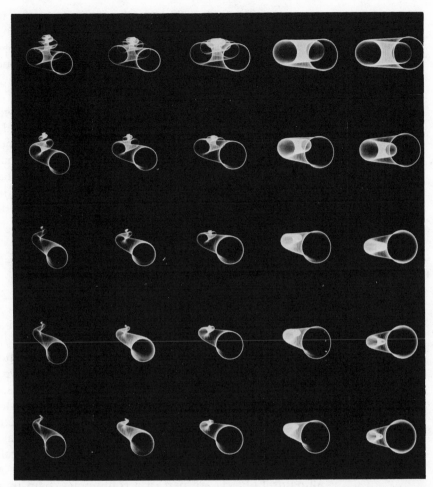

FIGURE 1–2. *Computer simulation of the forms of coiled shells. Rate of translation along the axis increases to the left while the rate of expansion of the generating curve increases from top to bottom. The shape of the generating curve and the distance between the generating curve and the coiling axis are the same in all.* [Photograph courtesy of D. M. Raup, from D. M. Raup and A. Michelson, Theoretical morphology of the coiled shell, *Science* 147:1294–1295, 1965. Copyright 1965 by the American Association for the Advancement of Science.]

puter analog simulations of actual shells; these are (1) cross-section shape of the generating curve, (2) rate of expansion of the generating curve with respect to revolution, (3) position and orientation of the generating curve relative to the axis, and (4) rate of movement of the generating curve down the axis. These simple parameters describe the shape to be generated, but they bear little relationship to the genetic program or the actual mechanisms by which organisms read out the genetic program for morphogenesis.

Although organisms obey the laws of chemistry and physics, there is an additional factor governing morphology, the evolutionary history of the organism. As elegantly pointed out by François Jacob, evolution operates by tinkering. New structures do not appear *de novo*; rather, evolution produces novelties by modifying already existing systems or structures. The first vertebrates, the fishlike Agnatha, were jawless. The origin of jaws, one of the great advances of vertebrate evolution, involved the transformation of an anterior pair of gill arches into primitive jaws. Similar conversions of preexisting structures have been shown in the evolution of specialized limbs, such as wings of pterodactyls, birds, and bats, or the origin of the bones of the inner ear of mammals from remnants of reptilian jaw articulation.

Because ontogenetic processes are highly integrated, they tend to be extremely conservative and stable. Thus, ontogeny and morphogenesis not only obey physical laws, as indeed they must, but also reflect the evolutionary history of the process. Historical accident and the necessity of maintaining integration clearly place limits on the kinds of evolutionary changes possible in the developmental process and thus constraints on morphological evolution.

Ontogeny, Phylogeny, and Recapitulation

In *Through the Looking-Glass* the White Queen informs Alice that "Why, sometimes I've believed as many as six impossible things before breakfast." To the modern reader the history of ideas about the relationship of ontogeny to evolution has very much the same flavor, yet the ideas we may currently recognize as absurd have had a profound effect on our perception of evolutionary mechanisms. How tenacious, after all, is the slogan "Ontogeny recapitulates phylogeny"?

To the transcendentalists of the early nineteenth century there was a fundamental unity of life, which was expressed as a parallel between the embryonic development of the individual and the scale of beings. The concept of a scale of beings, which was derived from Aristotle, held that all natural objects are part of a continuous chain that links inorganic creation with a series of increasingly complex living forms. Inanimate nature grades into plants, then simple animals such as sponges, to insects, fishes, birds, mammals, and finally, humans. This scheme was static and should not be construed as evolutionary in nature: It simply represented God's plan of creation. According to the law of parallelism, which is generally known as the Meckel-Serres law, after its two chief proponents, J. F. Meckel in Germany and Étienne Serres in France, a higher animal in its embryological development recapitulates the adult structures of animals below it on the scale of beings. Conversely, lower animals represent the permanent larval stages of more advanced forms.

Meckel, according to Russell, was a "timid believer in evolution," and indeed his final (1828) statement of the law of parallelism was cast in evolutionary terms:

> The development of the individual organism obeys the same laws as the development of the whole animal series; that is to say, the higher animal, in its gradual evolution [ontogeny], essentially passes through the permanent organic stages which lie below it; a circumstance which allows us to assume a close analogy between the differences which exist between the diverse stages of development, and between each of the animal classes.

However, as with the scale of beings, there was nothing inherently evolutionary about the law of parallelism. It too could be visualized as representing a divine plan of creation. This was the view of Louis Agassiz, who was to become a bitter opponent of Darwin. Agassiz, the discoverer of the Ice Age and the world's foremost authority on fossil fishes, extended the law of parallelism to the fossil record. By 1849 there were sufficient paleontological data available that Agassiz could demonstrate a three-fold parallelism such that a higher organism in its development passed through stages resembling not only the adults of a series of lower related forms, but also the progression of fossils of its class in the geological record. Agassiz, of course, unlike the transcendentalists, fully appreciated that Cuvier's classification had swept away a single scale of beings. Rather, there were, according to Cuvier (1812), four fundamentally different modes of body organization among animals: vertebrates, mollusks, articulates, and radiates. Recapitulation or parallelism could only exist within a class.

It was in the climate of the transcendentalist biology of the 1820s that von Baer carried on the studies of animal development that largely introduced embryology as a science. The magnitude of von Baer's achievements in embryology can be gauged by considering that he discovered the mammalian ovum and the notochord and that he formulated the theory of germ layers. His interpretations of his studies in comparative embryology led to a series of generalizations that made nonsense of the idea that animals in development recapitulate the scale of beings. Von Baer, like Cuvier, noted that instead of a single scale there were four basic plans of body organization. Development clearly reflects these basic plans. For example, the notochord and neural tube characteristic of vertebrates arise early in development, and thus, "The embryo of the vertebrate animal is from the very first a vertebrate animal, and at no time agrees with an invertebrate animal." Vertebrate embryos resemble only other vertebrate embryos, and von Baer denies resemblance to adults of anything else: "...the embryos of the Vertebrata pass in the course of their development through no (known) permanent forms of animals whatsoever."

Von Baer published the following major generalizations, his famous laws, in 1828:

1. *That the more general characters of a large group of animals appear earlier in their embryos than the more special characters.*
2. *From the most general forms the less general are developed, and so on, until finally the most special arises.*
3. *Every embryo of a given animal form, instead of passing through the other forms, becomes separated from them.*
4. *Fundamentally, therefore, the embryo of a higher form never resembles any other form, but only its embryo.*

These empirical laws retain their validity today and may be observed in operation in the development of any vertebrate, such as von Baer's favorite research organism, the chick. Early in development the chick embryo can be recognized only as a vertebrate, because early embryos of all classes of vertebrates are nearly identical; somewhat later it can be recognized as a bird; and only later still can it be recognized as a chick.

While von Baer's laws made the idea of recapitulation of the chain of being untenable, they were not, as pointed out by Ospovat as well as by Gould, really incompatible with a modified form of recapitulation, and in fact were to be eventually absorbed by Haeckel in his scheme of evolutionary recapitulation. The reason is not difficult to find. Von Baer's concept of development was progressive. Embryos pass from the general and simple to the specific and complex. Resemblances between embryos of higher forms and adults of lower forms exist, and are a necessary consequence, in von Baer's view, of two factors. Von Baer observed that the degrees of morphological complexity and differentiation, which characterize higher organisms as opposed to lower, coincided with the increasing histological and morphological complexity seen in the course of individual development. Thus, although von Baer realized that embryos of higher animals do not recapitulate the adult stages of lower forms, they do resemble them in complexity. To the modern reader this may sound contradictory to von Baer's fourth law, but von Baer himself explained that, "It is only because the least developed forms of animals are but little removed from the embryonic condition, that they retain a certain similarity to the embryos of higher forms of animals." The second factor is related. Von Baer held that primitive forms more closely resemble the hypothetical archetype, or idealized basic form, for a particular plan of body organization. Adult fishes are thus nearer the basic type than adult mammals: Both resemble the vertebrate archetype early in ontogeny, but mammalian development diverges farther from it than does that of the fish (Fig. 1–3).

While the concept of the archetype, which is a part of the transcendentalist approach to biology, would seem to be of little appeal to Darwin and his followers, it in part continued to exert a considerable influence on the interpretation of embryological facts. The importance of embryological data to late-nineteenth-century evolutionists was their phylogenetic content. The three-fold parallelism of Agassiz and the generalizations of von Baer were recast in evolutionary terms.

Darwin wrote in the 1859 edition of *On the Origin of Species* that "in the eyes of most naturalists, the structure of the embryo is even more important for classification than that of the adult. For the embryo is the animal in its less modified state; and in so far it reveals the structure of its progenitor." The archetype is here visible to Darwin as it was to von Baer. But of course Darwin made different use of the idea than did von Baer, who remained skeptical about evolution until his death in 1876. According to Darwin,

> In two groups of animals, however much they may at present differ from each other in structure and habits, if they pass through the same or similar embryonic stages, we may feel assured that they have both descended from the same or nearly similar parents, and are therefore in that degree closely related.
> Thus, community in embryonic structure reveals community of descent.

Darwin also suggested that an evolutionary rationale could also be made for the three-fold parallelism: "As the embryonic state of each species

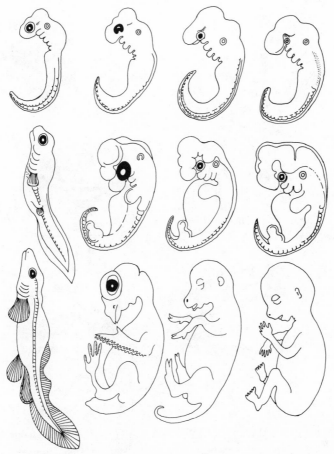

FIGURE 1–3. Embryos of fish, chick, calf, and human at various stages of development. The early stages (top row) resemble each other more closely than do the advanced stages in the bottom row. [Redrawn from Haeckel, 1879.]

and group of species partially shows us the structure of their less modified ancient progenitors, we can clearly see why ancient and extinct forms of life should resemble the embryos of their descendants—an existing species.''

The utility of such a principle to the working out of evolutionary relationships can be seen in the curious life history of barnacles. Barnacles are sessile, armor-plated filter feeders. Cuvier thought they were mollusks, but when their embryology was studied it became clear that barnacles are not mollusks at all, but crustaceans. Like shrimp, the earliest larval stage of barnacles is the nauplius. The nauplius larva, instead of developing into further larval stages leading to a shrimplike adult, becomes an ostracodlike cypris larva, which settles to a suitable substratum and attaches by means of cement glands at the base of the first antenna. The settled larva metamorphoses into a typical barnacle (Fig. 1–4).

The first attempt to provide a mechanism linking ontogeny with evolution was made by Fritz Müller in 1864 in a little book entitled *Für*

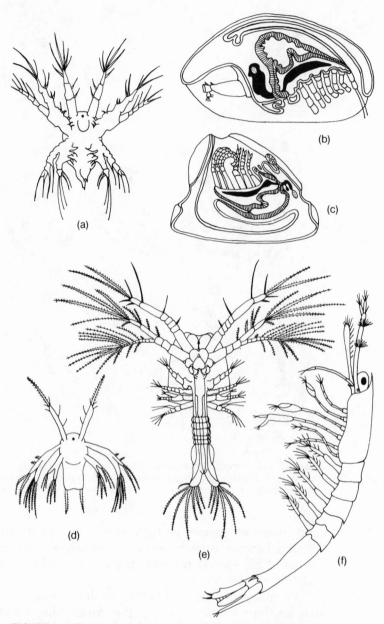

FIGURE 1–4. *Development of two crustaceans, barnacle and shrimp. (a) The nauplius of the barnacle* Balanus; *(b) section through the cipris larva of* Balanus; *and (c) section through the adult barnacle. (d) The nauplius of the shrimp* Penaeus; *(e) first protozoea larva; and (f) first postlarva. Barnacle and shrimp both possess similar nauplius larvae, but diverge in subsequent development.* [Redrawn from R. Bassindale, The developmental stages of three English barnacles, *Proc. Zool. Soc. Lond.* 1:57–74, 1936; Rees, 1970; and S. Dobkin, 1961.]

Darwin. Müller, from his studies of Crustacea from a Darwinian point of view, arrived at some momentous ideas. Müller proposed that "Descendants therefore reach a new goal, either by deviating sooner or later whilst still on the way towards the form of their parents, or by passing along this course without deviation, but then, instead of standing still, advance still further." Two modes of evolution are envisaged here. In the first, descendants will pass through only an early portion of ancestral development, and then diverge and exhibit a new pattern of later development. One might, for example, see barnacles diverging from other Crustacea in the manner. "In the second case the entire development of the progenitors is also passed through by the descendants, and therefore, so far as the production of a species depends upon this second mode of progress, the historical development of the species will be mirrored in its developmental history." Here the evolutionary mechanism is not the substitution of a new adult stage for the former one, but the addition of a new stage. The former adult stage is still retained but now as a step in development. The result, in terms of the developmental pattern of the descendant, is recapitulation.

Müller realized that the entire series of ancestral ontogenies in its full length and complexity could not be recapitulated. Stages must be compressed or lost. Thus "The historical record preserved in developmental history is gradually effaced as the development strikes into a constantly straighter course from the egg to the perfect animal, and it is frequently altered by the struggle for existence which the free-living larvae have to undergo."

Müller's ideas on recapitulation were seized upon and elaborated by Ernst Haeckel, who was to fuse embryology with evolution into an association that he believed would both provide infallible phylogenetic histories and explain the workings of development and evolution. Haeckel propounded his famous biogenetic law in his book *General Morphology of Organisms*, published in 1866, and reiterated it in subsequent books. The biogenetic law was Haeckel's term for his generalization that the ontogeny of an organism recapitulates its evolutionary history, or phylogeny. This concept was essentially an updated version of the transcendentalist law of Meckel-Serres, differing mainly from its more naive predecessor in that Haeckel recognized that evolution did not represent a single chain of beings, but many diverging lines of descent. Ironically, in this way the biogenetic law bears a superficial resemblance to the very generalizations with which von Baer thought he had laid recapitulation to rest.

Haeckel summarized his ideas in 1879 in his *Evolution of Man*:

> These two divisions of our science, Ontogeny, or the history of the germ, Phylogeny, or the history of the tribe, are most intimately connected, and the one cannot be understood without the other...Ontogeny is a recapitulation of Phylogeny; or, somewhat more explicitly; that the series of forms through which the individual organism passes during its progress from the egg cell to its fully developed state, is a brief, compressed reproduction of the long series of forms through which the animal ancestors of that organism...have passed from the earliest periods of so called organic creation down to the present time.

Although Haeckel called for an explanation of the connection between evolution and development based on physical and chemical laws, he never made clear what he meant by this. What he did say about the mechanical cause of development is vague but nonetheless rather startling:

> *The causal nature of the relation which connects the History of the Germ (Embryology, or Ontogeny) with that of the tribe (Phylogeny) is dependent on the phenomena of Heredity and Adaptation. When these are properly understood, and their fundamental importance in determining the forms of organisms recognized, we may go a step further, and say Phylogenesis is the mechanical cause of Ontogenesis.*

Evolutionary theorists in the late nineteenth century were bedeviled by their lack of understanding of heredity. Darwin, as well as others, fell back on Lamarck's theory that animals could somehow transmit useful characteristics acquired during their lifetimes to their descendants. This theory provided a mechanism for progressive evolution, and was also ideally suited to the biogenetic law. Development was recapitulationist because during evolution only the adult stages of the ancestors persisted long enough to acquire and pass on new characteristics. Embryonic stages were simply too fleeting in duration. As might be expected, Haeckel wholeheartedly embraced Lamarckian evolution. Haeckel held that there were three key factors in evolution: adaptation, heredity, and natural selection. He credited Lamarck as the real father of evolution because Lamarck discovered the role of the first two factors, adaptation and heredity. By adaptation, Haeckel meant the practice and habit that Lamarck believed led to the small but real improvements achieved by the individual. In his view, heredity was the transmission of these acquired traits, which led to the accumulation of improvements in generation after generation. Darwin of course contributed the third factor, natural selection.

Haeckel had no interest in embryology for its own sake: Embryology provided data for the working out of evolutionary histories, the construction of phylogenetic trees. Haeckel's influence was considerable, and he had no doubts about the validity of his approach or the former existence of ancestors reconstructed on the basis of the biogenetic law. Of course, this attitude of uncritical acceptance of recapitulation was bound to give rise to absurdities, and, for example, we find Darwin in the sixth edition (1872) of the *Origin of Species* suggesting that because various Crustacea all have a nauplius larva, the ancestral crustacean was naupliuslike. This interpretation goes beyond merely recognizing the crustacean nature of the barnacle from its larva. In fact the most ancient and primitive arthropods have numerous, relatively undifferentiated segments and in no way resemble the unsegmented nauplius (Fig. 1–5).

As had Müller, Haeckel recognized that phylogenetic history obtained from embryological data was not perfect. Stages might be missing from the sequence, but more seriously, there might be interpolations or new stages in development which represent embryonic adaptations or, in Haeckel's terms, caenogenetic processes. These, Haeckel claimed, have no evolutionary importance, but rather tend to falsify the record. He noted two other phenomena, one he called heterotopy, a change in place of origin of a structure, perhaps by means of a change in germ-

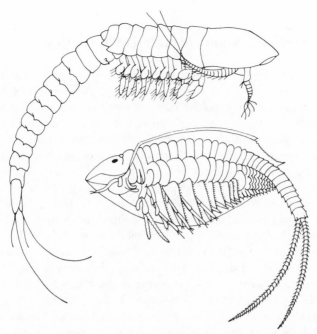

FIGURE 1–5. *Two primitive crustaceans, a cephalocaridan and a notostracan. These very primitive forms have a large number of segments, and bear little resemblance to the unsegmented nauplius.* [Redrawn from T. H. Waterman and F. A. Chace, Jr., General crustacean biology, in *The Physiology of Crustacea*, Vol. I: *Metabolism and Growth*, T. H. Waterman, ed., Academic Press, New York, 1960; and from W. T. Calman, Appendiculata, part 3: Crustacea, in *Treatise in Zoology*, Part VII, E. R. Lankester. ed. Adam and Charles Black, London, 1909.]

layer contribution to a particular organ or tissue. The second he called heterochrony, a change in timing or in sequence of appearance of organs from what would be expected from phylogenetic inference. Haeckel was unable to see that such phenomena provided potential mechanisms for significant evolutionary change. To him they were merely inconveniences to the tracing out of phylogenies using his biogenetic law.

Haeckel employed his biogenetic law not only to interpret juvenile stages of development, but the earliest events of development as well. The egg represented recapitulation of the original unicellular ancestor of all animals. The blastula represented a hypothetical ancient form with one layer of cells surrounding a hollow interior, the blastaea. The gastrula, resulting from invagination of the blastula to form a sac with two cell layers and an opening at one end, represented the gastraea with its simple mouth and two-layered structure. Haeckel visualized the coelenterates as living gastraea-stage animals.

Wide acceptance of the biogenetic law and Haeckel's interpretation meant that because even the earliest stages of embryological development were a direct consequence of phylogeny there was little point in looking for proximate causes in development. Instead, one should look for phylogenetic information. This point of view was to hamper the growth of embryology as an experimental science.

Developmental Mechanics and Mendelian Genetics

As the nineteenth century drew to a close, there was a growing tension between the two major philosophical approaches to biology—what Garland Allen calls the naturalist-experimentalist dichotomy. The naturalist tradition was concerned with the whole organism, its structure and its adaptation. Its method was observation. Following Darwin, workers in this tradition documented evolution and were deeply engaged in the unraveling of the evolutionary histories of living and extinct organisms. Morphological studies and observations of embryonic development were crucial parts of this program.

Experimentalists were less interested in the whole organism or its morphology, and focused on laboratory studies of isolated, analyzable aspects of function. There are two basic assumptions underlying the experimental approach to biology. The first is that the function of an isolated organ, cell, or enzyme in the laboratory can be extrapolated to the living organism. The second is that experimental perturbations of a system can yield information about its normal function. The experimentalists intended to make biology an exact science on the model of chemistry and physics. Physiology and biochemistry, exemplifying the experimental mode in biology, made profound advances in the late nineteenth century and would provide a model for embryology.

In the period dominated by Haeckel and his biogenetic law, embryology, naturalistic in tradition and a faithful trooper in the cause of phylogeny, was about to switch camps and become an experimental science with its own ends and methodology. The first real methodological challenge to the Haeckelian view came from Wilhelm His in 1874, who sought proximate, mechanistic causes for ontogeny in the physical characteristics of the protoplasm of the fertilized egg and in the environmental conditions of its development. In this he was strongly opposed and ridiculed by Haeckel and his followers, or ignored in the general enthusiasm to apply the biogenetic law. In 1888 His wrote in exasperation:

> This opposition to the application of the fundamental principles of science to embryological questions would hardly be intelligible if it had not a dogmatic background. No other explanation of living forms is allowed than heredity and any which is founded on another must be rejected—yet to think that heredity will build organic beings without mechanical means is a piece of unscientific mysticism.

Other embryologists were also beginning to perform experiments aimed at testing mechanistic hypotheses. In 1883 Pfluger studied the role of gravity in the determination of the plane of cleavage in the fertilized egg. His conclusion that the force of gravity did determine the plane of cleavage was incorrect, but that is of little concern here. What is important is that he used an experimental approach to isolate and study a particular mechanistic aspect of development. The pace of experimentation quickened when Chabry in 1887, using tunicates, and

Roux in 1888, using frogs, both published the results of experiments in which one of the blastomeres of a two-cell embryo was destroyed by puncture with a needle and the developmental capacity of the remaining blastomere observed.

Blastomeres were not simply martyrs to idle curiosity. The object of the puncture experiments was to test the proposal that the progressive and divergent specialization of cells in a developing embryo was caused by the unequal division of the chromosomes so that different cells in the embryo would come to differ because the hereditary particles they contained would differ. Roux thought he had demonstrated the validity of the hypothesis that development is strictly mosaic, but he was challenged by Driesch in 1892—also on the basis of experimental studies in which Driesch found that the separated blastomeres of sea urchin eggs could yield complete embryos.

By 1894 a generation of embryologists aware of the success of the experimental approach to physiology and biochemistry, and frustrated by the lack of exactness in phylogenetic speculation, was prepared to respond to Roux's call for a new science of Entwicklungsmechanik or developmental mechanics. In 1894 Roux published a very systematic prospectus of what this science was to accomplish in the introduction to the new journal *Archiv für Entwicklungsmechanik der Organismen* that he had founded for the reporting of studies in developmental mechanics. To Roux mechanics meant causation, thus "...the task of developmental mechanics would be the reduction of the formative processes of development to the natural laws which underlie them." He had in mind not only the basic chemistry and physics of the system, but also underlying biological mechanisms. He noted that "...all the extremely diverse structures of multicellular organisms may be traced back to the few *modi operandi* of cell-growth, of cell evanescence, cell division, cell migration, active cell-formation, cell-elimination, and the qualitative metamorphosis of cells." Roux's program called for studying the roles these processes play in specific developmental events, and for dissecting these cellular events themselves.

But the real revolution in embryology stemmed from Roux's insistence that although mechanisms may be inferred by observation, they can only be proved by experimentation. Individual components of a developing system would be interfered with by "isolating, transposing, destroying, weakening," and the effects on the normal process examined. Roux's Entwicklungsmechanik transformed embryology, and resulted in phylogenetics becoming increasingly irrelevant to the activities of embryologists pursuing the functional analysis of development. The mechanistic and reductionist approach promised real solutions to the problems of development in terms of molecular phenomena. By the 1890s many biologists were receptive to reductionism. It was, after all, in 1896 that Eduard Buchner published the experiments which showed that fermentation, thought to be a vital process inseparable from the living yeast cell, could be achieved outside the cell by isolated enzymes. Buchner's work was hardly obscure, and its importance at the time can be gauged by his being awarded the Nobel prize in chemistry in 1907. Enzymes, or ferments, provided a clear paradigm for visualizing life as a complex chemistry. Oppenheimer and Mitchel, for example, in their book *Ferments and Their Actions*, published in 1901, were able to present an extensive discussion of both the chemical nature and action of enzymes, and the various major classes of enzymes. They discussed incidentally the enzymes known to exist in embryos. Molecular mechanisms in embryology began to be visualized in the writings of Driesch (1894) and E. B. Wilson (1898 and 1904).

Heredity and recapitulation were replaced in the foreground of embryological thought by concern for the process by which the individual organism developed. The new attitude was nicely summed up by C. O. Whitman, one of the founders of American embryology and the first director of the Marine Biological Laboratory at Woods Hole, who wrote in 1895 that

> we have no longer any use for the "Ahnengallerie" [ancestor portrait gallery] of phylogeny....We are no better off for knowing that we have eyes because our ancestors had eyes. If our eyes resemble theirs it is not on account of genealogical connection, but because the molecular germinal basis is developed under similar conditions.

With the triumph of developmental mechanics the divorce of embryology from evolution was sudden and complete, and as we shall see, it contained the seeds of yet a second divorce—that of embryology from genetics. Curiously, the embryologists had not proved the biogenetic law incorrect, nor, in the period of Entwicklungsmechanik, did they really attempt to do so: Embryologists were excited by new problems unrelated to the biogenetic law. It was not until a generation later that Garstang and de Beer were to return to Haeckelian recapitulation and make it, as a universal mechanism of evolution, untenable on embryological grounds. Developmental mechanics had not denied the basis of the biogenetic law. In fact some aspects of recapitulation could be easily accounted for in a mechanistic fashion perfectly in accord with the new approach. The best example of this was Kleinenberg's 1886 proposal that those apparently functionless embryonic structures, such as the notochord or the tubular heart of vertebrates, which were taken as simple examples of recapitulation might be vital to the development of the embryo as participants in the formation of more advanced structures:

> From this point of view many rudimentary organs appear in a different light. Their obstinate reappearance throughout long phylogenetic series would be hard to understand were they really no more than reminiscenses of bygone and forgotten stages. Their significance in the process of individual development may in truth be far greater than is generally recognized....Through the stimulus or by the aid of these organs, now become rudimentary, the permanent parts of the embryo appear and are guided in their development; when these have attained a certain degree of independence, the intermediary organ, having played its part, may be placed upon the retired list.

And indeed Kleinenberg's idea is essentially correct: Such processes do exist; they have been studied experimentally; and they do account in large part for the apparently recapitulatory features of development.

The ultimately fatal weaknesses of the biogenetic law were its dependence on a Lamarckian theory of inheritance, and its demand that a new evolutionary stage can only be added on to the adult stage of the immediate ancestor. The rediscovery and development of Mendelian genetics at the turn of the century would reveal the biogenetic law for what it was—a mirage.

Mendel had performed his now familiar breeding experiments with the pea, *Pisum sativum*, and published his results in 1865. The scientific community, however, was not ready at that time for his theory of inheritance and his work was ignored. By the 1890s the widespread use of the microscope and its application in studies of cellular components and structure, most especially those of the nucleus and chromosomes by W. S. Sutton, Nettie Stevens, and E. B. Wilson, set the stage for a biological revolution. This revolution had as its first step the abovementioned rediscovery of Mendel by H. deVries, C. Correns, and E. von Tschermak in 1900.

All three had carried out breeding experiments similar to those of Mendel. Their results paralleled those he had reported 30 years previously. By utilizing different plant species they pointed out the essential correctness and general applicability of the Mendelian laws. Genes were discrete entities and behaved as if they were particulate. These entities were passed on from generation to generation in highly predictable and repeatable patterns, and most importantly, there was no blending of characters. Genes occurred in dominant and recessive forms and determined different and contrasting unit characters, or phenotypes. These attributes of genes did not seem to be affected by either environmental conditions or the association of different genes in hybrid individuals. A hidden recessive condition would reappear after several generations in the proper proportion of the progeny, exhibiting all of the characteristics it had prior to its hybridization.

The second step of the revolution was taken by W. S. Sutton and T. Boveri, who in 1903, in separate publications, pointed out the similarities in behavior between genes and chromosomes. This "chromosomal theory of inheritance" found a champion in T. H. Morgan who, after an initial resistance to the chromosome theory, became its most influential proponent, and established the American school of modern genetics. Morgan, whose training was in experimental embryology, brought the same mechanistic and experimental approach that characterized that field into the study of inheritance. The culmination of his studies came in 1915 with the publication of *The Mechanisms of Mendelian Heredity* by Morgan and his students. The general acceptance of a Mendelian view of heredity is, of course, inimical to Lamarckian thinking, and consequently, was in serious conflict with the biogenetic law.

A further event also served to lessen the credibility of recapitulation. In 1893 August Weismann published his *Theory of the Germ*. He had observed that early in the development of many embryos a group of cells is set aside that, in the adult, give rise to the reproductive tissues. These reproductive or germ cells are therefore separate from the body or soma, and it is these cells alone which pass on the determinants (genes) to the next generation. If the germ is to acquire characteristics to pass on to future generations in the Lamarckian manner it must therefore communicate with the soma. In 1909 W. E. Castle and J. C. Phillips designed an experiment to test this possibility. They had two strains of guinea pig; one black, one white. Both were true-breeding, and when crossed, produced the expected Mendelian ratios. These crosses also demonstrated black to be dominant to white. Castle and Phillips then transplanted ovaries from black females into white and ovaries from white females into black. At maturity they bred these recipient females to pure-breeding white males. What they found in the progeny was that the ovary type bred true. An ovary derived from a white female produced only white offspring, despite the fact it was held in the body of a black individual. Likewise, the ovary derived from the black donor produced only black progeny. The autonomy of the germ line coupled with the purity and constancy of the gene is, of course, contrary to the inheritance of acquired characters.

The final blow to the biogenetic law came when it was realized that morphology and morphological adaptations are not only important in the adult stages of an organism but during all phases of ontogeny. The work of de Beer, Garstang, and Huxley in the first half of the twentieth century was crucial in the development of this idea. If the form of the developing organism is just as important and perhaps more so than the adult, it becomes difficult to envision how this fits a Haeckelian scheme of evolution. Taken together, Mendelian genetics, the autonomy of the germ line, and the importance of morphological characters throughout development spelled the end of recapitulation *sensuo stricto.*

While experimental embryology had removed itself from evolutionary concerns, genetics on the other hand threw itself directly into the evolutionary fray. With the rise of Mendelian genetics a new explanation of Darwinian principles was seen. The experimental paradigm of the Morgan school was brought to bear on evolutionary questions and the school of population genetics, founded by R. A. Fischer, J. B. S. Haldane, and S. Wright, blossomed. These workers saw in the Mendelian laws and ratios a quantitative and mathematical approach to evolution. This school of thought deals with groups or populations of organisms in much the same manner as the Morgan school deals with individuals.

Developmental Genetics

There is little doubt that developmental genetics is presently one of the most active areas of biological thought and experimentation. However, during the first three decades of the twentieth century, when both genetics and developmental biology were enjoying the scientific limelight, there was little attempt to integrate the two. Embryologists concerned themselves with the mechanics of the ontogenetic process, while geneticists were involved with the elucidation of the laws of the transmission of traits. To a great extent, the two fields developed autonomously. Moreover, although the findings of the geneticists have played an important role in the development of neo–Darwinism, the same cannot be said for experimental embryology.

There were two reasons for this seemingly strange lack of synthesis. The first, discussed previously, was the repudiation of the biogenetic law by the experimental embryologists; the second was the break of embryology from genetics. Roux's founding of Entwicklungsmechanik was an attempt to define more precisely the mechanisms of development, that is, to find experimentally definable cause-and-effect relationships in ontogeny. This experimental, mechanical paradigm had a direct parallel in the founding and development of the American school of genetics by T. H. Morgan. Morgan's group incorporated many of the philosophical premises of the embryologists, especially the heavy emphasis on experimentation. However, the integration of genetics with embryology was delayed by the rejection of Mendelian genetics as an important component of ontogeny by embryologists. This rejection was most succinctly stated in 1928 by F. R. Lillie in an article entitled "The Gene and the Ontogenetic Process":

> The present postulate of genetics is that the genes are always the same in a given individual, in whatever place, at whatever time, within the life history of the individual, except for the occurrence of mutations or abnormal disjunctions, to which the same principles then apply. The essential problem of development is precisely that differentiation in relation to space and time within the life-history of the individual which genetics appears implicitly to ignore. The progress of genetics and physiology of development can only result in a sharper definition of the two fields, and any expectation of their reunion (in a Weismannian sense) is in my opinion doomed to disappointment. Those who desire to make genetics the basis of physiology of development will have to explain how an unchanging complex can direct the course of an ordered developmental stream.

The reasons for this categorical denial are three-fold. The early Mendelians viewed the gene as a *particle* transmitted to the offspring in the sperm and ova. It was these particulate genes or factors that caused the elaboration of the individual through ontogeny. This, in the minds of the experimental embryologists, smacked of preformation, a theory long since out of favor.

The second was the tacit assumption of the Mendelian school that in the somatic cell divisions the nuclear components, the chromosomes, and therefore, the genes, were faithfully replicated and identically partitioned to each and every cell. This seemed to fly in the face of the results of experimental embryology. It was well known that the ontogenetic process resulted in a sequential partitioning of the cytoplasm and a subsequent restriction of developmental potential. These two results were taken by embryologists to mean that the genes could not control ontogeny. It was believed that the cytoplasm, not the nucleus, was primary—a position represented in the quotation from Lillie.

Finally, we should note that a basic dichotomy existed between the Mendelians and the embryologists. Mendelian genetics had its primary interest in the study of the transmission of traits from generation to generation, while embryology was concerned with the elaboration of traits within a single generation. Both types of research progressed rapidly during the early twentieth century. The transmission genetics school of Morgan made giant strides, while simultaneously the American (Lillie, E. B. Wilson, Conklin, Harrison) and European (Spemann, Boveri, Hertwig) groups were doing equally well in experimental embryology. Each saw the value in the other's work, but unfortunately, the gap separating the two could not be bridged.

Although it is true that for the most part experimental embryologists had removed themselves from concerns with evolution and genetics, there were some who attempted a synthesis. The first of these was Driesch, who tried to reconcile the nucleus-versus-cytoplasm dichotomy. He formulated a model in 1894 which postulated that development was not all nuclear or cytoplasmic but was the result of an interplay between the two. His hypothesis sounds quite reasonable even today, nearly 90 years later; however, it seems to have been ignored by his contemporaries.

A second effort at synthesis was made several years later in 1932 by Morgan. His book *Embryology and Genetics* was intended to fulfill this role. Unfortunately, the book presents a series of chapters, some concerned with embryology, some with genetics, and there is very little correlation of the two.

Perhaps the most extensive attempt at a complete synthesis was by Richard Goldschmidt. Goldschmidt's early training was as an anatomist, and this naturalistic predilection followed him throughout his life and may account for some of the problems his ideas encountered later. He became interested not only in the transmission of traits but also in the physiology of genetics, that is, how inherited factors are translated into a phenotype, how genes work. A summary of these ideas appeared in 1938 in his book *Physiological Genetics*. The major contribution of this and his previous works is the concept that genes control the rates of developmental processes and thereby can dramatically influence dependent events during ontogeny. This postulation of "rate genes" is related to Huxley's idea of heterogonic growth in allometry. If a gene is involved in the rate of growth of some particular structure, it will control the size of that structure relative to the rest of the body. Moreover, rate genes can be envisioned to control the absolute time of appearance of any given structure. Ontogeny is the result of relational and interdependent processes; that is, the formation of a particular structure is dependent on the formation of other structures, both temporally and spatially. Thus, changes in the timing of a single developmental event can have profound effects by altering many subsequent dependent ontogenetic steps. Both Goldschmidt and Huxley recognized the importance of changes in timing of developmental events in evolution, especially as they related to neoteny, the presence of vestigial organs, and the production of large specialized structures.

Despite this conceptual advance, Goldschmidt was left with a problem. He found it difficult to envision how one could achieve major morphological change, most especially the evolution of novel structure through the selection of mutations in genes affecting minor components or small steps in ontogeny:

> Consider as an example a bird....The original species might have been a grain-eater, while the empty niche available is for a honeysucker. Adaptive radiation produces the type, which might be called a new genus. But how does such a complicated genetic change, leading, by accumulation of small mutational steps, to the perfect mechanism for honeysucking in the structure of bill and tongue, become available just when it has a chance for successful selection? If one tries to work out this idea in detail one soon comes to a point where it is evident that something besides the Neo-Darwinian tenets is needed to explain such macroevolutionary processes.

In order to overcome this problem Goldschmidt postulated two forms of evolutionary change, which he discussed in his book *The Material Basis of Evolution*. Microevolution was the changes in gene frequency observed and studied by the field of population genetics, and macroevolution was the production of large-scale morphological alterations, or what Goldschmidt liked to refer to as "hopeful monsters." Goldschmidt's insight into this basic problem in evolutionary theory was excellent, but his explanation of the two modes was not nearly as successful. As a matter of fact his explanation served to alienate the very groups he would have liked to convince. He contended that microevolution served only to increase fitness and variability within species. These kinds of small changes by gene mutation could not account for the morphological changes seen in the evolution of major groups of plants and animals. This conclusion made the work of the entire school of population genetics (e.g., Haldane, Fischer, Wright, and Dobzhansky) interesting but irrelevant.

Since Goldschmidt could find no explanation for major morphological change within the accepted doctrine of Mendelian genetics, he developed his own theory of inheritance. He took his cue from the newly discovered phenomenon of position effect, that is, the finding that in some cases the position of a gene in the chromosome can be shown to dramatically affect the expression of that gene. In order to account for far-reaching developmental changes in a highly complex interactive system, the developing embryo, he proposed an equally far-reaching global orientation within the nucleus. He hypothesized that macroevolution occurred by macromutation. It was the interactive "chromosome-as-a-whole" that changed, and an alteration in this whole changed the embryo as a whole. This hypothesis of course violated the pervasive belief in the particulate Mendelian gene. The experimental evidence supported the majority view, and Goldschmidt's thesis gained few adherents. Unfortunately, because of his inconoclastic explanation for the mechanism of macroevolution, Goldschmidt's belief that there was a difference between macro- and microevolution was not acceptable to the neo-Darwinians.

Why was it so difficult to produce a coherent modern synthesis? In order to demonstrate definitively that genes control ontogeny and, more importantly, *how* they control ontogeny, it was first necessary to understand how genes work and are controlled in their function. Such information was not truly available until the revolution of modern molecular biology was underway. Beadle and Tatum's one gene-one enzyme hypothesis, Watson and Crick's elucidation of the structure of DNA, Jacob and Monod's model of the operon, to name just a few, were absolutely necessary prerequisites to any real understanding of the genetic control of ontogeny. The conjoining of embryology and genetics is now not only possible but quite profitable. This can be seen most dramatically in the recent burgeoning of the schools founded in the 1930s and 1940s by C. H. Waddington in England, Curt Stern in the United States, and Ernst Hadorn in Germany. The proliferation of developmental genetics as an experimental science has parallelled the developmental mechanics of Roux, except that instead of perturbing the ontogenic process by physical means, mutations are used as the scalpel. That we have indeed come to the union of genetics and embryology is reflected in the final paragraph of Hadorn's book *Developmental Genetics and Lethal Factors* published in 1955:

> In the chromosomal substance of an organism, there are fixed places for thousands of mutable functional units or genes. Every change, or loss of a gene is a risk to the life of the developing organism. There can be no more impressive proof for the importance of the chromosomal factors than the knowledge that the loss of a single gene may lead to the breakdown of development while the fact that none of the many thousands of genes which are left can take over the task of the missing factor is evidence of the high degree of individuality in the structure and function of the gene. In addition, the process of development must make enormous demands on the harmonious cooperation of the numerous individual processes which are originated in the genetic substance of the chromosomes.

It is our belief that the time has come to take the final step in the modern synthesis: To fuse embryology with genetics *and* evolution.

Two

Paleobiology
and Evolutionary Theory:
Time and Change

...hearing far off great seas upon beaches that had long ago been washed away, and sea-birds crying whose race had perished from the earth.

J. R. R. Tolkien, The Fellowship of the Ring

Absolute and Relative Time

It should be clear at the outset that our attempt to explain morphological evolution in terms of developmental genetics will be hampered by a maddening peculiarity of the subject. Unlike physiological or developmental changes, which are obvious in the life of an individual and open to direct observation and experimentation, evolutionary changes in living organisms are elusive and limited in extent. As a consequence, most of our observations of morphological evolution have not been made on organisms at all, but on fossils which we are able to interpret as organisms by inferences drawn from biology. This is hardly meant to suggest that our only source of evolutionary data is the fossil record—merely that fossils yield a qualitatively different sort of evidence than do the biochemistry, embryology, and genetics that occupy most of this book. It is only by recourse to the fossil record that we gain a substantial view not only of the actual evolutionary histories of living organisms, but also of the evolution of long extinct lineages. Equally important, the geological record provides a measure of absolute time from which rates of evolution can be derived.

There are actually two geological time scales: relative and absolute. The relative scale was developed in the nineteenth century on the basis of the discovery by William Smith, an English engineer and surveyor, that certain characteristic fossils occurred in the same relative order in the rock strata wherever he found them. This insight was to provide geology with most of its basic methodologies. The two assumptions on which relative dating rests are simple. The first is that younger strata were deposited on top of older strata. The second is that individual

25

strata contain, as Smith observed, characteristic fossils. Among these are the remains of species that were short-lived in time and so are confined to a narrow range in the stratigraphic record. These provided index fossils, which allowed the correlation of rock strata over a wide geographical area, and the construction of a relative time scale.

The relative scale was in use well before Darwin published the *Origin of Species* in 1859, but its connection to absolute time was tenuous at best. It was only slowly realized in the eighteenth century that the chronology of the Bible, commonly accepted by scientists as well as laymen, was far too short to accommodate the vast changes that have occurred in the history of the earth, even if one accepted the catastrophist view that life has been annihilated again and again by devastating natural disasters. The flood of Noah was seen as only the last of these catastrophes that overwhelmed successive creations.

But, there was another way to regard the history of the earth and life. This had been provided by James Hutton in 1795 in his *Theory of the Earth*. Hutton realized that the present slow processes of erosion and uplift could, given sufficient time, completely alter the face of the earth. This theme was fully developed by Charles Lyell in his *Principles of Geology* first published in 1830. As the doctrine of uniformitarianism it came to dominate geology. No catastrophes or forces not observed on the earth today were needed: rain and frost and time could level mountains. An immensity of time was the key to the history of the earth. In the words of Hutton, "We find no vestige of a beginning, no prospect of an end."

Darwin's evolution by means of selection acting on small variations to produce gradual changes needed a prodigious length of time: Uniformitarian geology provided it. However, a rude shock was in the making. The physicist William Thompson, later Lord Kelvin, in papers written in 1862 and after, showed that uniformitarian geology violated the second law of thermodynamics. Unlimited time was impossible, because while the total energy of the universe remains constant, the amount of usable energy available is decreasing. Thus, the universe must run down. In this Thompson was absolutely correct. Only his estimate of the time available would prove to be far too short. The only mechanism known in the nineteenth century that could heat the sun was gravitational collapse. Thompson could demonstrate that this mechanism limited the sun, and thus the earth, to a life of less than 100 million years. This put a severe pinch on evolutionary time, since only the last fifth of earth history has any record of multicellular life.

The discrepancy between the span of time allowed by Lord Kelvin and the time felt to be necessary by evolutionists was only resolved after the discovery of radioactive decay, which revealed a new source of heat for the sun: a source that would allow the sun to burn for billions of years. The properties of the radioactive decay of certain heavy elements, such as uranium, also provided clocks that made possible the development of radioactive dating methods, and the establishment of an absolute time scale for the earth. Figure 2-1 shows the presently accepted correlation of relative and absolute time scales.

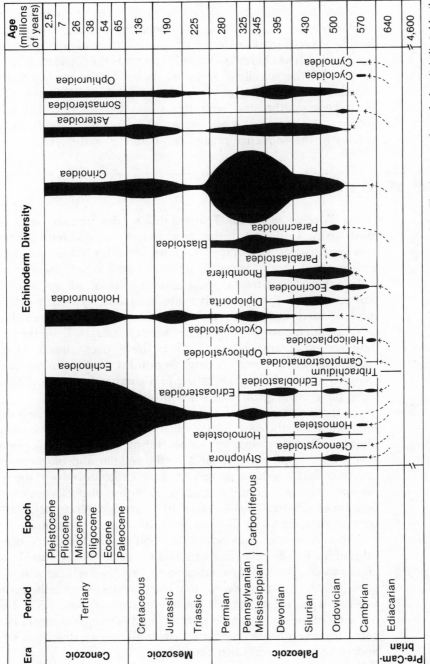

FIGURE 2–1. *Diversity of echinoderms through geological time. Each named group represents a class. The time range for each class is indicated by the length of the line representing it. Five major groups, Echinoidea (sea urchins), Holothuroidea (sea cucumbers), Crinoidea (sea lilies), Asteroidea (starfish), Somasteroidea (primitive starfish), and Ophiuroidea (brittle stars), survive to the present. Diversity in any period is indicated by the width of the group's line. Hypothetical relationships between classes are shown by dashed lines.* [Modified from C. R. C. Paul, Evolution of primitive echinoderms, in *Patterns of Evolution as Illustrated by the Fossil Record,* A. Hallan, ed., Elsevier Scientific Publishing Co., Amsterdam, 1977, p. 125.]

27

The theory and practice of radioactive dating lie outside the scope of this book, but one limitation is important: Most fossils cannot be directly dated. Dating methods can ony be applied to igneous rocks. Age ranges for fossils are generally obtained by finding situations in which fossil-bearing sedimentary rock layers are sandwiched between layers of datable igneous rocks. Under favorable circumstances quite accurate fossil ages can be obtained. A good example is the correlation by Gill and Cobban of the excellent series of index fossils found in the Late Cretaceous Pierre Shale of Wyoming with the absolute ages of volcanic ash layers interbedded with the shale. Evolutionary rates of long lineages of ammonites were estimated in this work, with the average species in the sequence persisting about 0.5×10^6 years.

Origins of Multicellular Life

In *The Phenomenon of Man*, Teilhard de Chardin remarked on one of the most vexing problems of the fossil record, the sudden appearance of new organisms: "Beginnings have an irritating but essential fragility, and one that should be taken to heart by all who occupy themselves with history." The fragile and elusive nature of origins is certainly in part caused by the destructive effects of time gradually effacing the record. However, it is becoming apparent that this is not the whole story. Sudden appearances are not simply artifacts. Evolution does not flow smoothly and serenely. There have been drastic changes in evolutionary rate, with many significant and even revolutionary changes in morphology occurring in a comparatively short time—and there have been qualitative shifts in evolution.

The fossil record of the first four-fifths of the history of life is distinctly different from the last fifth, which contains the record of teeming multicellular organisms. Evolution during much of the immensity of Precambrian time was primarily at the cellular and biochemical level. Unfortunately, we have no record of the earliest events of the origin of life and the most primitive organisms. Procaryotic cells appear to have become established by $3.4–3.0 \times 10^9$ years ago, since rocks of this age contain clear traces of life and the oldest fossil bacteria.

Procaryotes, cells that lack a membrane-bound nucleus, dominated most of the Precambrian Era. This was the age of bacteria and blue-green algae, metabolically active but monotonous. Nevertheless, it was from some progressive procaryotes that the first nucleated cells, eucaryotes, evolved. The oldest probable eucaryotes date to about 1.3×10^9 years, and consist mostly of simple spherical and filamentous algae. However, some striking macroscopic forms, ribbonlike algae, described by Walter and co-workers, were also present.

Defining the time of origin of eucaryotes from the fossil record is difficult because criteria by which the earliest eucaryotes might be distinguished from procaryotes are few, and in some cases open to argument. J. W. Schopf has listed a series of criteria based on size, shape, and morphological complexities of fossil cells. Structures interpretable as eucaryotes include branched filaments with internal cross walls, complex (e.g., flask-shaped) microfossils, large algal cysts, cells with internal dense bodies resembling the residues of eucaryotic organelles, and tetrads of cells or spores, possibly representing the products of meiosis. These tetrads have been interpreted by Schopf to mark the origin of eucaryotic sex by 0.9×10^9 years ago. Unfortunately, some of these presumably eucaryotic structures are dubious because, in experiments performed with cultures of living blue-green algae, Knoll and Barghoorn have shown that organellelike masses appear in degenerating blue-green algal cells. Brown and Bold, as well as Oehler and co-workers, have found blue-green algae that form nonmeiotic tetrads. A cetain degree of skepticism is not out of order in interpreting such difficult entities as fossil cells, but it probably is reasonable to accept that a diverse eucaryotic assemblage is present in the 0.9×10^9-year-old Bitter Springs Formation described by Schopf, and that the fundamental patterns of eucaryotic cellular organization were well established by that time.

One of the salient features of eucaryotic cell organization is the presence of membrane-bound organelles, the mitochondria and chloroplasts, which possess small DNA genomes and synthesize a limited number of their own proteins. These organellar genomes are vital to the assembly and function of the organelles and the survival of the cell, and evolved early in the history of eucaryotes. Raff and Mahler have proposed a mechanism by which these genomes may have evolved. The major consequences of evolution of organellar genomes by eucaryotes was that the presence of several genomes within a single cell required the evolution of mechanisms to govern and coordinate their functional interactions. Organellar genomes are controlled by the nuclear genome, and their activities are coordinated with nuclear gene activities. Nuclear genomes thus acquired the means to communicate with other associated genomes. This probably represented a crucial preadaptation to multicellular life that required coordination between the genomes of the cells of the organism.

Most of the genetic and molecular mechanisms required for development and differentiation of multicellular organisms were evolved by unicellular eucaryotes. That this preadaptation occurred is indicated by the multiple, independent attempts at multicellularity among eucaryotic groups. In his book *The Evolution of Development* J. T. Bonner listed at least 10 such attempts still represented by living organisms. The solutions vary. Cellular slime molds use aggregation of independent ameboid cells to produce a multicellular reproductive stage. *Volvox*, a green alga consisting of a few thousand cells, has a unique body plan, a complex pattern of development, and has achieved separate somatic and germ lines. Plants, fungi, sponges, and animals all independently evolved coordinated, highly differentiated multicellular forms.

The initial radiation of multicellular animals occurred during the late Precambrian Era. Remains of these first soft-bodied metazoa are preserved in rocks of about $0.7-0.6 \times 10^9$ years in age from Australia, Canada, England, and South Africa. It is for the best-preserved assemblage found at Ediacara, Australia that the fauna is named. The causes and timing of the metazoan radiation have proved to be fertile fields for speculation, because there are so few facts to restrain the imagination. Eucaryotes had been in existence for several hundred million years before the first known Metazoa appeared. It is possible that this time interval was necessary for the evolution of the necessary mechanisms for development of multicellularity. But it is equally likely that these mechanisms were in existence long before the evolution of Metazoa, and that the metazoan radiation was finally made possible by ecological changes near the end of the Precambrian Era.

One major possibility is that, as suggested by Berkner and Marshall, free oxygen levels only rose to a sufficient level to support metazoan life late in the Precambrian. Some of the biochemical consequences of this idea have been explored. Towe has suggested that until high oxygen levels were available, animals were able to produce little collagen, which requires molecular oxygen in its synthesis, and so remained soft-bodied and small in size. Raff and Raff showed that with low oxygen tensions primitive Metazoa dependent on diffusion for transport of oxygen to tissues must have been limited in thickness and complexity, and Cloud has noted that the Ediacara fauna is made up of animals consistent with this hypothesis. Some of the worms were quite large in area, but all were extremely thin and flimsy. Circulatory systems able to transport oxygen to the tissues became possible only when atmospheric oxygen levels became high enough for respiratory proteins with differing oxygen affinities to pass on oxygen efficiently via a chain, such as the hemoglobin-to-myoglobin-to-cytochrome system found in many animals. Once this occurred, more robust bodies and exoskeletons could replace the Ediacara wraiths.

Stanley, on the other hand, pointed out that the Precambrian was dominated by low-diversity, single-trophic-level ecosystems consisting mainly of blue-green algae. Diversity was limited because a few species of algae would best use the available space and resources to the exclusion of all other species. The evolution of the first herbivores meant that no longer was diversity controlled by competitive exclusion, and a greater diversity of producers became possible. This, in turn, increased the number of niches for herbivores, and made possible the evolution of carnivores, and thus several trophic levels. Stanley proposed that these ecological pioneers were protists but that the new variety of environments produced "explosive rates of evolution" that led to multicellularity.

The actual events are lost in the past, but one thing appears to be clear, and it is reflected in the models for metazoan origins. The radiation of multicellular animals occurred considerably after the origin of eucaryotes, but when it did it was rapid. The evolution of diverse and complex morphologies, developmental patterns, and all of the basic tissues took place within *at most* the 200 million years separating the unicellular eucaryotes of the 0.9×10^9-year-old Bitter Springs Formation from the 0.7×10^9-year-old metazoans of the Ediacara fauna.

The Ediacara fauna has been studied in detail by Glaessner: The fauna consists entirely of soft-bodied forms with coelenterates and annelids predominating, but with very primitive arthropods and a possible echinoderm also present. Altogether, seven classes belonging to four phyla are represented. Other coeval Metazoa include a little conical shell representing a possible mollusk, which has been described from the late Precambrian of California by Taylor, and two other animals of unknown relationship, which have been described from the late Precambrian of Russia by Zhuravleva. Whereas at least five phyla made their debuts by the end of the Precambrian, an extensive metazoan fossil record actually began in rocks of Lower Cambrian age when animals with readily preserved hard parts became common. Simpson has recorded 12 classes belonging to eight phyla in the Lower Cambrian, and Stanley has more recently listed over 18 classes in these same phyla. By the Middle Cambrian Period the number of phyla still extant today listed by Valentine was increased to 12. Conway Morris and Whittington have suggested that some other peculiar animals in the Middle Cambrian Burgess Shale of British Columbia belong to as many as 10 phyla no longer in existence, which may display unsuccessful body plans evolved during the great initial radiation of multicellular animals. Highly complex animals—echinoderms, trilobites, and other arthropods; articulate and inarticulate brachiopods; and several classes of mollusks, including cephalopods—all appear in the Cambrian Period in considerable diversity and without recognized ancestors.

Evidence of the rapidity with which profound evolutionary changes took place in the early history of the Metazoa can be had from examining the evolutionary history of the echinoderms, which possess a rich and well-studied fossil record, and serve well as a paradigm for the main issues of evolutionary theory raised by the fossil record. Living echinoderms comprise several quite distinct body plans, all exploiting pentaradial symmetry. These include the well-known starfish and sea urchins, and some less familiar creatures, such as sea cucumbers and crinoids. Altogether, there are five living classes of echinoderms. However, diversity among the ancient echinoderms was higher. Paul has recorded 15 classes known from Cambrian rocks, and 19 from the Ordovician. Figure 2–1 illustrates echinoderm diversity through time. One possible echinoderm, *Tribrachidium*, was present in the late Precambrian. Four classes occurred in the Lower Cambrian, and there was a rapid rise in diversity into the Ordovician. Classes then began to become extinct as they were excluded by the expansions of other, more efficient classes of echinoderms.

Three major evolutionary problems emerge from echinoderm history as we know it. The first is the lack of any identifiable ancestors for the phylum. Echinoderms appear in the record with all of the basic echinoderm patterns fully recognizable. Second, there are no transitional forms between classes. The relationships indicated by dashed lines in Figure 2–1 are largely hypothetical. The animals illustrated in Figure 2–2 possess basic structural similarities that lead to their

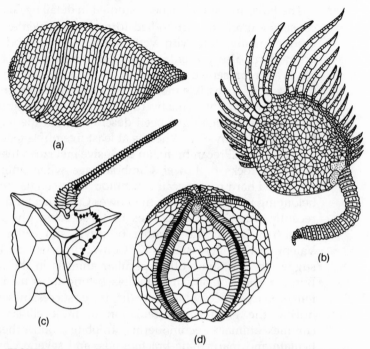

FIGURE 2–2. *Four Lower Paleozoic echinoderms:* (a) *helicoplacoid,* (b) *paracrinoid,* (c) *stylophoran, and* (d) *primitive echinoid. All are constructed of plates of calcite and have ambulacral systems. Symmetries and body plans are otherwise very dissimilar. There is a helically wound ambulacral tract present on the helicoplacoid, a pair of tracts bearing podia on the paracrinoid, a single tract on the "arm" of the stylophoran, and five tracts on the echinoid.* [Redrawn from J. W. Durham and K. E. Caster, Helicoplacoidea: A new class of echinoderms, *Science* **140**:820–822, 1963. Copyright 1963 by the American Association for the Advancement of Science; Parsley and Mintz, 1975; G. Ubaghs, Diversité et spécialisation des plus anciens échinoderms que l'on connaisse, *Biol. Rev. Cambridge Phil. Soc.* **46**:157–200, 1971, by permission of Cambridge University Press; MacBride and Spencer, 1938.]

inclusion in a common phylum, but beyond that even the earliest echinoderm classes are quite distinct morphologically from one another.

Early echinoderms included classes with body plans unlike those of any living forms. All are constructed of characteristic plates of calcite, and show evidence of possessing the unique water vascular system found in living echinoderms since ambulacra are present. The ambulacra were sites of surface appendages hydraulically linked to the water vascular system and functioning in collection of food, locomotion, and respiration, as do the podia occupying the ambulacra of living echinoderms. However, there family likeness ends. We are accustomed to having our echinoderms pentaradial, because all living and most fossil forms show this symmetry, but some of the ancients, such as the helicoplacoids, ctenocystoids, and paracrinoids, although not otherwise similar, show an asymmetrical disposition of parts superimposed on a primitively bilateral symmetry. Helicoplacoids are constructed of plates arranged in a spiral manner—presumably embedded in a tough integument producing a sort of flexible, helical mail armor. A single, forked ambulacrum winds around the body. Other very ancient echinoderms are also nonradial, and in some cases, quite asymmetric. Of these, the stylophorans are perhaps the most enigmatic. Stylophorans have a solid test of large plates arranged in a distinctly asymmetrical manner, and an apparently movable spinelike protuberance which Ubaghs called an aulacophore and has proposed as a feeding structure at the oral end of the animal. Jefferies, on the other hand, because of several peculiarities of internal and external structure, proposed that stylophorans are actually a subclass of chordates, which he called calcichordates, possessing a complex brain but having echinoderm affinities. By this interpretation the stylophoran aulacophore with an ambulacrum becomes the calcichordate tail with a notochord. The homosteleans and one or two other classes also have no obvious symmetry. The remainder of known classes are radial, with most conventionally pentaradial. An exception is the egregious *Tribrachidium*, which is triradial instead of pentaradial. This curiosity and its bearing on the evolution of echinoderm symmetry are discussed in detail in Chapter 5.

The third problem is the question of rates of evolution. The living classes of echinoderms have been very conservative in their evolution. Crinoids, holothurians, starfishes, and sea urchins recognizably quite similar to those living today had their origins in the Ordovician Period, which ended about 450 million years ago. According to the tabulations in *The Fossil Record*, most living orders of crinoids appeared in the Triassic and Jurassic Periods, as have most living families of regular sea urchins, and thus have persisted for nearly 200 million years. A few families of irregular urchins date to the Jurassic, but most are younger, appearing in the Cretaceous and Tertiary. The other classes of echinoderms include even more ancient surviving orders and families. Roughly a third of families of sea cucumbers (class Holothuroidea) originated in the Devonian and Mississippian Periods (about 350–400 million years ago), and the rest from the Jurassic. The sole remaining order of the Somasteroidea has survived since the Lower Ordovician, almost 500 million years. Some orders of starfishes (class Asteroidea) date to the Ordovician, with most of the rest from the Lower Jurassic. The brittle stars (class Ophiuroidea) include living suborders appearing in the Ordovician, Silurian, and Devonian Periods, and in the Jurassic.

And yet we have before us classes with radically different body plans appearing in the first radiation of echinoderms. Did the echinoderms have an immensely long history, of which we have no fossil record, extending back hundreds of millions of years into the Precambrian, so that echinoderms are as ancient as the first fossil unicellular eucaryotes? This seems an unlikely proposition. More reasonably, the first echinoderms, in the late Precambrian, evolved very rapidly to achieve a spectrum of more-or-less successful echinoderm body plans. Subsequent evolution has involved consolidation and diversification within body plan.

Gaps, Missing Links, and Evolutionary Mechanisms

The lack of ancestral or intermediate forms between fossil species is not a bizarre peculiarity of early metazoan history. Gaps are general and prevalent throughout the fossil record. Darwin was troubled by the lack of a continuum of evolutionary intermediates in the fossil record, since such transitional forms were to be expected from his theory: "...so must the number of intermediate varieties, which have formerly existed, be truly enormous. Why then is not every geological formation and every stratum full of such intermediate links?" In the *Origin of Species*, Darwin expended a good deal of effort in attempting to answer and rationalize this question. He proposed three main solutions. The first was the incompleteness of the fossil record. Gaps were caused by wholesale destruction of fossils by erosion and other processes, or the nonpreservation of once-existing intermediates.

The second was distortion of the record by taxonomic artifacts introduced by the investigators themselves. Intergrading forms might be given specific names, thus obscuring their transitional nature. Taxonomic artifacts are avoidable by modern statistical and stratigraphic methodology. However, one important example is worth mentioning here. Until recently, Cretaceous angiosperms were placed in modern genera on the basis of leaf shape, making the origin of modern genera of flowering plants appear to be more nearly instantaneous than newer studies of the morphology of fossil leaves and flowers, such as those discussed by Dilcher, indicate it actually was.

Darwin's third suggestion was that gaps may be a consequence of the nature of the evolutionary process itself. Evolutionary transitions, according to Darwin, might commonly occur in small, geographically limited populations followed by the rapid spread of the new form into the larger range of the ancestral species, or, "...the period during which each species underwent modifications, though long as measured by years, was probably short in comparison with that during which it remained without undergoing any change." Local or rapidly evolving populations would be rarely preserved in the fossil record.

The fossil record is highly biased. Certain environments, such as shallow marine basins, are more likely to leave fossil-bearing deposits than others, such as mountain ridges. Some organisms are more readily preserved than others. Mollusks, vertebrates, and echinoderms for example, have rich fossil records; insects have a rather poor record; whereas planarians and nematodes, which are extremely widespread in living faunas, have left essentially no record. But gaps occur even within those phyla with excellent histories.

G. G. Simpson, writing in 1959 for the Darwin Centenary, suggested that in spite of the intensive discovery and study of fossil forms since the publication of the *Origin of Species* the known fossil record remains a very small and imperfect sample of past life. Simpson presented the results of a very interesting "paper experiment" to test the consequences of an imperfect record. The experiment entailed the taking of a random sample containing 10% of the species of an hypothetical phylogenetic tree, consisting of a few families divided into a number of genera that in turn were subdivided into a large number of species. The species included in the sample correspond to an incomplete fossil record. The recovered species were classified, as would be actual fossils, into genera and families. As might be expected for a small random sample, most of the species in the sample were scattered, so that in only a few cases were a series of contiguous species representing a direct ancestor-descendant sequence recovered. The mean length of gaps (number of "unknown" species missing between recovered species) increased with taxonomic rank, such that there were gaps of many species between all families recovered. This situation resembles the actual record in which gaps between higher taxonomic levels are general and large.

Because of the close correspondence of the nature of the gaps found in his experiment and the gaps actually found in the fossil record, Simpson concluded that the fossil record represents a small, random sample of originally continuous phylogenetic sequences without gaps. From that postulate, Simpson predicted that:

1. Fossils falling into a major gap should very occasionally be discovered. *Archaeopteryx*, which nicely filled the gap between reptiles and birds, is the best known such case.
2. Sequences of genera should be more frequent than sequences of species. This is because sampling favors large, widespread species; however, unlike the paper experiment, in which all species had equal probabilities of being sampled, on the average small, localized species may be more frequent in real genera. But the larger population species, although more likely to leave a fossil record, are no more likely to give rise to a new genus than are the small population species. In such a case, successive genera will be known, but from species not in themselves directly successive.
3. Some major taxa should reappear in the record after long absences. Coelacanth fishes are the best-known examples of this phenomenon. The last fossil coelacanths are Cretaceous in age, yet living coelacanths still thrive off Madagascar. Other cryptic Methuselahs are *Neopilina*, which is the only living representative of the Monoplacophora, a class of mollusks last observed in the fossil record in Silurian rocks, and *Platasterias*, a member of the echinoderm subclass Somasteroidea last found as a fossil in the Devonian. That classes should turn up after absences of 300 or 400 million years suggests something about the degree of imperfection of the fossil record, and appears to validate Simpson's view that gaps can result from sampling and biases of preservation.

In 1972 Eldredge and Gould proposed that imperfections in the fossil record could not account for all of the troublesome gaps, because even species with long durations in the record typically show no gradual and sustained evolutionary trends, and remain essentially unchanged throughout their histories. Descendant species often appear suddenly and with marked discontinuity. They suggested that in general, "If new species arise very rapidly in small, peripherally isolated local populations, then the great expectation of insensibly graded fossil sequences is a chimera. A new species does not evolve in the area of its ancestors; it does not arise from the slow transformation of all its forbears. Many breaks in the fossil record are real." Thus, "Most evolutionary changes in morphology occur in a short period of time relative to the total duration of species." Whereas to Darwin this was only one of a number of possibilities, to Eldredge and Gould it is the major reason for gaps. The whole species population does not gradually evolve. Rather a small, peripheral population, isolated from the main population, changes rapidly. This evolutionary change is accompanied by speciation. At some subsequent time the new species may expand and replace the main population of the ancestral species over the entire species range. To someone examining the fossil record this sequence of events would appear as a sudden break—the descendant species arising without any sign of evolutionary transition from its ancestor. This does not mean that transitional forms did not exist, that evolution is by saltation, only that transitional populations were small in number, short in duration, and geographically limited to a small area at the edge of the main ancestral population. The chance of their preservation as fossils was therefore low.

By this proposal Eldredge and Gould were able to tie the gaps that are so prevalent in the fossil record to modern concepts which recognize that allopatric speciation involving small, local populations is a common mode of speciation, and that these events can be extremely rapid. Rates and modes of speciation are considered in Chapter 3, after we have more thoroughly dealt with the spectrum of kinds of evolutionary rates and their mechanistic interrelationships.

While the Eldredge and Gould model is intellectually satisfying and consistent with the existence of gaps, it is by no means proven by the presence of gaps or of stable species between the gaps in the fossil record. Punctuation has a certain "whodunit" quality about it. All of the interesting events happen out of sight, off stage in elusive peripheral populations. Actual examples representing episodes of punctuational evolution are needed. This is particularly true because even if this model is correct and is the prevalent or even a common mode of evolution the fossil record is nonetheless imperfect, and the consequences of that imperfection pointed out by Simpson remain in force. A probable example of punctuational evolution was provided by Ovcharenko (1969) who studied the evolution of two common and widespread species of Jurassic brachiopods: *Kutchithyris acutiplicata* and its descendant *K. euryptycha*. Stratigraphically, *K. euryptycha* occurs above *K. acutiplicata*. In one limited locality, Ovcharenko found a rock unit about 1.0–1.5 m thick, which contained only *K. acutiplicata* in the lower part of the unit and only *K. euryptycha* in the upper part. However, in between was a thin (10 cm thick) band in which both species occurred along with intermediate forms. In their 1972 paper, Eldredge and Gould provided two examples from their own work in support of their hypothesis. These examples are of interest not only because they represent actual fossil lineages interpreted in a punctuational manner, but also because it is possible to comment on the possible developmental-genetic bases of the observed evolutionary changes.

The first, studied by Gould, is the evolution of *Poecilozonites bermudensis zonatus*, a terrestrial snail living during the last 300,000 years of the Pleistocene of Bermuda. The fossil subspecies are extremely well preserved, and there is a living subspecies available for comparison. Both persistent eastern and western populations of *P.b. zonatus* gave rise to paedomorphic offshoots. Paedomorphosis refers to the retention of traits characteristic of the juvenile ancestral form by the sexually mature descendant form. The mature shells of the paedomorphic subspecies resemble the immature shells of the parent subspecies in color pattern, general shape of the spire, thickness of shell, and shape of the apertural lip. Gould proposed, on the basis of detailed study of geographic, stratigraphic, and morphological characteristics of the paedomorphic subspecies, that they do not represent a persistent paedomorphic lineage; rather, he suggested that the basic populations of *P.b. zonatus* gave rise to several successive paedomorphic offshoots in response to recurrent changes in environmental conditions, resulting in lime-poor soils that favored thin shells because calcium was less available. The offshoots arose rapidly, with paedomorphosis representing the most readily available path to thin shells.

If Gould's interpretation of the record is correct, it is interesting to note that the genetic changes required to modify the development of an organism to result in paedomorphosis appear to be minimal. In the best-understood cases of paedomorphosis, in salamanders of the genus *Ambystoma*, the genetic basis for the developmental decision between metamorphosis and paedomorphosis resides in a pair of alleles of a single gene (see Chapter 6). Assuming that an analogous control may apply to the development of snails, the initial genetic change toward the paedomorphic condition might have been extremely rapid in a small population.

The second example presented by Eldredge and Gould was Eldredge's study of the evolution of subspecies of the Middle Devonian trilobite *Phacops rana* (Fig. 2–3). The conspicuous eyes of *Phacops* consist of a discrete number of large lenses arranged in dorsoventral files. The major evolutionary changes observed in the *P. rana* subspecies are reduction in the number of dorsoventral files of lenses from 18 in the oldest subspecies to 17 and then to 15 in younger subspecies. The individual subspecies were generally stable in number of files of lenses, although populations with variable numbers have been found. Changes in file numbers between subspecies appear as sudden discontinuities, with a new stable file number replacing the former. Eldredge interpreted these events as having occurred allopatrically in peripheral populations. The subsequent spread of the new subspecies with files of lenses stabilized at a new number was instantaneous in terms of resolution in the fossil record.

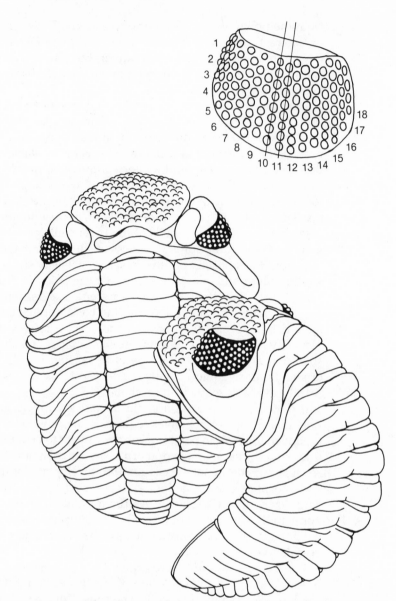

FIGURE 2–3. *The Devonian trilobite* Phacops rana *and the organization of the files of lenses of the eye. In this subspecies,* P. rana crassituberculata, *there are 18 dorsoventral files of lenses.* [Drawn from photographs published by R. Levi-Setti, *Trilobites,* 1975, by permission of the University of Chicago Press. Copyright 1975 by the University of Chicago.]

This interpretation poses no genetic problems because although one might suspect that gene control of the development of as complex a character as number of rows of lenses in the eye might also be complex, this is not the case. Direct investigation of the genetics of trilobites is unfortunately out of the question, but examination of the control of meristic traits, such as bristle number in *Drosophila* or the number of toes in mammals, indicates that control of such traits involves only a small number of genes (loci) (see Chapter 5). Substitution of a few alleles in guinea pigs, for example, causes discontinuous changes in the number of toes. If the number of lens files in *Phacops* was under similar control, substitution of alleles and thus changes in the number of lens files could have resulted from founder effect in a small population. The genetic change required was minor and could have occurred in a geological instant. The modest extent of genetic alteration probably required to produce the morphological transformations noted in *Poecilozonites* and *Phacops* should not be taken to mean that such events are trivial. On the contrary, "easy" gene changes may give rapid initial impetus to further genetic changes in an evolving population.

In 1977, Gould and Eldredge reconsidered punctuational evolution in light of paleontological studies done since the publication of their model in 1972. They were able to point to several cases in which a punctuational explanation seems most likely in radiolaria, ammonites, trilobites, and even hominids. Only one study, that of Ozawa on the evolution of a Permian foraminifer, emerged in their view as an unquestionably good case of gradual evolution of a whole population. Bookstein, Gingerich, and Kluge have hotly disputed some of Gould and Eldredge's examples, and in their own reanalysis of Gingerich's data on the evolution of some Eocene mammals concluded that in the data examined there was evidence for twelve instances of gradualistic change, four of punctuation, and one of stasis. Similarly, Cronin et al. have recently analyzed the available data for hominid evolution, and have concluded that human evolution is more reasonably interpreted as gradualistic, with some periods in which evolution was accelerated and some during which the rate slowed. A punctuational interpretation remains possible for the basic divergence of hominids from the apes. Molecular distances between humans and chimpanzees support a rapid divergence, but there is almost no fossil evidence currently available.

Assurance that a lineage of organisms has evolved in a punctuational mode requires a long-term, continuous fossil record in which both stasis and rapid evolution can be recognized, and it requires adequate absolute dating of short intervals within the sequence. These conditions were met in a recent detailed study by P. G. Williamson of the evolutionary histories of a number of late Cenozoic freshwater mollusks from the east African Turkana Basin. He was able to examine thousands of fossils representing 13 lineages of snails and clams from a thick sedimentary sequence interbedded with accurately dated volcanic tuffs. Many of the species remained unchanged over several million years, and no lineage exhibits gradualistic changes in morphology. Instead, new species arose relatively rapidly in times of environmental stress resulting from regressions in the lake they inhabited. At such times of stress the mollusks were probably isolated from their conspecifics in other lakes. Williamson's stratigraphy was fine enough, and his collection of fossils sufficiently large that he was able to document the transitional events in his lineages. He found intermediate populations to be more variable in morphology than established species. He proposed that this variance resulted from a partial breakdown in developmental homeostasis, which leads to greater phenotypic variability and rapid morphological evolution.

There is certainly no question but that the Turkana mollusks exhibit the hallmarks of punctuated evolution—long stasis with occasional episodes of relatively rapid evolution. Just how rapid is relatively rapid evolution? According to Williamson, the evolutionary episodes took place within a span of 5,000–50,000 years. Indeed, this is a short interval for the resolution generally available in the fossil record. However, as J. S. Jones has suggested in his discussion of Williamson's observations from a geneticist's point of view, 5,000–50,000 years is not a short time for the organisms concerned. Their living relatives have generation times on the order of six months to one year, indicating an average of 20,000 generations to complete the observed morphological transitions. As Jones has observed, this is equivalent to a 1,000-year experiment with *Drosophila*, or 6,000 years with a mouse selection experiment, or 40,000 years in the breeding of dogs or other domestic animals. Conventional selection experiments have produced striking morphological changes and even reproductive isolation in as little as 20–50 generations. Williamson's mollusks have not evolved particularly rapidly, but they do elegantly illustrate punctuated equilibrium.

The paleontological broadsides accompanying this issue are entertaining, but the hypothesis of punctuation appears correct, at least in some instances. As shown by Harper, punctuational evolution and phyletic gradualism are end-members of a spectrum of possibilities: Both appear to have occurred. The real question is not, of course, whether punctuation and gradualism are mutually exclusive, but rather, has one resulted in quantitatively more significant or qualitatively different morphological evolution, and what do the possible modes of evolution as observed by paleobiologists signify in terms of the developmental-genetic processes underlying morphological evolution?

The ammonites, extinct cephalopods with chambered shells resembling that of the living *Nautilus*, provide examples supporting both modes. The sequence of ammonites shown in Figure 2–4 evolved over an interval of about 3×10^6 years and show several trends: a gradual increase in size, then gradual decrease; gradual enrollment from loose to tight apposition of living chamber; an increase in complexity of the suture pattern followed by a decrease in complexity and some qualitative changes in shape of the sutures; and finally, a gradual increase, then a decrease in the coarseness of ribbing. These trends resulted in only minor changes in any direction, and some were even reversed during the time encompassed by this lineage.

A very different set of evolutionary events is pictured in Figure 2–5, which presents the proposed evolutionary origin of heteromorph ammonites from normally coiled progenitors during a short time interval of the Late Triassic. The length of the entire Norian Stage is approximately $5–10 \times 10^6$ years. The very profound changes recorded

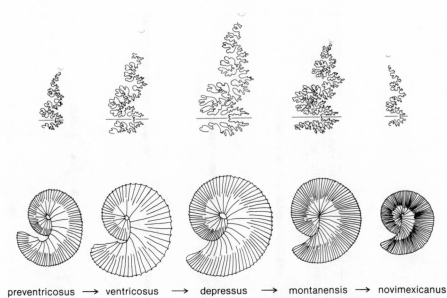

preventricosus → ventricosus → depressus → montanensis → novimexicanus

FIGURE 2–4. An evolutionary sequence of ammonities showing changes in size, shape, ribbing, and suture pattern over a period of approximately 3 million years. The suture patterns, which define the attachment of the walls of the air chambers to the inside shell, are drawn beside the corresponding species. The lineage of Upper Cretaceous ammonites extends from the oldest species Scaphites preventricosus, *through* S. ventricosus, S. depressus, *to* Clioscaphites montanensis, *to* C. novimexicanus. [Redrawn from W. A. Cobban, 1951.]

in Figure 2–5 occupied only a limited interval and were far more rapid than the minor, gradual changes seen in the lineage shown in Figure 2–4. Other groups of Jurassic and Cretaceous ammonites also gave rise to heteromorphs. Presumably, an adaptive zone very different from those occupied by normally coiled ammonites was available for exploitation by heteromorphs. Evolution of heteromorphs, with their radically different modes of shell coiling, required considerable modification of developmental patterns, yet appears to have been achieved with dispatch.

Evolutionary rates can be studied without reference to particular modes (phyletic transformation of an entire population versus speciation involving small peripheral populations). Such an approach can be very informative, as may be seen by considering the treatment given the question of variable rates of evolution by Simpson in *The Major Features of Evolution* written in 1953. Simpson proposed that evolutionary rates can be assigned to three broad categories: horotely, bradytely, and tachytely. Horotely includes that groups of rates more or less average for evolution in a particular group of organisms. While horotely might be equivalent to Eldredge and Gould's gradualism, it is not necessarily so, because a horotelic rate could result from a long-term averaging of punctuational events.

Bradytely refers to rates of morphological evolution approaching zero. The concept of a "living fossil" is a familiar one, denoting an organism whose lineage has persisted with little obvious morphological change for millions of years, while related forms have undergone significant

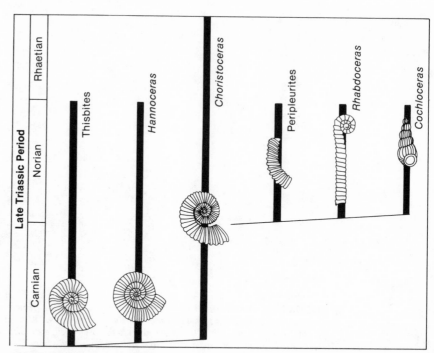

FIGURE 2–5. *Apparently rapid evolution of heteromorph ammonities from normally coiled ancestors during the Late Triassic. The lengths of the bars indicate the time ranges for each genus. Ammonities, which were swimmers, are drawn in life orientation. Typical ammonities were planispiral. Evolution of heteromorphs required changes in growth patterns to produce uncoiled shells, such as that of* Rhabdoceras, *or shells with helical whorls, such as that of* Cochloceras. [Redrawn from J. Weidmann. The heteromorphs and ammonoid extinction, *Biol. Rev. Cambridge Phil. Soc.* **44**:563–602, 1969, by permission of Cambridge University Press.]

evolution. Living primates, for example, include among their number some very primitive prosimians that are not terribly different from their (and our) Eocene ancestors. There are also a variety of specialized prosimians; a diverse collection of monkeys, some more advanced than others; the apes; and most diverged from the basic prosimian pattern, humans. Clearly, primitive living prosimians have evolved far less rapidly in morphology, and presumably in behavior, than has the lineage of hominids that boasts our species as its current ornament.

Evolutionary rates can change drastically within a single lineage. An excellent example is provided by the work of Westoll on the rate of evolution of lungfishes. Westoll's data, replotted from Simpson, are presented in Figure 2–6. Fossil lungfishes were scored for the degree to which they approach living lungfishes in "modernization" of morphological characteristics. A score of 100 was assigned to modern characters and zero to the most primitive. The plot of score versus age indicates the acquisition of modern characters. A relative rate of evolution was estimated by plotting change in score per 10^6 years as a function of age. Lungfish evolution was relatively rapid (for lungfishes) during the first 50×10^6 years of the group's history, and has been

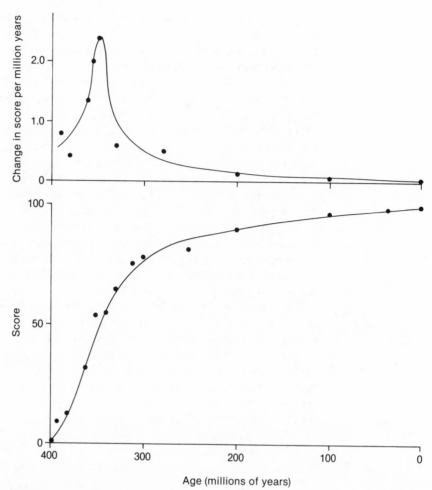

FIGURE 2–6. Evolution of morphological characteristics of lungfish. The lower panel illustrates the acquisition of "modern" characters (score of 100). The upper panel presents the rate of lungfish evolution. [Modified from Simpson, 1953.]

extremely slow since: so slow, in comparison with other fishes, that we regard lungfishes as living fossils.

Bradytelic rates do not imply that no mutation or selection is occurring, only that selection favors little or no net morphological change. Analysis of the degree of protein polymorphism and genetic heterozygosity has been made for one living fossil, *Limulus polyphemus*, the horseshoe crab, by Selander and co-workers in 1970. The subclass Xiphosura to which *Limulus* belongs has a fossil record beginning in the Middle Cambrian. Although fossil horseshoe crabs have exhibited variations in habitat, including marine, brackish, and freshwater forms, suggesting evolutionary flexibility in physiology, the Xiphosura have been conservative in their morphological evolution. Upper Paleozoic, Mesozoic, and living forms are alike in their general features. *Limulus* itself is not well known from the fossil record, which is vanishingly poor in Tertiary Xiphosura. However, living horseshoe crabs are quite similar to the Jurassic *Mesolimulus*, which like *Limulus* was a marine form, indicating a relatively slow rate of evolution in this lineage. Protein polymorphism was estimated from electrophoretic variation in 24 different proteins encoded by 25 genetic loci. The proportion of polymorphic loci per population of *Limulus* was found to be similar to the proportion in mice, *Drosophila*, and humans. Individuals of *Limulus* also had as high a proportion of heterozygous loci as the species with which they were compared. Thus *Limulus*, a bradytelic organism, has as high a degree of genetic variability as members of more rapidly evolving groups. This variability extends to other bradytelic organisms. Opossums, which have shown little evolution since the Cretaceous Period, have populations as variable in morphology as other mammals. Furthermore, bradytelic organisms possess a significant store of variability to speciate, some profusely. *Selaginella*, for instance, which has over 200 living species, is a genus of a plant that, according to Phillips and Leisman, is little changed in basic structure from the Pennsylvanian lycopods *Paurodendron* and *Selaginellites*.

Simpson suggested that bradytely is maintained in organisms occupying an adaptive zone persisting for an exceptionally long time. Such a zone might be narrow and held by a tenacious occupant, such as the lungfish; or it might be broad and occupied by a consumate generalist, such as the opossum. Significant change appears to be checked by normalizing selection—maintaining a well-adapted form. On occasion, bradytelic organisms have given rise to rapidly evolving forms. This implies that when subjected to strong selection away from the norm bradytelic organisms have the genetic flexibility to respond.

The fastest evolutionary rates fall in the last, and perhaps most interesting category, tachytely. Simpson stated that, "It is my opinion that tachytely is a usual element in the origin of higher categories and that it helps to explain systematic deficiencies of the paleontological record." Tachytely resembles the punctuation of Eldredge and Gould in that both rely on exceptionally high rates of evolution. However, while Eldredge and Gould focused on a speciation model deriving from population genetics, Simpson regarded tachytely from a different, complementary point of view. He suggested that the primary concomitant of tachytely is a shift of a population from one major adaptive zone to another.

The crossing of a threshold from one zone to another implies that the shifting population will find itself in a metastable zone during transition. Survival in such circumstances is probably only possible in the absence of competition, so that a temporarily poorly adapted form can hold on long enough to begin to achieve a satisfactory degree of adaptation to the new zone. Thus tachytely is possible during early radiations of new groups expanding into vacant adaptive zones. During the rapid radiation all lineages are relatively poorly adapted and not mutually competitive. The result, as we have seen with the echinoderms, is the production of diverse lines that quickly become extinct as other lines consolidate their positions in the adaptive zone at the expense of their less-efficient cousins. The difficulty in achieving the rapid changes in developmental processes required in tachytelic evolution comes from the fact that no matter what the ecological factors that encourage or allow a tachytelic population to make the transition between zones, no matter how loosely adapted a transitional organism is permitted to be to its external environment, developmental and functional integration must continue to be maintained if the organism is to be able to exist and reproduce at all. Or, as Frazzetta aptly put it in his *Complex Adaptations in Evolving Populations*, "The evolutionary problem is, in a real sense, the gradual improvement of a machine while it is running."

Rates of Evolution

Up to this point we have dealt with rates in a general way, and have not attempted to define just how much change in morphology per million years corresponds to horotelic or tachytelic rates. However, concrete evolutionary data do exist in the fossil record that allow, in combination with sufficiently accurate absolute time measurements, some quantitative assessments of rates of evolution. Three broad categories of evolutionary rates can be gleaned from the fossil record: rates of taxonomic change, rates of size change, and rates of shape change. These are, of course, not independent of one another, but operationally they can be dealt with separately.

Rates of taxonomic change are the most subjective of these because individual taxonomists working on related organisms may differ in the criteria they use in establishing categories, such as genera and families. A "lumper" might assign ten related species to one genus while his "splitter" colleague who sees them as belonging to three genera looks on disapprovingly. More seriously, taxonomists working in very different groups must necessarily use very different morphological characteristics, terms, and criteria in their work. It is difficult to decide if taxonomic categories, such as genera or families, have the same evolutionary significance in comparisons of organisms belonging to different classes or phyla with unique body plans and evolutionary histories. Nevertheless, bearing this difficulty in mind, it is possible to use the vast amount of taxonomic data obtained from the fossil record to estimate evolutionary rates, particularly among related organisms, or to estimate changes in evolutionary rate within a single lineage. Comparisons between unrelated groups are more tenuous, but taxonomy is not completely arbitrary. The same general principles are applied by workers in all groups in attempting to arrive at hierarchical classifications that express the evolutionary relationships between and within groups of organisms.

Thus, taxonomic categories express the summation of the taxonomist's estimation of the degree of evolutionary separation, and reflect the extent of the morphological differences between the organisms being classified. If absolute dates are available, as they are in some instances, the time required to achieve significant degrees of morphological change can be estimated. These estimates may allow us to determine the degree of time resolution necessary to study punctuational events, and to assess how significant horotelic rates might in fact be.

In 1953 Simpson pointed out that evolutionary rates can be derived from taxonomic data in several ways. Phylogenetic rates are rates at which taxonomic units, such as species or genera, evolve in a particular lineage. Ideally, rates can be determined for an evolving lineage in which species or genera originate at known times from known ancestors and disappear not through extinction, but by giving rise to known descendants. Although reliably dated ancestor-descendant sequences are not common, there are some cases in which this direct approach has been possible. The evolving sequence of scaphitid ammonites shown in Figure 2-4 provides a good example for the determination of direct phylogenetic rates. The species in this lineage had durations of 0.5–1.0 $\times$ 10^6 years and terminated in the origin of sequential species. Similar evolutionary series of species of Upper Cretaceous ammonites of the genus *Baculites* studied by Gill and Cobban have yielded average species durations of 0.5×10^6 years. Rates are the reciprocals of durations—for these ammonites, one to two species per 10^6 years. These rates for cephalopods are similar to the rates of phylogenetic change in a very different group of animals, mammals. The evolution of primitive Lower Eocene mammals analyzed by Gingerich and by Bookstein, Gingerich, and Kluge spanned about 4 million years. One simple lineage consists of four species of the primate *Pelycodus*, each of which persisted for about 1 $\times$ 10^6 years. Evolution of the condylarth *Hyopsodus* in this interval was more complex, with several species evolving in a branching phylogeny, apparently including both gradualistic and punctuational events. Species of *Hyopsodus* persisted for $0.3–0.7 \times 10^6$ years, yielding rates of 1.5–3 species per 10^6 years. Equivalent calculations can also be made for rates of generic change; for example, Simpson estimated that for a lineage of eight successive genera of horses encompassing a span of about 60×10^6 years the average rate was 0.13 genera per 10^6 years.

While the ammonites and mammals are considered to have evolved at similar average rates, other lineages have exhibited different, far slower average rates of change. Long species durations have emerged for Miocene to present-day scallops of the genus *Argopecten* studied by Waller. The span of this genus is about 19×10^6 years with speciation by branching, as well as by possible gradual transformation of species in linear sequence continuing to the present. The mean duration of extinct species of *Argopecten* was approximately 5×10^6 years, or about 0.2 species per 10^6 years. On the face of it, clams evolve, as they live, at a more leisurely pace than mammals.

The marked difference in rate of evolution of mammals and clams has been confirmed using another approach to determine rates of taxonomic evolution, the survivorship curve. In this approach, the average longevity of all species or other categories in a group is estimated. A knowledge of direct evolutionary lineages is not required. Survivorship curves are obtained by plotting the percentage of genera that first appeared in the fossil record at a given time in the past and survive today. Not surprisingly, fewer genera with origins in the remote past survive than genera with recent origins. Survivorship curves for extinct genera are generated by plotting the percentage of genera surviving a given length of time, using the durations from the first appearance to the last appearance in the fossil record for each genus. Simpson first applied this method in 1953 and used it to compare the durations of clam and mammalian genera, as shown in Figure 2–7. From these curves, Simpson estimated that the duration of an average clam genus was 10 times that of the average mammalian genus, and suggested that mammalian genera may have evolved as much as 10 times faster than clam genera.

A similar result has been reported at the species level. Stanley (1976, 1977) plotted survivorship curves for Pleistocene mammal and clam species, and arrived at a mean duration of about 1.2×10^6 years for the average mammalian species, whereas his estimate for clam species was 7×10^6 years. Thus, evolution of both mammalian genera and species appears, as we have previously noted from phylogenetic rates, to be 5–10 times as fast as for evolution of clam genera and species.

This conclusion has been questioned in a cogent way by T. J. M. Schopf and his collaborators, who have suggested that the degree of taxonomic change perceived in evolving lineages may merely be a function of their overall morphological complexity. More complex organisms having more parts to change will thus appear to evolve more rapidly than less complex organisms. Do clams only appear to evolve more slowly than mammals or does their less complex morphology

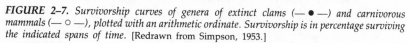

FIGURE 2–7. Survivorship curves of genera of extinct clams (— ● —) and carnivorous mammals (— ○ —), plotted with an arithmetic ordinate. Survivorship is in percentage surviving the indicated spans of time. [Redrawn from Simpson, 1953.]

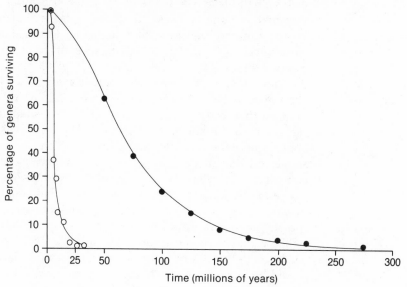

conceal an equally rapid rate of as yet poorly defined genomic evolution?

Schopf et al. tested this proposal by estimating morphological complexity from the number of morphological terms in use for various groups. Clams and mammals each have about 3,000 genera, but there are only 300 morphological terms used for clams, whereas there are about 1,000 used for mammals. Mammals thus appear to be morphologically more complex than clams. While this test is consistent with the idea that rates of morphological evolution might be an artifact of complexity, this explanation clearly fails. Living fossils, such as the opossum, do not appear to be appreciably less complex than their more rapidly evolving cousins, and rates of evolution can be rapid in relatively "simple" groups, as can be clearly seen from data on evolutionary rates collected by Van Valen in 1973.

Van Valen, using more recent data for durations of fossil organisms than were available to Simpson in 1953, prepared survivorship curves for genera and families of a large number of groups. Unlike Simpson, Van Valen used a logarithmic ordinate. This turns out to be a more informative way to plot survivorship than with an arithmetic ordinate, because if the probability of extinction is constant for the members of a group, that is, if at any time any genus is as likely to become extinct as any other, regardless of time of origin, then the survivorship curve plotted with a logarithmic ordinate will be linear. Van Valen found that survivorship curves so plotted were, in fact, linear. The curves do not distinguish between pseudoextinction (a genus disappearing by evolving into another) and actual termination of lineages. In Figure 2–8 three of Van Valen's curves are replotted for genera of extinct mammals, genera of extinct clams, and genera of rudists (a specialized extinct group of clams). Comparison of the curves for clam and mammal genera yields a half-life for clams of 35×10^6 years, or a mean duration of genera of 70×10^6 years, whereas the much steeper curve for mammals yields a half-life of 3×10^6 years, or a mean duration of 6×10^6 years for mammalian genera. These durations indicate that mammalian genera appear and become extinct at a rate about 10 times that of clam genera. However, it is obvious that the rudists evolved more rapidly than other clams, exhibiting a survival half-life of 10×10^6 years for a mean generic duration of 20×10^6 years. The rudists arose in the Upper Jurassic and became extinct at the end of the Cretaceous. In morphology rudists were unlike other clams in that one valve was shaped like a cone and cemented to the substrate at the tip. The other valve served as a lid. Some species were enormous, up to 2 m in length. Stanley has suggested that most clams evolve slowly because of a general lack of competition between them for resources. Rudists lived in tightly packed agglomerations, and even formed reefs. Their relatively rapid rate of evolution may have been driven by competition for space. Since the rudists are not morphologically more complex than other clams, it is unlikely that their more rapid evolution is an artifact.

Although there is no doubt that on the whole clams evolve slowly, it would be illusory to believe that they are unable to attain high rates of

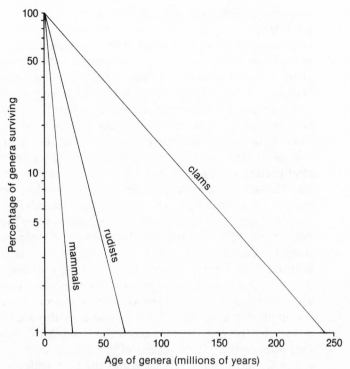

FIGURE 2–8. Survivorship curves of genera of extinct clams, rudists (a group of specialized Mesozoic clams), and mammals, plotted using a logarithmic ordinate. [Redrawn from Van Valen, 1973.]

evolution. E. G. Kauffman, in a study of evolutionary rates among Cretaceous clams, found that evolutionary rates were tied to such factors as trophic strategy and levels of environmental stress. Some clams may have evolved as rapidly as mammals, exhibiting average species durations as short as 1.25×10^6 years. Like the horseshoe crab, clams are not limited in evolutionary rate by some innate genomic peculiarity. The genomes of clams can respond to produce rapid morphological evolution when the opportunity arises.

By definition the rate of taxonomic change derived from survivorship data for the members of a large group such as clams corresponds to horotely, the average rate of evolution for the group in question. It is interesting to note that the rates estimated from survivorship data, even the relatively fast rates observed for mammals, are not fast enough to account for the suddenness of appearance of new forms in the fossil record. Thus, the radiation that produced most of the modern mammalian orders occurred in the $10–15 \times 10^6$-year span of the Paleocene Epoch. Gradualistic evolution within lineages of species exhibiting the mean 1.2×10^6-year durations calculated for Pleistocene mammal species by Stanley would hardly account for this spectacular radiation.

The same conclusion has been reached for evolution of mammalian genera in the Plio-Pleistocene by use of a different approach, determination of taxonomic frequency rates. These include rates of change in total frequency, such as in number of genera, and the two rates that control total frequency: rate of origination and rate of extinction. Rate of origination is defined as number of first appearances in the fossil record per million years, and rate of extinction as last appearances per million years. These rates are the most readily obtained of evolutionary rates, because no knowledge of evolutionary lineages within the group in question are needed, and taxonomic identifications and stratigraphic occurrences are the prevalent data in the paleontological literature. Thus, to determine the rate of origination of genera in a particular family one need only tally the number of new genera appearing in a dated stratigraphic interval. In 1977 Gingerich presented rates of origination of Plio-Pleistocene genera of rodents, artiodactyles, carnivores, and primates. All of these groups underwent considerable radiation of new genera during this time. Rates of origination were high—as high as 145 genera per 10^6 years for artiodactyls and 222 genera per 10^6 years for rodents. Pseudo-origination, the evolution of one genus into another without branching, accounts for only 5–20% of the rates. The average duration of a rodent genus is 5.9×10^6 years, with about half persisting approximately 2×10^6 years. Gradualistic evolution cannot account for the explosive radiation of new genera in the roughly 3×10^6 years available. The only way to accommodate high rates of origination and long survivorship is by means of branching, with new species originating in the punctuational mode and then persisting for a relatively long period without further modification.

Rates of Size Change

Change in size is one of the most frequent of evolutionary phenomena. In general, size increase is a prevalent trend. Groups as different as foraminifera and dinosaurs have produced giants from small progenitors. But bigger is not always better, and large animals like elephants have on occasion given rise to pygmy forms. Rates of size change in evolution are easily determined. The dimensions of homologous structures such as shells, bones, or teeth from evolutionarily related organisms can be measured with precision, and rates obtained from the length of time elapsed over which the size change occurred. Such measurements provide the simplest quantitative measure of evolution, and can be made independently of transformations in shape, which are much more difficult to quantify, thus providing a measure of evolutionary change without some of the problems of objectivity that appear in rates of taxonomic change.

Comparisons of absolute size changes are usually futile, because it is generally necessary to compare organisms of different sizes to start with. Thus, a measure of fractional or percentage change in size in a standard time span is needed. Such a relative measure was devised by Haldane in 1949. For example, if in an interval of time, t, the mean length of a bone or other structure has increased from x_1 cm to x_2 cm, then the proportional rate of change can be expressed as

$$\frac{1}{x}\frac{dx}{dt} = \frac{d(\ln x)}{dt} = \frac{\ln x_2 - \ln x_1}{t}$$

Haldane used this equation to calculate relative size increases, and suggested that the term darwin be used to designate a unit of evolutionary size change equal to a factor of $e/10^6$ years. Haldane suggested that as a practical approximation 1 darwin is equivalent to a change of 1/1,000 per 1,000 years, which yields a change in size by two-fold in 10^6 years.

Size rates can vary greatly in magnitude and duration. Simpson has estimated, for example, that horse teeth in the Eocene-Oligocene lineage *Hyracotherium* (*Eohippus*) to *Mesohippus* increased in height with an average rate of about 25 millidarwins. The rate from the Lower Oligocene *Mesohippus* to the Miocene *Hypohippus* increased slightly, to 45 millidarwins. These forms were all browsers. The living *Equus* is a plains animal, a grazer. The grazing horses diverged from the browsers in the Miocene with the lineage *Mesohippus* to *Merychippus*, which exhibited an acceleration in rate of increase of tooth height to 80 millidarwins, producing a four-fold increase in tooth height in about 20×10^6 years. The modestly higher rate of tooth size increase was only a part, although a significant one, of the evolution of grazing horses. There were modifications in skull shape, improvements in the brain, and profound changes in feet, limbs, and skeleton for rapid running. Radical structural changes accompanied greater tooth height: an increase in the number of cusps and a change from a relatively simple tooth with enamel covering a dentine core to one consisting of very high enamel ridges with spaces in between filled by a hard material, cement, to produce an efficient and durable grinding structure. Tooth size evolved far less radically than shape in *Merychippus*, although the shape changes could not have occurred without concomitant increase in tooth height. Details of these changes can be found in a charming book by Simpson called *Horses*.

The rates of size increase in the molars of horses are not unusual for mammals, nor for many other groups of organisms. Van Valen has tabulated evolutionary rates of size change for a variety of protists and invertebrates, and found an average rate of 40 millidarwins, with a range of 3–300 millidarwins. Relatively fast rates of size increase are also known for mammals. The rate of increase in crown height in Plio-Pleistocene mammoth teeth calculated from data presented by Maglio for the lineage *Mammuthus africanus* to *M. meridionalis* to *M. armenicus* was 300 millidarwins sustained over about 2×10^6 years, resulting in a 1.8-fold net increase. Hallam has also recorded a wide range of rates of size increase for Jurassic clams and ammonites. The range for clams was 6–546 millidarwins, with a mean rate of 109 millidarwins. In some very long lineages, such as the clams *Gervillela lanceolata* to *G. aviculoides*, a moderate average rate of 55 millidarwins maintained over almost 40×10^6 years resulted in a four-fold overall increase in shell size. Trends were constant in some, but varied in others. *Gryphaea* showed a pattern of increase, then decrease, and finally again an increase in size. Other species slowed or accelerated their rates of size increase at various times. Ammonites, consistent with their faster rates of evolution by taxonomic criteria, exhibited faster rates of size increase. The range was 64 millidarwins to 3.7 darwins, with a mean of 584 millidarwins. Hallam noted one particularly interesting phenomenon. Size increase might be considerable within a species or short sequence of species; however, younger species often appear in the fossil record that are appreciably smaller than their progenitors, but without discernible size intermediates. Hallam interpreted his observations to suggest that two modes of size evolution occurred in his Jurassic mollusks. The first is a more or less gradualistic increase in size, which he proposed generally leads to an evolutionary dead end. The other mode is a relatively sudden diminution in size, which may lead to changes in morphology and speciation.

Extremely fast rates of dwarfing are known for late Pleistocene mammals from Australia, Eurasia, and North America. The ecological reasons for rapid dwarfing are still not entirely clear, but pressure to maintain an adequate population size in situations in which resources become more limiting appears to be likely. The phenomenon of geologically instantaneous size reduction is well documented. Kurten calculated rates of dwarfing of European mammals in late glacial and postglacial times. In spans as short as 5,000–15,000 years, marten, bear, wildcat, wolverine, and other animals exhibited marked reductions in size. Rates of dwarfing ranged from 3.7 to 43 darwins, with a mean rate of 12.6 darwins, a rate that, if sustained, would result in a halving in size in only 80,000 years. Similar, very rapid rates of dwarfing occurred in Australian marsupials, such as kangaroos and dasyures, in the interval between 30,000 and 20,000 years ago. Rates calculated by Marshall and Corruccini for dwarfing of marsupials ranged from 9 to 26 darwins.

Rapid dwarfing in the mammals discussed thus far was only sustained long enough to achieve 10–35% size reductions. However, more drastic shrinking did occur in some Pleistocene mammals. The seemingly most improbable example is that of pygmy elephants. The large European elephant E. namadicus gave rise to a series of dwarf elephants living on various Mediterranean islands during the late Pleistocene. E. falconeri, the smallest, was the size of a pony. Unfortunately, insufficient stratigraphic data exist to allow an accurate determination of rates of dwarfing, but Maglio suggested that the process took place in a relatively short time—a few hundred thousand years at most. The island elephants may have conformed to the behavior of populations undergoing evolution in the punctuational mode. They were isolated from the main species population of E. namadicus; they inhabited limited geographical areas; and they were probably limited in population. Did they evolve by punctuation? If so, is it necessary to postulate unusually rapid rates for the reduction from E. namadicus to E. falconeri? If we assume that dwarfing occurred over 100,000 years, a rate of about 16 darwins would have been sufficient. This is a high rate, but even higher rates were found for the dwarfing of other mammals. At the maximum rate of 43 darwins recorded by Kurten, only 40,000 years—a geological instant—would have been required.

Both Stanley and Hallam have suggested that, genetically, the most available route for the rapid dwarfing of organisms as different as vertebrates and ammonites is paedomorphosis, in which small, morphologically juvenile forms achieve sexual maturity. This process may represent an important means for the origination of new morphologies in a punctuational manner. Paedomorphosis may also have played a role in some aspects of the dwarfing of Australian marsupials, but paedomorphosis is only one aspect of the genetics of size change. The rapid rates of size reduction observed in Pleistocene mammals are easily accounted for when one considers the basis for dwarfing in living mammals. Dwarfs have been observed and studied in several families: horses, cattle, sheep, pigs, dogs, humans, and quaintly enough, mice. McKusick has categorized two basic types of dwarfing. One type, with disproportionately short limbs, is best exemplified by achondroplastic dwarfing, and is inherited as a simple autosomal dominant. This sort of dwarfing is probably inadaptive. The other type, ateliotic dwarfing, results in a well-proportioned miniature version of the normal-sized animal. In humans, this type of dwarf can result from three basic causes, all involved in the production or utilization of pituitary growth hormone. Type I dwarfs lack growth hormone, whereas type III dwarfs lack all anterior pituitary secretions. Type II individuals lack the growth hormone receptor on the target tissue, and despite their higher than normal growth hormone levels, are dwarfed. Both types II and III show a simple autosomal recessive pattern of inheritance with a major locus subject to the action of modifier genes. Thus, a single gene change resulting in an alteration of a simple humeral factor can have striking developmental results. Under conditions of strong selection or isolated population, rapid rates of dwarfing are quite plausible. Such dwarfing has occurred in human populations, and in the case of the Ituri pygmys of the Congo, who average about 4 feet in height, an entire population has become dwarfed apparently because the population has fixed an allele, resulting in the production of a defective growth hormone.

The Evolution of Shape

Because the morphologies of actual organisms or their parts can be extremely complex, quantitative determinations of rates of change in shape are more difficult to obtain than are taxonomic or size rates. This is unfortunate because taxonomic decisions are often made on the basis of complex characters whose evolutionary changes may be subtle or at least difficult to describe in any quantitative fashion. Rates of morphological change are reflected in the taxonomic rates we considered earlier. If, as maintained by the proponents of the punctuation hypothesis, major evolutionary steps are always accompanied by a speciation event, then taxonomic rates may indeed measure the pulse of evolution. Nevertheless, taxonomic rates only indirectly approach the problem of shape change.

The most significant morphological changes in evolution produce novelties, structures qualitatively different from what existed before, which make possible new modes of life. The origins of a large number of novelties have been documented in the fossil record. Among these are amphibian limbs, the amniotic egg, mammalian jaw articulation, and wings. The origin of other novelties, such as homeothermy or the mammmary gland, which involve soft tissues or physiological functions, can only be inferred from the study of living organisms. Novelties do not arise *de novo*. Preexisting structures are modified through changes in the developmental processes involved in morphogenesis to yield the new structure. The rates at which this happened have varied. The transition from a reptilian to a mammalian type of jaw articulation took place gradually over many millions of years, and is well documented in the extensive fossil record of advanced mammal-like reptiles. As for the rate of acquisition of homeothermy, we can only guess. Other novelties seem to have evolved rapidly, but documentation is all but absent. Bats, for example, which invaded a completely new adaptive zone for mammals, appear quite suddenly in the fossil record early in the Eocene Epoch. The paucity of the record makes any real assessment of the rate of this truly profound reorganization of the forelimb into wing impossible, but it most probably occurred during the 10–12×10^6-year Paleocene radiation of placental mammals.

Evolutionary transformations of structures, such as generalized mammalian forelegs into the wings of bats, require only changes in developmental programs so that the same components become arranged in a different configuration. The bat's wing contains all the same bones as are present in the forelegs of other mammals, and the development of a wing begins with a limb bud. The nature of evolutionary modifications of developmental programs concerns us at length in later chapters. For our purposes here, we only need ask if evolutionary processes that yield structural novelties are qualitatively different from those that yield more modest sorts of evolutionary changes in morphology. Our working assumption is that they do not. Evolutionary changes in morphogenesis all involve similar genetic factors controlling gradients, pattern formation, rates of cell division, inductive interactions, and the other processes that in detail establish a differentiated structure in development. In the fossil record we can recognize two kinds of readily quantifiable changes in morphology that result from evolutionary changes in the genes controlling these processes. These are changes in meristic traits, and changes in allometric relationships.

Meristic traits are those that involve numbers of identical or similar parts, such as ambulacra in echinoderms, eye files in phacopinid trilobites, bristles in insects, ribs on brachiopod and mollusk shells, or vertebrae and toes in vertebrates. Rates of change in number of parts can also be expressed in the darwin units of Haldane. Van Valen has listed several rates of change in meristic traits calculated from the fossil record. Rates of change in the number of chambers in certain fossil foraminifera range from 70 to 120 millidarwins. Ribs on brachiopod shells have changed in number with rates of from zero to 100 millidarwins, and the number of ribs on scallop shells with rates of from 6 to 190 millidarwins. In 1973 Maglio calculated rates in darwins for several meristic traits measurable in elephant teeth. The change in number of enamel ridges or plates that form the shearing surfaces of elephant molars is plotted in Figure 2–9. Three lineages are shown, all stemming from *Primelephas gomphotheroides*, with primitive molars possessing only a small number of plates. *Loxodonta*, the genus of the existing African elephant, increased molar plate number only slowly, whereas *Elephas*, now represented by the Indian elephant, and *Mammuthus*, the mammoths, increased plate numbers rapidly. In *Elephas* the average rate was about 200 millidarwins, and in the later mammoths it reached a burst of about 600 millidarwins. Both lineages increased their molar plate numbers from 7 to 23 plates in about 5.5×10^6 years. In combination with a complex set of other changes in thickness of enamel, height of crowns, and shape of the plates, the increase in number of plates resulted in the evolution of highly efficient shearing molars.

Generally, merely changing the number of similar parts results only in minor modifications in morphology. Fewer or greater numbers of bristles on flies or ribs on brachiopod shells are not profound evolution-

FIGURE 2–9. *Evolution of enamel ridges on the molars of three lineages of elephants. Lineages are those of the African elephant,* Loxodonta *(— ○ —); the Indian elephant,* Elephas *(— ● —); and the mammoth,* Mammuthus *(—△—).* [Elements of this figure adapted from V. J. Maglio, 1973.]

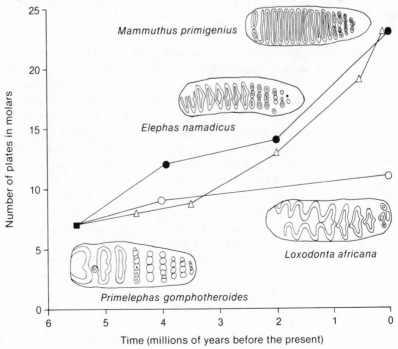

ary events. But in certain cases changes in meristic traits have played key roles in evolution; a striking example is provided by snakes, which may have up to 400 vertebrae and almost as many pairs of ribs, making possible an efficient, if specialized, mode of locomotion, and in the case of constrictors such as pythons and boas, a unique method of subduing prey.

So far we have considered changes in size as occurring independently of changes in shape. Yet this is seldom the case. Large animals are generally not merely enlarged versions of their smaller ancestors, nor are adults merely enlarged juveniles. A major portion of changes in shape that take place during growth of an individual or in an evolutionary lineage involves changes in the relative dimensions of parts of the body. In both cases such modifications in proportions are the result of changes in relative growth during development. This is allometry. Consideration of allometric relationships allows one to sort out which shape changes are caused by growth and which are the result of modifications of the developmental program. We can distinguish three kinds of allometric series: those that are a function of growth in ontogeny of a species, relationships among related species of different sizes, and finally, allometric relationships in an evolving lineage. Allometric relationships in dimensions between two structures often obey the simple formula devised by Huxley in 1932:

$$y = bx^\alpha$$

where y represents the size of a particular structure and x the size of the whole body or another structure with which y is to be compared. The term b is a scale factor while α is the ratio of specific growth rates of y and x. The equation may be written in the form

$$\log y = \log b + \alpha \log x$$

The dimensions of x and y are generally plotted on double logarithmic axes. A linear plot with slope α and intercept $\log b$ results. In cases in which $\alpha = 1$, the relative sizes of the structures represented by x and y are constant regardless of size; that is, they remain in the same proportion and growth is isometric. Isometric growth is a special case of a more general spectrum of allometric relationships. In most cases $\alpha \neq 1$, and proportions change with change in size.

A particularly interesting example of developmental allometry discussed by Huxley in his 1932 book *Problems of Relative Growth* is shown in Figure 2–10. Worker ants are polymorphic in some species, with the biggest workers having exaggeratedly large heads and mandibles and serving as soldiers. It is probably advantageous to an ant colony to be able to field a variety of workers, each most efficient at a particular range of tasks. Such a series of workers is shown in Figure 2–10, along with a plot of the relationship of head size (x) to body size (y) for workers of a species of ant. Within a species all of the different sizes of workers fall on a single allometry curve. This means that although the larger workers look different from the smaller in having monstrous heads and mandibles, the whole series represents the expression of a single, genetically determined growth law.

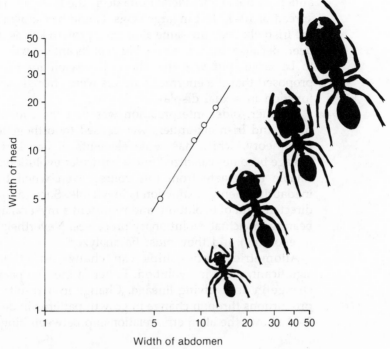

FIGURE 2-10. *Allometric relationship between size of head and body in the ant* Pheidole instabilis. [Modified from J. S. Huxley, 1932.]

We have so far only considered allometric relationships expressed by individuals during growth. However, allometric curves can also be drawn to compare the adults of successive species of an evolutionary lineage. Evolutionary size increase is, as we have seen, extremely common. In some cases size increases have had very interesting consequences. When allometric trends characteristic of the smaller, ancestral species are maintained, if α ≠ 1 for a trait, the paradoxical result is that by retaining the ancestral growth pattern the larger descendant attains a different shape from its ancestor. This Alice in Wonderland mode underlies the morphological changes seen in some famous evolutionary lineages. One of these, the titanotheres, a group of herbivorous mammals, culminated in the Oligocene in huge forms with a pair of massive blunt horns on their noses, fully half as long as the skull. Their Eocene ancestors were smaller and either lacked or had very small horns. Horn size increased radically with the evolutionary increase in titanothere body size during the Oligocene. Did the evolution of these large horns represent the acquisition of a new morphogenetic pattern? In 1934 Hersh provided the surprising answer that this was not the case. In a double logarithmic plot of horn length (y) against skull length (x), all species of titanotheres fall on the same allometric curve with a very high value of α. Thus, the growth law governing ontogeny of the titanothere skull was such that with overall increase in size the nasal region enlarged drastically. Evidently, both large body size and horns were selectively advantageous and so the allometric trend was maintained. A similar case is that of the enormous

antlers of the extinct Pleistocene deer, the Irish elk. These antlers had a
spread of 10–12 feet in large stags. Gould has shown that the antlers of
the Irish elk obey the same allometric growth law as the antlers of other
deer. Because this was a very big deer its antlers would also be expected
to be large. But was this the only reason for giant antlers? Gould
proposed that the enormous antlers were strongly selected for by their
function in sexual display.

Another, older interpretation was that the evolution of titanothere
horns and Irish elk antlers was caused by orthogenesis. According to
this theory, which has some elements of Greek tragedy about it, a
lineage becomes canalized into a particular evolutionary direction and is
unable to deviate from this course even when the trend becomes
inadaptive; hence, extinction is inevitable. Such ideas imply a finalistic,
directed kind of evolution, and represent a mystical approach with little
bearing on actual evolutionary processes. Nevertheless, trends do exist
in evolution and they must be analyzed.

Allometric relationships can change in other, perhaps more
significant, ways in evolution. Either of the two parameters α or b can
change in an evolving lineage. Change in α results in modified body
proportions through change in growth patterns in development. Figure
2–11 shows the allometric relationship between hinge length and shell

FIGURE 2–11. *Shift in the allometric relationship between hinge length and shell perimeter in a*
lineage of fossil clams of the genus Myalina. *The lineage exhibited gradual size increase, along*
with a change in allometry, during the Pennsylvanian and Permian. [Modified from N. D.
Newell, 1942, 1949.]

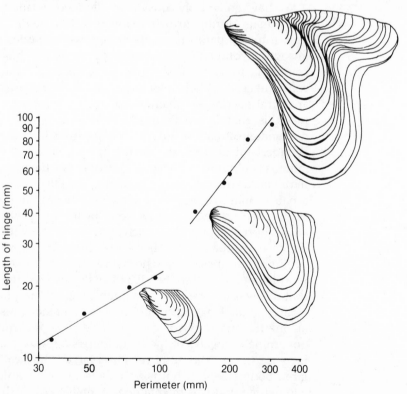

perimeter in a lineage of fossil clams of the genus *Myalina*, exhibiting gradual size increase. The earlier species fall on one curve, the later on a curve with a different, higher α, and culminate in a large form with a shape strikingly different from that of the ancestral forms. The overall course of these events occupied about 50×10^6 years, but the change in allometry took place within 10×10^6 years or less. A similar sort of change has been recorded by Pilbeam and Gould for the allometry of brain size as a function of body weight in human evolution. Apes and our extinct hominid cousins in the genus *Australopithicus* have an α-value of 0.34, whereas living and extinct representatives of the genus *Homo* have an α-value of 1.73. Because recent discoveries made in East Africa by Johanson and the Leakeys suggest a divergence time for *Homo* from *Australopithicus* of about 3.5×10^6 years ago, it may become possible to estimate how rapidly this particularly momentous shift in allometry progressed as more fossils from this period are collected.

Allometric growth during ontogeny often has a value of α different from that obtained from an allometry plot for a group of related, adult forms. Such a situation is diagrammed in Figure 2–12. The dashed line gives the allometric relationship between brain weight and body weight for a group of related species of insectivores of mean adult size. The solid lines give the ontogenetic allometries of each of the species. In this example, the slopes of the ontogenetic curves are lower than the interspecific slope, but the reverse is just as likely because values of α > 1 are often present in development. Note that the α characteristic of ontogeny differs from the α for adults in interspecific comparison, but is

FIGURE 2–12. *Allometric relationship between brain and body weight for related species of insectivorous mammals from Madagascar (----). Ontogenetic allometries within each species are given as solid lines.* [Redrawn from S. J. Gould, Geometric similarity in allometric growth: A contribution to the problem of scaling in the evolution of size, *Am. Naturalist* **105**:113–136, 1971, by permission of the University of Chicago Press. Copyright 1971 by the University of Chicago.]

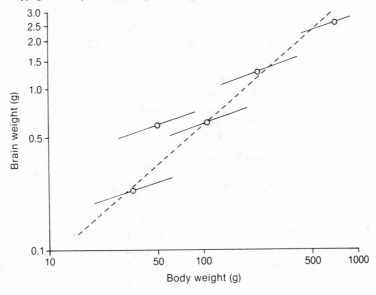

the same for each of the species. What differs from species to species is the value of b.

If the interspecific $\alpha = 1$, then the larger-species adults are scaled-up versions of their smaller relatives (or ancestors). Gould has suggested that this can be achieved in evolution by the larger-descendant species retaining the α of its smaller ancestors but starting the allometric growth of the particular structure in question from a larger rudiment, thus, a larger value of b. A change in time of onset, an acceleration, or a delay of rudiment growth is required to produce a change in size. This provides an alternative way in which scaling-up might be achieved in evolution of larger size without requiring a change in the growth law. If organisms retain the ontogenetic growth curve of their ancestors and the growth allometry is very different from $\alpha = 1$, then body proportions may change drastically with size. By retaining allometry but starting at a different size, drastic changes in body proportions are obviated.

A full mechanistic explanation of regulation underlying isometric or allometric growth is still not possible, but some interesting aspects of this control have begun to emerge. Goss has pointed out that two possibilities are open to related organisms of greatly differing body size. As long as the basic plan of organization remains similar, component parts must either change in size or in number. The general rule is that small functional units, such as cells in an organ or lenses in a compound eye, change in number whereas large functional units—organs, limbs, or eyes—are modified in size. Within limits, increase in size results in more effective organs. The brain is a case in point: A larger overall size permits more neurons and more interconnections, yielding a greater functional potential. An important consequence is that during growth, proportionality between body parts will be largely a function of relative rates of cell division.

Evidence for size-controlling factors dates back to the elegant organ graft experiments performed in the 1920s and 1930s, reviewed by Twitty in 1940. For example, grafts of eyes between young and old salamander (*Ambystoma*) larvae resulted in the retardation of the growth of large eyes transplanted to small hosts, and conversely, the acceleration of the growth of small eyes transplanted to large hosts, so that their eventual sizes were properly proportioned with respect to the size of the host.

A feedback system involving circulating regulatory substances apparently exists, and probably involves both agents that stimulate cell division in specific tissues as well as inhibitors of cell division. One well-known tissue-specific stimulatory regulatory substance is erythropoietin, which is produced in the kidneys in response to blood loss and stimulates the production of red blood cells. Positive regulators involved in morphological development also exist. The androgens and estrogens that begin to be produced in adolescence interact with target tissues and are responsible for the development of striking morphological traits—the secondary sexual characteristics, breasts, beard, and bodily proportions of the mature adult. These regulatory substances are humeral and act on a target tissue at a distance from the source. But another class of positive regulation exists in which substances produced by a tissue stimulate the growth and, in not a few cases, differentiation and pattern formation of an adjacent tissue. This is the classical induction of embryology, which is discussed in Chapter 5.

Specific inhibitors of cell division, called chalones by Bullough, also exist, although they are as yet poorly characterized. These substances are produced in the target tissue and inhibit its growth. Experimental removal or damage to part of an organ, such as the liver, reduces the circulatory level of specific chalone and induces compensatory growth. As with the growth hormone system, the various elements of these controls—production of regulators, structure of regulators, and number and specificity of receptor sites—are accessible to genetic change in evolution of new morphological proportions.

Changes in allometry can occur rather gradually, as recorded for hyenas and other Pleistocene mammals by Kurten, or extremely rapidly, as in the breeds of domestic dogs developed within a span of centuries. The powerful borzoi, bred to pursue wolves, and the pug-faced Pekingese, bred (one can only wonder why) to sit on laps, differ in size and proportions. Genes are known in dogs that modify body form in such aspects as length of legs, length of snout, and size. These traits have been avidly exploited by breeders. Rapid rates of allometric change are also known for naturally evolving lineages, such as some of the Hawaiian *Drosophila* to be discussed in Chapter 3. The fossil record is limited in what evidence it can provide on mechanisms of evolution. Preservation of an entire species range for a sufficient span of time in the life of the species to catch peripheral isolates petrified in the act of speciation is not impossible, but the number of examples will be vanishingly few. Nor are the genetic systems of extinct organisms available to us, although inferences can certainly be made from their living relatives. Some fossil ontogenies are well known; larval stages of a few trilobites, the record of a lifetime of shell growth in the spiral shells of ammonites, and the eggs, juveniles, and adults of the famous Mongolian dinosaur *Protoceratops*. There are other examples, but on the whole the fossil nursery is not crowded. The fossil record does reveal that rates of evolution are highly variable, whether measured by changes in taxonomy, size, or allometry. Aside from the wonderful documentation of past life, of a multitude of lost worlds, provided by the fossil record, this is probably the chief contribution of paleobiology to our problem. We are freed from the concept of evolution by gradual nucleotide substitution to produce new genes, and forced to seek mechanisms of evolution at the level of gene organization and gene expression in ontogeny that will account for rapid and profound changes in morphology.

Three

Morphological and Molecular Evolution

I believe that our Heavenly Father invented man because he was disappointed in the monkey.

Mark Twain

Varieties of Molecular Evolution

A tacit assumption of our discussion of the fossil record and the rates of morphological evolution derived from it is that these rates reflect genomic changes. The relationship of genomic to morphological evolution is a generally accepted tenet, although in a sense the term genomic evolution is a tautology because all evolutionary changes require genetic changes—thus, genomic evolution.

But the real difficulty in dealing with genomic evolution in eucaryotes is that eucaryotes are not simply *E. coli* writ large. Eucaryotic genomes are exceedingly complex and contain a multiplicity of kinds of genetic elements. This complexity is reflected in attempts to reconcile genomic and morphological evolution. For instance, in their study concerning the influence of complexity on rates of morphological evolution, Schopf et al. suggested that external morphological complexity may not in fact be an accurate index of the extent of genomic evolution. Thus, although forms with complex morphologies, such as brachiopods, might show a great deal more evolutionary change in morphology than, say, bacteria with their limited morphological repertoire, the bacteria might indeed have experienced more genomic evolution in the same span of time. Or, as these authors concluded, " 'Rates of evolution' as customarily reported by paleontologists may therefore be a poor indication of evolutionary changes in the underlying genome." This conclusion is partially correct. Bacterial groups possess a wonderful spectrum of metabolic pathways: Their adaptations are biochemical rather than morphological. On the other hand, eucaryotes lack the metabolic virtuosity of bacteria. With a few exceptions Metazoa all employ the same metabolic pathways, and their adaptations are primarily morphological. The changes recorded in the fossil record thus reflect the

62

organization and function of the portion of the genome that governs morphogenesis.

It is difficult to assess the relative degree of genomic evolution required for the evolution of groups of greatly differing morphological organization and complexity. An approach to this conundrum has been made by studies of the amino acid sequences of homologous proteins from different organisms. The significant result of these studies is that proteins of quite different organisms have related amino acid sequences; that is, changes in amino acid sequences of proteins or nucleotide sequences of DNAs provide molecular criteria of evolutionary relatedness independent of morphology or taxonomic assignment. Nevertheless, the problem of deciding which portion of molecular evolution is pertinent to morphological evolution remains.

Most studies of evolution at the molecular level have dealt with structural genes because they are most readily accessible. Structural genes are those that are transcribed to yield RNAs that either function as messenger RNAs (mRNAs) and are translated to produce the amino acid sequences of proteins, or have a functional role, such as ribosomal (rRNAs) or transfer RNAs (tRNAs). Most of the structural genes currently under intensive study are those that encode proteins produced in large amounts by specialized cells, for example, globins, ovalbumin, actin, and histones. These proteins do not function in the direct regulation of gene action; thus, their genes are classified as structural. However, the distinction between structural and regulatory genes is in some respects arbitrary. A protein such as the *lac*-repressor protein of *E. coli* is the product of a structural gene in the sense that a protein results from the transcription and translation of the gene; however, its function is solely regulatory in that the *lac*-repressor protein acts directly to control the expression of a specific set of structural genes coding for enzymes. Other genomic elements important in the regulation of gene action or in maintaining chromosome organization may not be transcribed at all in order to perform their functions.

Because of the complexity of eucaryotic genomes a profusion of evolutionary events comprise genomic evolution. Genomic changes have a number of possible outcomes. A by no means exhaustive list of events and their consequences that have occurred in the evolution of the genomes of multicellular animals is presented in Table 3–1.

The first set of events embraces most of classical molecular evolution, that is, modifications of the coding regions of structural genes. Such events involve changes in nucleotide sequences and in many cases result in a changed amino acid sequence in a protein. The degree of modification expressed by the protein can range from minimal to rather drastic changes, the extremes of which are loss of function or gain of new functions. A substantial proportion of nucleotide substitutions in structural genes will only be detectable at the DNA sequence level because the genetic code is degenerate, and substitutions at the third position in most cases produce a synonymous codon. Thus, no amino acid substitution results. Some substitutions will be conservative: They will result in the substitution of an amino acid chemically similar to the

TABLE 3–1. Varieties of Genomic Evolution

Event	Consequence: DNA Structure	Consequence: Protein Structure	Consequence: Phenotype
In structural genes:			
Nucleotide substitution, silent	Change in base sequence	No change in amino acid	None or little
Nucleotide substitution, conservative	Change in base sequence	Substitution of similar amino acid	None or little
Nucleotide substitution, other	Change in base sequence	Substitution of amino acid	None to loss or change of function
Deletion	Loss of base(s)	Deletion of amino acid(s), nonsense protein, or premature termination	Little to loss of function
Duplication followed by nucleotide substitution in duplicate gene	Duplication of sequence; change in sequence of duplicate	New, related amino acid sequence	New function with old function retained
Gene fusion	Loss of intervening bases	Fused protein	None, loss of function, or new function
In noncoding sequences:			
Nucleotide substitution in highly repetitive satellite DNA	Change in base sequence	None	?
Nucleotide substitution in spacer sequences between genes	Change in base sequence	None	None
Nucleotide substitution in noncoding moderate repetitive sequences	Change in base sequence	None	?
Nucleotide substitution in noncoding single copy sequence	Change in base sequence	None	?
Nucleotide substitution in introns	Change in base sequence	None to insertion of amino acids	None to loss or change of function
Nucleotide substitutions in promoters or other regulators	Change in base sequence	None	Change in level or time of expression
In sequence frequency:			
Change in frequency of satellite sequence	Change in number of copies of existing sequence	None	?
Change in frequency of moderate repeat sequence	Change in number of copies of existing sequence	None	?

TABLE 3–1 (continued)

Event	Consequence: DNA Structure	Consequence: Protein Structure	Consequence: Phenotype
In sequence frequency (cont.)			
Change in ploidy	Most or all sequences multiplied equally	None	None, or increase in size; isolating mechanism
In movement of sequences to new locations in genome:			
Insertion of intron into structural gene	Preexisting sequence in new location	None to insertion of amino acids	None to loss or change in function
Transposition of *cis*-acting regulator	Preexisting sequence in new location	None	Change in level or time of expression
Movement of blocks of satellite DNA between chromosomes	Preexisting sequence in new location	None	?
In higher order:			
Inversions and translocations	Preexisting sequence in new location	None	Generally none or little; some selective advantage in maintenance of blocks of genes
In transfer of genes between species:			
Horizontal transfer of genes between unrelated species	Introduction of new sequence	Introduction of new protein	None to introduction of novel function

one replaced. For example, substitution of valine, a hydrophobic amino acid, for leucine with a very similar hydrophobic side chain can be detected by amino acid sequence analysis of the mutant protein, but will probably not have any effect at the phenotypic level.

Evolution of structural genes is not limited to nucleotide substitution: A variety of other events, such as deletions and gene fusion, have also occurred. The most signifiicant changes in the evolution of new proteins are duplications of an existing gene followed by divergent evolution of one of the duplicated sequences to yield a related protein. The original gene is retained so that the net result is the augmentation of the organism's biochemical abilities by the addition of a novel protein. On the phenotypic level the changes run a corresponding gamut, with the most interesting changes leading to the acquisition of novel functions.

Noncoding DNA in the genome presents a more enigmatic problem. These DNA sequences do not code for proteins, although in some cases they are transcribed in association with structural genes. Empirically, noncoding DNA falls into four categories. The first group includes the best-understood noncoding DNA sequences, which are those that serve as spacers between structural genes. Spacers appear to be more tolerant to nucleotide substitution than the structural genes they separate. A second, recently discovered, and as yet poorly understood group of noncoding sequences are intervening sequences, which are also called introns. Introns are DNA sequences inserted into, and interrupting, the coding regions of structural genes. When the gene is transcribed the initial transcript includes both coding and intron sequences. The intron sequences are removed by special RNA-processing enzymes that convert the primary transcripts into mRNA containing an uninterrupted coding sequence. Introns are widely distributed in eucaryotic genes, being present in both nuclear and organellar genes, but missing from the genes of procaryotes. Suprisingly, in some cases the intron sequences are considerably longer than the coding sequences they interrupt. It is not known what effect mutations within introns might have, but any mutations that interfere with correct removal of intron sequences from primary RNA transcripts will have serious consequences. The third group of noncoding sequences includes non-transcribed regulatory regions, such as the promoters to which the enzyme RNA polymerase, which carries out transcription, must attach in initiation of transcription of an adjacent structural gene. Mutations in these regions produce no change in the amino acid sequence of the proteins produced, but may have profound effects on the amount or time of expression of the gene. The final group contains sequences of no apparent function. Mutations in this DNA result in changes in nucleotide sequence, but have no known phenotypic effect.

Although most structural genes exist in one copy per haploid genome, changes in frequency of particular sequences have been a common evolutionary occurrence. Most of the sequences that are present in eucaryotes in more than one copy per haploid genome are not structural genes. Thus, change in frequency of such sequences has no effect on the amino acid sequence of any protein. Regulatory roles have been proposed for multiple-copy DNA sequences, but none has been demonstrated at this point. Phenotypic effects of frequency changes are unknown.

The class of genomic events involving movement of already existing sequences to new locations within the genome will not in general be detected by conventional methods for studying molecular evolution. Yet movement of regulatory sequences so that a new regulator of different specificity is inserted adjacent to a structural gene and replaces the former may, as shown by Berg, result in phenotypically dramatic changes and provide a potentially rapid means for morphological evolution without requiring any nucleotide substitution at all. The overall content of base sequences is unchanged—no protein is modified—and yet a phenotypic change results. Only direct nucleotide sequencing of the region of insertion can reveal the event at the molecular level.

Finally, there are large-scale rearrangements of chromosomes in which large blocks of DNA containing many genes are inverted or moved to entirely new chromosome locations. This is not molecular evolution *per se*; however, chromosome rearrangements are frequently observed in metazoan evolution. M. J. D. White, in his book *Animal Cytology and Evolution* in fact assigns chromosome rearrangements a central role in evolution.

The events discussed above represent changes occurring within an existing genome. Recent studies by Busslinger, Rusconi, and Birnstiel have indicated that in rare instances horizontal transfer of genes can occur between distantly related species by heterodox mechanisms, perhaps involving retroviruses which can cross species boundaries. In the case studied by Busslinger and his collaborators, a histone gene cluster apparently has been recently transferred between two families of sea urchins that diverged about 65 million years ago, and, except for the transferred gene cluster, possess quite distinct histone genes. The genes of the transferred cluster are expressed to produce functional proteins. The significance of events of this kind to evolution is unknown.

A number of important aspects of genomic evolution outlined in Table 3–1, such as introns, moderately repeated and satellite DNA, and the organization and function of various types of regulators, are discussed in detail in later chapters. In the present chapter we focus on molecular evolution in the more restricted sense of nucleotide substitutions in DNA and amino acid substitutions in proteins. Because most of our understanding of evolutionary events at the genomic level has resulted from studies of structural genes and their products, there has been a distinct temptation to extrapolate modes and rates of structural gene evolution to those genes involved in morphogenesis and morphological evolution. However, the studies of A. C. Wilson and his colleagues discussed later in this chapter make it clear that evolution by nucleotide substitution in structural genes is largely uncoupled from morphological evolution. Nevertheless, evolutionary data at the molecular level provide an invaluable tool for tracing relationships among morphologically dissimilar organisms, and rates of molecular evolution provide a clock against which other rates may be compared.

Genes, Proteins, and "Molecular Clocks"

The majority of studies in molecular evolution have concentrated on changes in structural genes as expressed in the sequence of amino acids in the proteins encoded. A large number of protein sequences have been determined and collected into a very useful, growing compendium, the *Atlas of Protein Sequence and Structure* edited by M. O. Dayhoff. The few hundred sequences so far determined represent only a fraction of the vast number of interesting and potentially available proteins. Unfortunately, phylogenetic representation is very uneven: Mammals with their mere 4,060 living species (Anderson and Jones) are represented by over 350 sequenced proteins, whereas insects, of which almost a million species have been described (Daly, Doyen, and Ehrlich) are represented by a paltry 11 sequenced proteins. The number of sequences known for other major phyla, such as mollusks and echinoderms, are also disproportionately low. Nevertheless enough sequence data are available so that rates of structural gene evolution can be estimated, phylogenetic inferences drawn, and the relationships of structural gene evolution to morphological evolution assessed. It should be noted that unlike the fossil record, evolutionary data from protein sequences pertain only to lineages that survive today. So whereas the fossil record allows us to see extinct and abandoned morphologies, sequence data can never reveal the special features of the proteins of extinct groups that have left no descendants.

Biochemistry is highly conservative. Pathways and even protein sequences are maintained through vast stretches of geological time. This provides one of the unique values of sequence data: Sequences are independent of morphology. Thus, relationships can be readily detected in sequences of conservative proteins, such as cytochrome c, across phylum and even kingdom boundaries. Protein sequence data are quantifiable, with each amino acid position in each sequenced protein a potential variable. Since there are 20 amino acids possible at any position, independent origin or convergence of similar proteins in two organisms is unlikely. For instance, human, chimpanzee, and gorilla hemoglobin α-chains, 141 amino acids in length, are identical in sequence. The possible number of different sequences of this length is 20^{141}. An independent origin for the globins in apes and humans is, to say the least, improbable. Close correspondence of sequence most probably indicates close evolutionary relationship; this rule thus serves as the basis for the construction of quantitative phylogenetic trees for proteins. Because the absolute times of morphological divergence of the organisms from which the sequenced proteins were isolated can be obtained from the geological record, rates of amino acid substitution can then be calculated.

When quantitative comparisons of protein sequences first became possible, the promise of this new approach to evolutionary relationships was enthusiastically welcomed. Zuckerkandl wrote in 1962: "Thanks to the knowledge that has been gained recently about the relationship of proteins and genes, the study of amino acid sequence in proteins is now able to achieve the most precise and the least ambiguous insight into evolutionary relationships and into some of the fundamental mechanisms of evolution." And according to Dayhoff and Eck in 1969: "One of the grand biochemical ideals is to be able to work out the complete, detailed, quantitative phylogenetic tree—the history of the origin of all living species, back to the very beginning. Biologists have had this hope for a long time; biochemistry now has the actual capacities of accomplishing it." A rather Haeckelian goal indeed.

The major working assumption in the construction of phylogenetic trees from molecular sequence data is that rates of nucleotide, and therefore, amino acid substitution are constant within any particular set of homologous sequences, such as cytochrome c. This is a gradualistic hypothesis with the interesting corollary that rates of substitution behave as molecular clocks that may run independently of morphological rates.

The possibility that protein sequence evolution and morphological evolution are not coupled was recognized by Margoliash in 1963. Margoliash pointed out that if elapsed time determines the number of substitutions accumulated in a protein, then protein sequence evolution could serve as a clock to measure the divergence time of any two species. He made the prescient suggestion that a "...useful test of the importance of time as the main factor in the collection of variation in cytochrome c would be the comparison of the amino acid sequences of the homologous proteins from species known not to have evolved morphologically for long periods with those from species which have changed rapidly..." The use of molecular clocks to dissect the relationship between evolution of structural genes and morphological evolution has yielded some very interesting insights into the aspects of genomic evolution responsible for morphological change. An excellent introduction to protein clocks was written by R. E. Dickerson in 1971, and a recent and thorough assessment was published by Wilson, Carlson, and White in 1977.

Before discussing the relationship of molecular clocks to morphological evolution we should examine the validity and limitations of such clocks.

The basis for accepting the hypothesis of a uniform and characteristic rate of evolution for any particular protein is illustrated in Figure 3–1. The number of mutational steps, which is estimated from the number of amino acid differences observed between the sequences of homologous proteins, is plotted as a function of divergence time for the organisms from which the proteins were extracted. The divergence time is taken as the number of years before the present that two organisms shared a common ancestor. Taking as an example the case of the comparison between mammalian and reptile cytochrome c, the fossil record indicates that the mammallike reptiles diverged from other reptiles about 300×10^6 years ago. About 15 substitutions per 100 amino acids separate the cytochromes of living mammals from those of living reptiles. Thus, in this comparison a 15% difference required 300×10^6 years, or 20×10^6 years for a 1% divergence. The time required for a 1% divergence in any protein has been defined as a unit evolutionary period (UEP) by Dickerson. The UEP for cytochrome c is 20×10^6 years. Other proteins similarly exhibit constant average rates of evolution, but the absolute rates are not the same for all proteins. For the examples illustrated in Figure 3–1 the UEP for hemoglobin is 5.8×10^6 years; that for fibrinopeptides is a short 1.1×10^6 years.

Differences in UEP apparently reflect differing degrees of selection to which various proteins are subject. The constraints limiting the rate of amino acid substitution in cytochrome c probably stem from its intimate

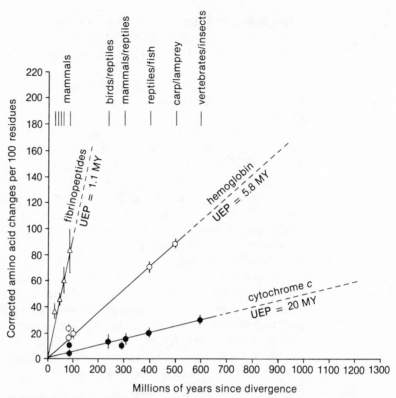

FIGURE 3–1. *Rates of evolution of three proteins: fibrinopeptides, hemoglobin, and cytochrome c. The rate of amino acid substitution is plotted as a function of the time elapsed since the species compared shared a common ancestry. The unit evolutionary period (UEP) is the time required for a 1% divergence.* [Redrawn from Dickerson, 1971.]

association with other proteins of the mitochondrial electron transport chain. Globins too are functional proteins that interact with both small molecules and other globin subunits. Conversely, fibrinopeptides have no known function except to be cut out of a longer protein, fibrinogen, when it is converted to fibrin during formation of a blood clot.

Fitch and Langley have tested the validity of the molecular clock hypothesis by considering an aggregate rate for seven different proteins for which extensive data on amino acid sequence evolution were available. Although the structural genes for each of the proteins had their own characteristic rate of acceptance of nucleotide substitutions, a plot of the number of substitutions for the aggregate of the seven proteins as a function of organismal divergence time yielded a straight line with a slope of 0.47×10^{-9} substitutions per nucleotide pair per year. Only the values for primate-derived sequences deviated. This deviation may be a result of variations in the rates of protein evolution in primates, or more likely, of misestimations of divergence times among primates because of the poor fossil record of this group. The average rate of nucleotide substitution determined by Fitch and Langley refers only to those substitutions that have resulted in an amino acid change. Using the sequences for hemoglobin RNA published by Salser et al. and Forget et al., Fitch and Langley concluded that silent mutations, that is, base changes that do not result in the replacement of one amino acid by another, may occur five times more often than those that do result in amino acid substitutions. Thus, the overall rate of nucleotide substitution in structural genes could be as high as 2.8×10^{-9} per nucleotide pair per year. It should be noted at this point that the underlying assumption that the clock is linear has been recently questioned by Corruccini et al., and existing data are not precise enough to decide if divergence with time is linear or nonlinear.

Whereas protein sequences provide an assay for a very specific aspect of genomic evolution, the coding sequence portions of structural genes, direct investigations of DNA have made it possible to quantitatively assess rates of nucleotide substitution in both codogenic and noncodogenic genomic elements. Obviously the most direct way to obtain such data would be through the determination of the nucleotide sequences of DNA in a manner analogous to that used for protein sequences. The methods that make this approach possible have been recently developed and a large number of DNA sequences will be forthcoming. Most of our presently available quantitative information on DNA evolution comes from nucleic acid hybridization experiments, which do not depend on direct knowledge of the DNA sequences in question.

The principles of hybridization are simple. The double helix consists of two DNA strands that are associated by hydrogen bond-pairing of complementary bases: adenine pairs to thymine, guanine to cytosine. In native DNA there is perfect pair-matching between the strands. Double-stranded DNA can be dissociated to produce individual, unpaired single strands. Under appropriate conditions of salt concentration and temperature, complementary single strands can reassociate to reform double-stranded DNA. Several procedures exist for separating double-stranded from single-stranded DNA so that the progress of reannealing can be readily monitored.

The genomes of procaryotes are organized such that each gene or nucleic acid sequence is generally present in only one copy per haploid genome. The situation in eucaryotes is more complicated. Most of the genome consists of single-copy genes, but much (e.g., 25% in *Drosophila* and 40% in mouse) of the genome consists of sequences repeated from 10^2 to 10^6 times per haploid genome. The most highly repeated sequences, such as the 10% of the mouse genome that is repeated 10^6 times, are satellite DNA sequences, which consist of simple tandem repeats located in discrete blocks in certain regions of chromosomes. The moderately repeated sequences present in a few hundred to several thousand copies are interspersed among the single-copy sequences of the genome.

Hybridization experiments can be used to observe the evolutionary changes that have occurred in both unique and repeat-copy sequences. Two kinds of information can be obtained. By examination of hybridization kinetics, the number of copies, and thus evolutionary change in representation of a sequence, can be determined. Further, hybridization experiments are not limited to reannealing complementary DNA strands from the same organism. Dissociated DNAs from a pair of different organisms can be mixed and allowed to anneal. If there is a sufficient relationship, hybrid double-stranded DNAs will form. The degree of divergence in the related but not identical strands in such hybrids can be readily estimated because any divergence in nucleotide sequence means that hybrid DNAs will contain some bases that cannot pair. Nonpaired bases lower the overall stability of the hybrid. The reduction in stability can be measured by the consequent depression in melting temperature of the DNA, that is, the temperature at which the two strands separate. A 1.5% mismatch results in a lowering of the melting temperature from that of the native DNA by 1°C. This is a powerful tool: The degree of nucleotide divergence in the DNA of any two organisms can be determined independently of protein or direct DNA sequencing.

Because with only a few exceptions, such as the genes coding for histones, structural genes are found in the unique-sequence portion of the genome, rates of nucleotide substitution in single-copy DNA have been of particular interest. In 1969 Laird, McConaughy, and McCarthy found that the unique sequences of artiodactyls evolved with a rate of about 2.5×10^{-9} substitutions per nucleotide pair per year, whereas Kohne and his colleagues observed rates of 1×10^{-9} to $3.6 \times ^{-9}$ substitutions per nucleotide pair per year for primates. Because the imperfections of the primate fossil record make the dating of interesting divergence points uncertain, including that between humans and apes, the average substitution rate of about 2×10^{-9} per nucleotide pair per year probably can be applied to all primates. Similar uncertainties about estimates of divergence time apply to studies of single-copy DNA evolution in sea urchins by Angerer, Davidson, and Britten, and in frogs by Galau and his co-workers, who arrived at substitution rates of 1×10^{-9} to 3×10^{-9} per nucleotide pair per year.

Although the rates for single-copy DNA evolution are in agreement with each other and with the average rate of nucleotide substitution derived from protein evolution, the rate of substitution in some rodents was found by Laird and his collaborators to be 10 times faster. Similarly, Sarich has found that the immunological distance between rat and mouse albumins is an order of magnitude greater than that between human and chimpanzee albumins. Part of this discrepancy may once again arise from estimates of divergence times from the fossil record for a group with a poor record. Laird et al. used a date of 10×10^6 years ago for the divergence of rats from mice, but Sarich has suggested a divergence time of about 30×10^6 years on the basis of molecular clock data. However, Jacobs and Pilbeam have pointed out that new fossil evidence strongly supports a divergence between rat and mouse in a period of $8–14 \times 10^6$ years ago. This implies that the rodent clock runs faster than that for other organisms. Further, the recent work of Hake on molecular evolution in corn indicates that in at least some plants the clock may run several orders of magnitude faster than is commonly observed in animals. Conversely, some groups of organisms, such as the warblers studied by Avise et al., may exhibit very slow molecular clocks. It is possible that the clock, as some have proposed, should be geared to generation time and reproductive strategies rather than absolute number of years elapsed.

The evolution of moderately repeated DNA sequences is more complex than that of single-copy DNA because it includes events that produce new repeat families in an apparently saltatory manner, as well as subsequent divergence of sequences by nucleotide substitution. Repeated sequences probably arise by the amplification of a previously existing unique sequence. These events have occurred in many lineages to produce a large number of repeat families. The origins of repeat-sequence families in Old World primates are diagrammed in Figure 3–2, which is taken from the work of Gillespie. Gillespie compared, by hybridization, the repeated DNAs of the higher primates, and found that some repeat families were shared by several lineages, whereas others were restricted to one particular group. Thus gibbons, chimpanzees, and humans all share a common repeat family, whereas a second family is shared only by chimpanzees and humans, and a third is found only in humans. The events that gave rise to these families are plotted as events 1–3 in Figure 3–2. Similar events are plotted for other groups, such as the baboons and their relatives. The macaque, the baboons, and

FIGURE 3–2. *Times before the present when repeated DNA sequence families arose in primates. Saltatory replication events that produced particular repeat families are numbered. Thus, the repeat family produced by event 1 is shared by humans, chimpanzees, and gibbons, whereas the family produced by event 2 is present only in humans and chimpanzees. Some events have occurred recently enough to have produced families that appear only in a single group (e.g., event 3).* [Redrawn from D. Gillespie, Newly evolved repeated DNA sequences in primates, Science 196:889–891, 1977. Copyright 1977 by the American Association for the Advancement of Science.]

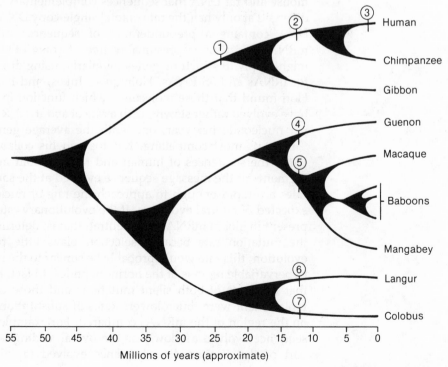

the mangabey all share a family that originated in event 5. The guenon lacks this family of repeated DNA, but possesses its own repeat family that originated after the divergence of the guenon lineage from the baboon lineage. All of the primates shown in Figure 3–2 share even older families that originated prior to the branching of these groups.

Although several organisms may share a repeat family, melting curves of hybrids of the related repeats from the two organisms show a depression in melting temperature. This result indicates that after the origination of a repeat family the member sequences begin to diverge by accumulation of nucleotide substitutions. Interestingly, the moderately repeated DNAs studied by Gillespie have evolved at the same rate as unique DNA sequences. This conformity to the clock rate is also true for a very different, but specific family of repeat sequences, the structural genes for the histones of sea urchins studied by Weinberg and his collaborators.

The overall impression from the relative uniformity of rates of nucleotide substitution into DNA is that there is indeed a "genomic metronome" that allows an average, relatively constant rate of nucleotide substitution in genomic DNA independent of either phylogenetic position or rate of morphological evolution. But does the rate of nucleotide substitution represent a long-term average of many classes of rates of sequence evolution, subject to varying degrees of selection, or does it mainly represent a selectively neutral process?

Comparison of evolutionary rates of several kinds of nucleotide sequences indicates that the former possibility is the correct one. Rosbash, Campo, and Gummerson found by hybridization studies in mouse and rat DNA that sequences complementary to total mRNA had diverged at only half the rate of total single-copy DNA. Total single-copy DNA contains a preponderance of sequences never expressed as mRNA, and thus is presumably free of some of the constraints that might apply to structural genes. Similarly, using direct sequencing data for tRNAs and 5S RNAs, Holmquist, Jukes, and Pangburm as well as Hori found that these molecules, which function in protein synthesis, have evolved rather slowly, with a rate of about 0.2×10^{-9} substitutions per nucleotide per year, one-tenth the average genomic rate.

Perhaps most compelling, Kafatos and his collaborators have compared the sequences of human and rabbit globin mRNAs to see if all segments of the message sequence evolved at the same rate, and if any rates are rapid enough to approach the rate of nucleotide substitution expected of neutral evolution. If the evolutionary rate for the sequences present in globin mRNA were neutral, that is, determined essentially by the mutation rate because selection plays little part in neutralistic evolution, this rate would probably be similar to the rate observed in the hypervariable regions of the fibrinopeptides. In fact, Kafatos et al. found that the rates of both silent mutations and those causing amino acid replacement were much lower. Rates of substitution varied depending on the region of the mRNAs compared. For example, the 5'-noncoding sequence evolved as slowly as the overall coding sequences, whereas part of the 3'-noncoding sequence evolved rapidly. Regions of the coding sequence giving rise to critical sections of the protein involved in heme interactions, Bohr effect, and α-β contact sites have been free of amino acid replacements and very low in silent base substitutions in contrast to regions that have accepted amino acid substitutions. Clearly, silent substitutions are not necessarily neutral.

The conclusion that clock rates are not determined by any single theme and are instead the average of several rates reflecting a variety of levels of selection does not impair their usefulness in the construction of molecular phylogenies. Two examples of protein evolutionary trees, for cytochrome c and myoglobin, are presented in Figures 3–3 and 3–4 Because the average rates of evolution of these two proteins are different, they are useful in mapping evolutionary events that occupied quite different time scales. Myoglobin has evolved with sufficient rapidity to have produced significantly divergent myoglobins among the various orders of placental mammals during their evolution since the late Cretaceous. Thus, myoglobin is an ideal protein for the construction of a molecular phylogeny for mammals. Cytochrome c, with its slower rate of evolution, can be used to trace a far broader and more ancient set of relationships—those between eucaryotic kingdoms, phyla, and classes. Protein phylogenetic trees are constructed by determining the smallest possible number of nucleotide changes required to achieve the observed evolutionary separation of related amino acid sequences. The branches connecting any two sequences are drawn to a length proportional to the number of mutational events separating the two sequences.

FIGURE 3–3. *Phylogenetic tree for the cytochrome* c *proteins of eucaryotes. The species represented by sequences are* (1) Tetrahymena pyriformis, (2) Crithidia fasciculata, (3) C. oncopelti, (4) Euglena gracilis, (5) *smut fungus*, (6) *baker's yeast*, (7) Candida *sp.*, (8) *bonito*, (9) *chicken*, (10) *human*, (11) Drosophila *sp.*, (12) *prawn*, (13) *garden snail*, (14) *starfish*, (15) *brandling worm*, (16) Ginkgo biloba, (17) *elder, and* (18) *wheat*. [Redrawn from Schwartz and Dayhoff, 1978.]

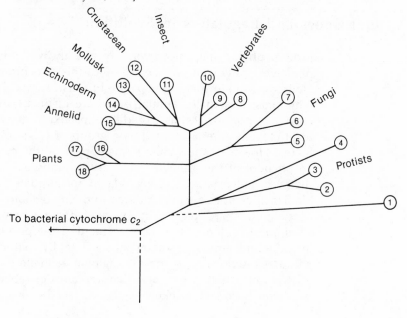

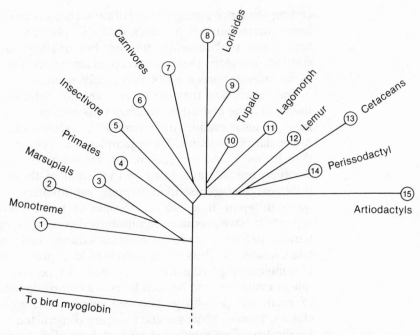

FIGURE 3–4. *Phylogenetic tree for the myoglobins of mammals. The species represented by sequences are* (1) *platypus;* (2) *kangaroo;* (3) *opossum;* (4) *human, baboon, and marmoset;* (5) *hedgehog;* (6) *dog and badger;* (7) *sea lion and harbor seal;* (8) *galago;* (9) *slow loris;* (10) *treeshrew;* (11) *rabbit;* (12) *sportive lemur;* (13) *porpoises and whales;* (14) *horse; and* (15) *cow, pig, and sheep.* [Redrawn from Hunt, Hurst-Calderone, and Dayhoff, 1978.]

On the whole these protein phylogenies are reasonably consistent with conventional phylogenetic trees that are based on the more classical methods of comparative anatomy, embryology, and paleontology.

Structural Genes and Regulators in Evolution

Protein phylogenies, however, are not always congruent with morphological phylogenies. For example, the cytochrome *c* sequences in Figure 3–3 fall into neatly separated protist, fungus, plant, and animal kingdoms; and major animal phyla are distinct. Annelids, mollusks, crustacea, and insects, as would be expected from classical approaches to phylogeny, are associated as a group of phyla distinct from vertebrates. However, the echinoderms appear on the cytochrome *c* tree as an offshoot of the annelids, contrary to embryology, which indicates that echinoderms are closely related to chordates (see Figure 4–1). Similar difficulties are encountered in the myoglobin tree in Figure 3–4. Most groups are placed in reasonable agreement with fossil and anatomical evidence, but the placement of the primitive primates, the lorisides and lemurs, is somewhat startling. By anatomy these forms are primates. According to their myoglobin sequences they are no more related to the higher primates than are dogs or rabbits. This extraordinary conclusion is unlikely to be correct because the conventional phylogeny is based on far more characters than a single gene product. The general correspondence of molecular and organismal phylogenetic trees is probably a result of long-term averaging of both morphological and molecular rates. Noncorrespondence can result either from variations in the rate of evolution of a protein or from variations in the rate of morphological evolution in a particular lineage.

Variations in protein rates are apparently greatest during the evolution of new functions. Proteins with well-established functions evolve in a clocklike manner, and are then useful for determination of molecular phylogenies. However, while a new protein is evolving its function, evolution apparently deviates from the clock mode. If the rate of globin evolution is extrapolated backward, as is done in Figure 3–1, one finds the divergence of globins projected into the late Precambrian, well before the first evidence of Metazoa in the fossil record. Thus lamprey globin is estimated to have diverged from insect globin over $1,000 \times 10^6$ years ago, lamprey globin from vertebrate hemoglobins 800×10^6 years ago, and hemoglobin from myoglobin 900×10^6 years ago. Such extrapolations are possibly serious overestimates. In accord with the fossil record, Goodman, Moore, and Matsuda have assumed more recent divergence times: lamprey from insect globins at about 700×10^6 years ago, lamprey globin from vertebrate hemoglobins at about 500×10^6 years ago, and hemoglobins from myoglobins also at 500×10^6 years. Divergence of α- from β-hemoglobins was at about 450×10^6 years. These divergence times appear reasonable because the oldest primitive chordate is known from the Middle Cambrian (about 550×10^6 years ago), and the remains of the oldest vertebrates date to the Late Cambrian (about 500×10^6 years ago). A consequence of using these revised estimates of divergence times is that the rate of globin evolution in the period from 500 to 400×10^6 years ago was very much more rapid than has been the case subsequently.

Other examples also exist. The best is perhaps that of α-lactalbumin, a subunit of the lactose synthetase of the mammary gland. α-Lactalbumin is similar in amino acid sequence to lysozyme and probably evolved from lysozyme during the evolution of early mammals. According to the *Atlas of Protein Sequence and Structure* the UEP for α-lactalbumin is about 2.3×10^6 years, whereas for lysozyme the UEP is about 5×10^6 years. If the rates had been constant through the history of both proteins, the amino acid differences between α-lactalbumin and mammalian lysozymes require that α-lactalbumin arose 300×10^6 years ago, approximately 100×10^6 years before the appearance of the first mammals in the Late Triassic. A more reasonable alternative is that α-lactalbumin arose as a part of the mammalian character complex during the Triassic, and was subject to a period of rapid evolution early in its history.

Another and particularly intriguing example has been found by Hennig, who has described a testis-specific cytochrome c from mouse that differs in sequence by 13 amino acids from the cytochrome c found in all other mouse tissues. He noted that if these 13 substitutions were located on the three-dimensional model for cytochrome c, they were concentrated on one region of the surface. Otherwise the remainder of the sequence is identical to the principal cytochrome c of the mouse and other rodents (Carlson et al.). Hennig proposed from the number of substitutions that the divergence of the testis cytochrome c occurred early in tetrapod evolution (or even earlier), and that the remainder of the molecule evolved in parallel to the principal cytochrome c. However, it seems more likely that the testis species represents a relatively recent gene duplication event within the early rodents with rapid subsequent evolution of the specialized protein.

In these examples, accelerated rates of change occurred during the period in which new functions were evolved, but acceptable substitutions became limited to noncrucial portions of the molecule once functional adaptation was attained. This is apparently the general case. Deviations from the clock behavior of protein evolution do not inhibit our use of protein clocks for evolutionary intervals in which a reasonably constant rate can be demonstrated. In cases for which the divergence times cannot be well established in the fossil record, or in lineages in which the rate of protein evolution may not be constant, caution is obviously desirable.

Noncorrespondence between morphological and molecular phylogenies in instances in which the proteins appear to be evolving in a clocklike manner can only be caused by nonconstancy of morphological rates. Apparently this is not an unusual relationship between protein and morphological evolution, and the uncoupling of morphological evolution from the molecular clock is of great significance. It is largely to the studies of A. C. Wilson and his collaborators that we owe the experimental demonstration that morphological evolution involves a different portion of the genome than that measured by molecular clocks.

The most spectacular paradigm for this hypothesis is the relationship between human and chimpanzee. These two species are classified as members of two separate families on the basis of morphological differences, yet as shown by King and Wilson, they are very closely related on a molecular basis. The degree of closeness is striking. The amino acid sequences of 12 rather varied proteins differ in only 7.2/1,000 amino acid sites; that is, chimpanzee and human protein sequences are greater than 99% identical. King and Wilson also used a second method for estimation of protein similarities by making electrophoretic comparisons of 44 intracellular and serum proteins. Most of these proteins have allelic variants detectable by electrophoresis. The proportion of alleles electrophoretically identical at a particular locus in both humans and chimpanzees was found to be 0.52. Sequence differences for the 44 loci examined in this way were calculated by determining the proportion of amino acid substitutions detectable by electrophoresis, and then using a Poisson distribution to estimate the total number of amino acid substitutions that have accumulated. This calculation yields a value of 8.2 substitutions per 1,000 amino acid sites. Again human and chimpanzee proteins emerge 99% identical.

Finally, King and Wilson estimated, from the data of Kohne and Hoyer and their collaborators on the thermal stability of human-chimpanzee DNA hybrids representing only single-copy sequences, that for an average sequence of DNA equal in length to 3,000 bases (thus equivalent to 1,000 amino acids) there are 33 nucleotide substitutions between humans and chimpanzees. This is higher than predicted from the rate of substitution observed in proteins, but the discrepancy is not difficult to rationalize. Both silent mutations, which result in nucleotide substitutions without changes in amino acid sequence, and mutations in noncoding regions will produce such a result. The molecular distances estimated by all of these methods can be related to taxonomic distances in other organisms for which good molecular comparisons also exist. When King and Wilson carried out an analysis of genetic distance from electrophoretic and DNA hybridization data they found the surprising result that humans and chimpanzees are as close as sibling species of *Drosophila* or other mammals. This similarity has again been emphasized by a detailed study of the structure of human and chimpanzee chromosomes by Yunis et al. who found near-identity in the chromosome-banding patterns of the two species.

Of course one might argue that our placement of chimpanzees into a different family from humans might be the result of a long-standing cultural prejudice that would like to see our species stand distinct from our benighted nearest relatives. Benjamin Disraeli's heartfelt conviction that given a choice of man as ape or angel he was on the side of the angels is still shared by many. That being the case, to a nonhuman taxonomist humans and chimpanzees might indeed appear to be sibling species in a single genus. Such an argument has been advanced by Merrell, and tested by Cherry, Case, and Wilson. Quantitative comparisons of morphological features used by taxonomists to discriminate between the body shapes of frogs were applied to humans and chimpanzees as well as to frogs. The interesting result that humans and chimpanzees differ by these criteria to a somewhat greater extent than frogs belonging to separate suborders does not support an exaggerated taxonomic splitting of humans and chimpanzees by human taxonomists. Instead it appears that in the approximately 5–10×10^6 years since the divergence of these two lineages, homonids have experienced extremely rapid morphological evolution but standard rates of molecular evolution.

In 1974 Wilson, Maxon, and Sarich proposed that, "There may be two major types of molecular evolution. One is the process of protein evolution, which goes on at about the same rate in all species. The other is a process whose rate is variable and which is responsible for evolutionary changes in anatomy and way of life. We propose that evolutionary change in regulatory systems accounts for evolution at and beyond the anatomical level." Wilson and his co-workers have tested this hypothesis by contrasting molecular distances between organisms with indexes of regulatory distance between organisms. Thus they made use of an unlikely comparison of frogs with placental mammals.

In their diversification since the end of the Cretaceous, the placental mammals have undergone rapid evolution. They range in their morphological diversity from bats to whales, from elephants to humans. This is an anatomical and adaptive spectrum far beyond any achieved by frogs which throughout their 150×10^6-year history have remained pretty much alike in anatomy and way of life. Despite the morphological conservativeness of frogs, their proteins have undergone considerable sequence evolution, and their molecular clocks keep the same time as the homologous clocks of mammals. Many pairs of mammal and frog species have been tested for their ability to produce viable cross-species hybrids (i.e., hybrids capable of development to adulthood although not necessarily fertile). These interspecific hybrids provided Wilson, Maxson, and Sarich with an assay for the relative role of protein sequence evolution in governing organismal changes. If the protein-sequence relationship is crucial, cross-species hybrids should only be possible between species with closely related sequences. Such a result might indicate either that closely related protein subunits are required for correct assembly of functional protein complexes, or that protein similarities serve as indicators that the two genomes combined in the hybrid have diverged sufficiently little that they remain compatible to support embryonic development.

In a study in some ways reminiscent of a survey of the passengers aboard Noah's ark, the serum albumins of each of the members of 31 pairs of mammalian species and 50 pairs of frog species that produce viable cross-species hybrids were compared by quantitative microcomplement fixation, a measure of the degree of immunological similarity of proteins. The results were given in immunological distance units, which are a direct function of amino acid sequence difference. This approach provides a molecular clock for albumins, which were used because albumins are easily purified and are potent antigens. Immunological distances for albumins correspond well with the overall rate of the molecular clock, because the immunological distances of the albumins of pairs of primates, other mammals, and salamanders examined by Wilson et al., by Sarich and Cronin, and by Maxson and Wilson are directly proportional to the total single-copy nucleotide sequence differences between DNAs of the same sets of organisms.

The results of the comparison of albumin distance and ability to produce viable hybrid progeny indicated a striking difference between frogs and mammals. Pairs of mammal species from such groups as primates, carnivores, perissodactyls, and artiodactyls yielding viable hybrids are very close in terms of albumin immunological distance. The range of immunological distances was from 0 to 10 units, with a mean of 3. This rather narrow range was in contrast to the frogs, among which species with albumins as far apart as 90 immunological distance units still formed viable hybrids. The average distance was 37 units. If mammals with immunological distances as great as those observed between frogs forming interspecific hybrids could produce viable offspring, then such crosses as human and monkey, dog and seal, or sheep and giraffe should be observed. Wilson et al. proposed that such crosses are not possible because mammals, in contrast to frogs, have been subject to rapid evolutionary changes in systems regulating the expression of genes in development. Because the albumin molecular clock has changed at the same rate in both frogs and mammals, the rate of evolution of developmental regulatory systems must be 10 times faster in mammals than frogs.

Although this is an attractive hypothesis, proof is still lacking. The major possible complication stems from the fact that mammalian embryos interact directly with the mother via the placenta. Thus hybrids producing proteins different from those of the maternal species might suffer immunological rejection. Frogs and birds, which Prager and Wilson have shown to be able, like frogs, to produce viable hybrids between species with considerable protein divergence, differ from mammals in that they develop from an egg entirely isolated from the mother's immune system. Experiments in which the maternal immune system was suppressed have not resulted in any improvement in hybrid viability in mammals, and so this objection too remains moot.

The lack of correlation between molecular and morphological evolution also applies to other organisms. For instance, Avise has found that minnows, which have speciated rapidly, have had the same rate of protein evolution as sunfish, which have speciated slowly.

How might regulatory evolution function independently of nucleotide substitution in structural genes? Wilson and his collaborators have suggested that gene rearrangements rather than point mutations drive morphological evolution. A variety of processes, some very difficult to detect, are encompassed by the term gene rearrangement. Karyotypic events, such as changes in number of chromosomes or chromosomal arms, reflect fission of existing chromosomes to produce a larger number of chromosomes, fusion of chromosomes to yield a smaller number, or other sorts of events, including inversions or changes in the amount of heterochromatin. Wilson, Bush, Case, and King regard these changes as providing "crude manifestations of the phenomenon of gene rearrangement."

It remains unclear just how closely gross karyotypic changes reflect the gene rearrangements hypothesized to be important in morphological evolution. Wilson and Bush and their collaborators have attempted to test the relationship of chromosomal changes to morphological evolution by comparing rates of karyotypic change to morphological change in placental mammals, reptiles, amphibians, and fish. Although rapid morphological evolution proceeds independently of molecular evolution, it is strongly correlated with rapid changes in chromosome number. The rate of change in chromosome number for mammals is up to 10–20 times faster than that calculated for the morphologically more conservative amphibians, reptiles, and fish. Interestingly, there is a similar disparity between mammals and mollusks with their generally low rate of morphological change. This provides a sort of overall correlation, which suggests that rates of chromosome change are higher in more rapidly evolving groups. Bush et al. have refined these measurements to show that in fact the rate of chromosome evolution is strongly correlated with rates of speciation. These results have been taken to suggest that chromosome evolution might in fact be the direct mechanism whereby genomic rearrangements important in morphological evolution may occur. However, it must be noted that there are exceptions to this generalization; two recent examples make this quite clear. Comparisons made by Gold of the rates of karyotypic evolution among minnows show, quite contrary to expectation, that chromosomal changes are slower in the rapidly speciating genus *Notropis* than in less rapidly speciating genera in the same family. Rates of karyotypic evolution can also vary greatly among mammals. Baker and Bickham have found that although bats are a morphologically conservative group, they have undergone very disparate rates of chromosome change. Some have undergone no changes in relation to the proposed primitive pattern for their families, whereas others have undergone changes as rapid as any recorded for any animal group. Rates of karyotypic evolution do not behave in a clocklike manner; nor do they necessarily correspond to rates of morphological evolution. Our own view is that genomic reorganization is crucial to morphological evolution. However, these changes are achieved by mechanisms more subtle than gross chromosomal rearrangement, and gross changes are not a necessary component of speciation and morphological change.

Patterns of Speciation

Our discussions up to this point have largely focused on rates of molecular and morphological evolution, and we have been able to arrive at reasonable metrics for several evolutionary processes. However, such measures as rate of size change in darwins or rates of DNA evolution in nucleotide substitutions per year may create an illusion of continuity, a gradualism, even in cases in which a series of punctuational events have actually occurred. If evolution commonly occurs by the punctuational mode, it becomes necessary to determine the nature of the process which produces a rapid and possibly radical evolutionary change. The most strongly held view, which derives from population biology and genetics, holds that speciation is the critical process.

Within this framework species is defined as a group of interbreeding organisms that share a common gene pool. Inherent in this definition is the process and mechanism by which speciation must occur. If members of a species share a common gene pool, then the events that separate a single species into two must segregate portions of that gene pool and prevent the flow of genetic information between the two separate populations. What we need to discover is what is the nature of the mechanisms that separate incipient species—the nature and amount of genetic alteration necessary to the process and the minimum time required. However, before our headlong rush into virgin territory two caveats are warranted. Bush has quite correctly pointed out that no one has ever observed the event of speciation from beginning to end in nature and as such the study of speciation is an "ad hoc science." What we see in nature are time points, a series of stop frames in an ongoing process and from these glimpses we of necessity must infer the remainder of the process and its underlying mechanism. It is a bit like

Sherlock Holmes astounding Dr. Watson by deducing the past history of a prospective client from his limp and the kind of cigar he smokes. As we shall see, the process of speciation is not uniform and constant in all organisms and universal statements about it are hard-won. In fairness we should point out that several laboratory "speciation" events have been recorded in the literature. These for the most part have resulted from the experimental or serendipitous erection of barriers to reproduction among individuals of the same species. What remains to be determined, however, is the relationship of these laboratory events to those that occur in nature.

The second caveat relates to the difference between adaptive changes within a population and the kinds of isolating events resulting in a split or cladogenesis. Every natural population has a certain amount of variability, be it chromosomal, morphological, or in enzyme polymorphisms. It can also be shown that these karyotypic, morphological, and enzymic attributes change with time in seasonal fluctuations or spatially; for example, with altitude. The classical example of this type of adaptive change is industrial melanization in the moth *Biston betularia*. In this particular case a black-color morph became predominant to the gray in populations in the English midlands during the nineteenth century because of a change in the habitat caused by coal-fired industrialization. These moths rest on tree trunks during the day and black soot on trees rendered the gray morph more visible to birds and thus more subject to predation than the black morph. It is true that this is a change in the population; however, there has been no splitting into two separate reproductively isolated groups. Black and gray moths will still mate and produce viable fertile offspring. Another case is the chromosomal inversion polymorphisms in *Drosophila pseudoobscura* analyzed so elegantly by Dobzhansky and his students. The third chromosome of these flies has many different gene sequences rearranged relative to an arbitrary standard sequence. In many local populations several of these inverted sequences are present. The frequency of any given sequence, however, changes in the population throughout the growing season for these insects. There are also altitudinal clines in inversion frequency such that different inverted sequences are predominant at different sites as one proceeds from lower to higher elevations. Again the flies carrying the different chromosomal sequences are all interfertile and therefore not different species. It appears that changes in gene frequency or chromosome arrangement within a population and not accompanied by a speciation event may be important in maintaining the adaptiveness of the population, but may result in little evolutionary change.

Species, particularly in animals, are defined by their reproductive isolation. This separation is maintained by a variety of isolating mechanisms that can be divided into two broad categories. These are pre- and postzygotic, referring to whether the passage of genetic information is inhibited either before or after fertilization. Prezygotic barriers serve to prevent the union of the gametes and can be as simple as ecological differences between the two prospective mating organisms. If two groups of animals are ecologically isolated either by actual physical distance or by occupying sufficiently different niches in the same general area there is little likelihood of their mating. A second type of observed prezygotic isolation is temporal. If animals exhibit different diurnal rhythms or if plants flower during different seasons they will not have the chance to exchange genetic information. A third type of prezygotic isolation is specific to the sexual process itself. Many animals have developed elaborate courtship rituals that must be culminated in a very precise manner before mating can be accomplished and syngamy achieved. In some cases this ritual includes not only an auditory and visual give and take but the production of specific pheromones or sex attractants by the male, female, or both. The fourth prezygotic mechanism is physical incompatibility. This mechanism relates to the size and shape of the male and female genitalia. For example, in animals with internal fertilization the male intromittent organ must be compatible with the anatomy of the female in order for sperm to be transferred. In plants this incompatibility comes about through the utilization of specific insect pollinators by certain species and is the result of the size, shape, color, and odor of the flower and its ability to attract one type of insect.

Finally there is gametic incompatibility. The gametes released by the organism in order to achieve syngamy, or fusion of male and female pronuclei to form the diploid nucleus of the zygote, must recognize each other and for this purpose possess specific recognition cues. In animals this is seen most clearly in those species having external fertilization and in which the gametes are shed into the surrounding environment, usually an aqueous one. Sea urchins and other echinoderms have this type of fertilization and protection against transspecific gene flow. There is a certain morbid fascination in imagining the chaos that would result from the random conjugation of freely shed gametes in the oceanic environment. In plants the gametic incompatibility is most often seen in the inability of transspecific pollen to grow in a foreign style, thus preventing the male pronucleus from reaching the ovum. This form of sterility is also the apparent mechanism by which many monoecious plants prevent self-fertilization.

The second major class, the postzygotic isolating mechanisms, include those mechanisms that act after syngamy has been accomplished. The first of these is hybrid lethality. Hybrids, at least in the laboratory, can be formed but they die at some point in development. This can occur either soon after fertilization or quite late in development. The cause is generally an inability of either the paternally supplied genome to be maintained and/or function in the maternal cytoplasm of the egg, or an incompatibility of the paternal and maternal genomes. These mechanisms have been demonstrated by Denis and Brachet in their investigations of the cause of lethality in a transspecific cross between two echinoderms, *Paracentrotus lividus* and *Arabacia lixula*. Eggs of *P. lividus* can be fertilized by sperm of *A. lixula*, and embryonic development is initiated. However, the hybrid embryos die before gastrulation. The cause of this arrest in development may be the loss of DNA of paternal species origin from the cleavage-stage embryo as a result of elimination of paternal chromosomes during these cellular divisions. Moreover, the normal increase in RNA synthesis, which should take place at or just preceding gastrulation, does not occur. A somewhat different hybrid lethality occurs in transspecific crosses between another pair of echinoderms, *Dendraster excentricus* and *Strongylocentrotus purpuratus*, analyzed by Whiteley and Whiteley. In this cross, development proceeds through gastrulation but a normal larval stage is not achieved. The probable cause is the failure of expression of the genome of paternal origin because no paternal species proteins are produced.

A second form of postzygotic isolation is hybrid sterility. Transspecific hybrids are viable but themselves produce no offspring. The classic case is the mule. The cause of the sterility falls into two basic categories, chromosomal and genic. Chromosomal sterility often results from the inability of the hybridizing chromosomes of paternal and maternal origin to pair and disjoin properly in meiosis, thereby producing massive "nondisjunction" at the first meiotic division. Figure 3–5 shows the normal course of meiosis for a hypothetical cell on the left and a case of nondisjunction on the right.

The lesson of this hypothetical example is that the resultant abnormal distribution of the genetic material in meiosis results in the production of chromosomally and therefore genically unbalanced or aneuploid gametes. This occurs despite the fact that the hybrid individual is quite normal in all other respects. However, the resultant aneuploid reproductive cells are incapable of uniting either with each other or with normal gametes to produce a normal pattern of development, and the hybrid is consequently sterile. An example of this sterility was shown by Clausen and Goodspeed in crosses between two species of tobacco. *Nicotiana tabaccum* has a diploid chromosome number of 48, whereas in *N. glutinosa* the diploid number is 24. Viable hybrids can be formed between these two species with a diploid chromosome number of 36. These plants are sterile and produce no seeds because of the fact that the 12 *N. glutinosa* and 24 *N. tabaccum* chromosomes are apparently unable to pair and disjoin in a normal pattern during meiosis.

Another kind of chromosomal sterility can result from two closely related species possessing different chromosomal constitutions. This is most easily seen if one species has a rearrangement of its genome such that different chromosome arms are associated with each other relative to another related species, that is, if there has been a translocation or fusion of arms. If a hybrid is formed between two such species the result will be the production of aneuploid gametes because of the abnormal segregation of the two different arrangements of the genetic information. A hybrid individual heterozygous for one translocation element will have its fertility reduced by half, and further rearrangements will drop fertility even further. In a less-obvious fashion chromosomal inversions can and do have a similar effect on the fertility of intraspecific hybrids. Therefore, chromosome numbers and rearrangements can be seen to block the flow of genetic information and can produce isolation and contribute to species formation.

What should be made clear, however, is that care must be taken to

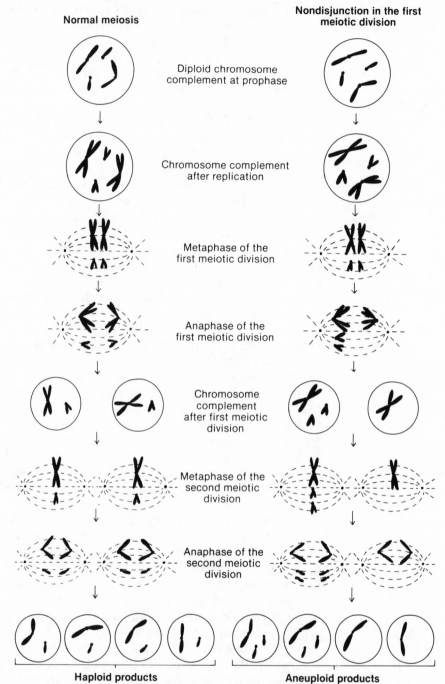

Normal meiosis

Nondisjunction in the first meiotic division

Diploid chromosome complement at prophase

Chromosome complement after replication

Metaphase of the first meiotic division

Anaphase of the first meiotic division

Chromosome complement after first meiotic division

Metaphase of the second meiotic division

Anaphase of the second meiotic division

Haploid products

Aneuploid products

FIGURE 3–5. Normal meiosis and abnormal meiosis, resulting in nondisjunction. In normal meiosis all gametes contain an equivalent haploid chromosome complement. In nondisjunction, meiosis fails to correctly separate chromosomes with the result that gametes contain unbalanced chromosome complements.

avoid confusing cause and effect when relating genomic alterations (especially gross visible ones) to speciation events. Chromosomal alterations in some cases certainly function in the maintenance of isolation rather than as the cause of speciation. That this is the case can be supported by two examples. One is that, as we shall see, speciation has occurred quite frequently among the Hawaiian Drosophilidae without any concommitant chromosomal rearrangements. Moreover, among these same drosophilids there are a large number of inversion polymorphisms that are maintained by stabilizing selection within populations of the same species, with no resultant cladogenesis. Inversion polymorphisms may be best interpreted as representing blocks of linked genes that undergo normal segregation as balanced polymorphisms in the population. They apparently function as supergenes that confer heterosis because most cross-overs within an inversion are not incorporated into the oocyte, but are eliminated in the polar bodies during meiosis in females. *Drosophila* males generally show no crossing-over in meiosis. Therefore, inversions do not reduce fertility. In this respect they differ from translocations. Chromosomal rearrangements are not always associated with speciation, nor is their presence necessarily an indication of a speciation event.

In genic sterility one or both sexes of a hybrid are sterile, usually because of abnormal gametogenesis. The time of action can be either pre- or postmeiotic. An example of this combined with an interesting pattern of hybrid lethality is seen in crosses between the two sibling species *Drosophila melanogaster* and *D. simulans*. These two species are morphologically indistinguishable, except for the external genitalia of the males. Their metaphase karyotype is identical and an examination of their polytene chromosomes shows one major inversion in chromosome 3 and five or six smaller ones scattered through the rest of the genome. A. H. Sturtevant found that if he crossed *D. melanogaster* females to *D. simulans* males only female offspring were produced. In the reciprocal cross of *simulans* females with *melanogaster* males only sons were produced. In both cases the viable progeny were completely sterile when crossed either to each other or back to either parental species. The hybrids were found to have small and incompletely developed gonads that did not produce gametes.

The pattern of lethality observed was interpreted by Sturtevant to indicate that the presence of a *simulans* X chromosome was necessary for survival of the hybrid progeny. However, even the presence of this X is not sufficient to recover viability if a *melanogaster* X is also present in *simulans* cytoplasm. A clue to the proximate cause of this lethality as well as a testament to the correctness of Sturtevant's hypothesis has come from more recent studies on this phenomenon. D. S. Durica and H. M. Krider have been able to demonstrate that in the hybrid genotypes one of the nucleolus organizers (the site of ribosomal RNA synthesis) is suppressed, presumably resulting in a lowered efficiency in protein synthesis. Interestingly, T. Takamura and T. K. Watanabe have recently found a naturally occurring dominant variant in *D. simulans* that when present rescues both lethal classes in the reciprocal interspecific crosses.

It is fair to speculate that this autosomal locus on the second chromosome is responsible for the activation or inactivation of the ribosomal genes in the transspecific hybrids. Further, genetic investigations carried out some years ago by H. J. Muller and G. Pontecorvo suggest that the number of genes controlling hybrid sterility is greater than 1, but is not large. Although a single gene on the second chromosome apparently is responsible for the lethal interactions, the precise number of loci seemingly dispersed throughout the genome that are responsible for the sterility of the viable hybrid progeny remain to be shown. A clearer estimate of the genetic differentiation between two sibling species is covered later in this chapter.

The final type of postzygotic isolation is referred to as hybrid breakdown or dysgenesis. Hybrid individuals are formed and are fertile; however, the progeny of these hybrid individuals possess a variety of developmental defects ranging from lethality to decreased vigor and sterility. An example of this phenomenon can be seen in a cross between *Zea mays* and *Zea mexicana*. Mangelsdorf produced hybrid *Z. mays/Z. mexicana* plants and then back-crossed these individuals to the *Z. mays* parent line. The progeny resulting from this cross were seen to be highly "mutable" and produced offspring with defects in endosperm, stature, and other characteristics affecting the vigor of the plants. This type of effect is not unique to plants but also has been observed and analyzed recently in *D. melanogaster*. In this case the dysgenic activity can be seen in crosses between wild caught flies and laboratory-bred stocks that have been isolated from the wild for only a few years. The results of the dysgenesis can be seen in chromosomal breakage and elimination during the mitotic cell division of the developing larvae and in the production of a high mutation rate. Genetic analyses of this process have shown that it is controlled by one or a few genes and may be correlated with the movement (transposition) of small segments of "nomadic" DNA. It would appear therefore that hybrid breakdown is caused by genes that affect the maintenance of the genome. Indeed, it may be this type of mechanism that causes the hybrid lethality observed in the echinoderm example presented earlier in this discussion. What must be stressed finally is that reproductive isolation rarely results from a single cause; as in the example of the *D. melanogaster/D. simulans* hybrids it may result from several blocks to gene flow. Some of these may be pre- and some postzygotic in their action. The particular combination and nature of the mechanism is dependent on the species involved and the peculiarities of their reproductive physiologies and behaviors.

The mechanisms of isolation of gene pools do not in themselves indicate how the isolation arose in the first place. Bush has outlined four possible patterns of speciation that would result in cladogenesis if they were to occur. These four patterns are presented diagramatically in Figure 3–6. The first is generally referred to as classical or allopatric speciation (type Ia in Figure 3–6). The initial state of the population is large and panmictic (freely interbreeding). A change in geography or climate splits this large population into two separate elements and

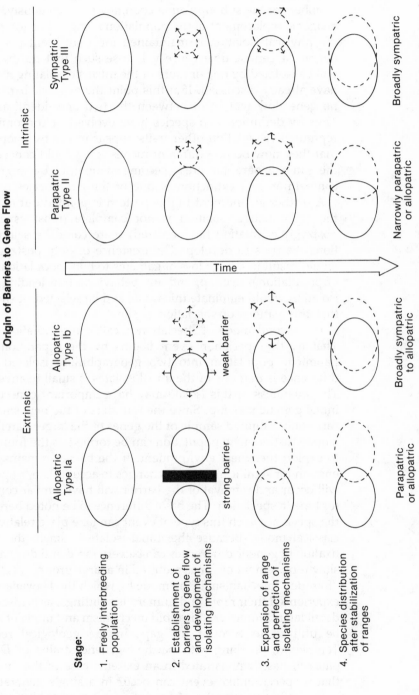

FIGURE 3–6. *Patterns of speciation in animals.* [Redrawn from G. L. Bush, Modes of animal speciation. Reproduced, with permission, from the *Annual Review of Ecology and Systematics,* Volume 6. Copyright © 1975 by Annual Reviews Inc.]

eliminates the possibility of genic exchange by virtue of a physical barrier between populations. Selective pressures or simply random drift of genetic elements separates the two once-identical populations into two genetically discrete groups. This change does not result from all members of the subpopulation changing simultaneously, but by a slow anagenic replacement of the population's original genotypic makeup. If the physical isolation is maintained for a sufficient period of time, a degree of genetic divergence will arise such that the two populations will be isolated by one or more of the intrinsic isolating mechanisms we have already discussed. If at this point the extrinsic barrier is removed no gene flow will occur between the two once-identical populations: Thus by definition two species have evolved. If concomitant with the reproductive isolation other traits appear in the two populations such that they now occupy different niches, they could conceivably cohabit the same geographic area (become sympatric) by migration. If not, competition and extinction of one of the two species could result.

A further amendment to this pattern is pointed out in Figure 3–6. If the reproductive isolation is not complete (most especially if only postzygotic isolation has occurred) there could be selection for additional barriers to develop. The existence of only postzygotic isolation would result in a large loss of gametes to both populations at any point of populational overlap, and any behavioral trait leading to prezygotic isolation would eliminate this waste and be selected. It should be noted that this pattern of speciation is slow.

The second mode of speciation is called type Ib allopatric by Bush. That is, the speciation event begins by the separation of an initial panmictic population into two geographically isolated groups. The difference is that one of the initial isolates is small relative to the other. The smallness of this isolate may have important consequences to its initial genetic makeup. Since the founders of the new small population carry only a limited sample of the genes of the larger parent population, a quite distinct new population can be formed as the founders multiply to exploit their new environment. If the barrier remains in place long enough for significant genetic changes to accumulate, a speciation event will occur and removal of the barrier will have similar consequences as for type Ia speciation. The basic difference to be noted here, however, is the speed at which this type of event can take place relative to the more classical mode. Because the initial isolate is small, the selection and fixation of genetic differences can be very rapid and does not require the slow replacement of genes required in a large group. It is likely that this "founder" mechanism is the mode by which the Hawaiian drosophilids experienced their rapid and numerous splittings, as is discussed in more detail later. Moreover, this rapid mechanism and mode of speciation can be utilized to account for gaps in the geological record. Indeed Templeton, working with parthenogenetic strains of Drosophila mercatorum, has demonstrated in an extreme case of the "founder" effect that a speciationlike event can occur in a single generation.

These first two modes of speciation are reasonably well accepted by evolutionary biologists; however, the last two modes diagrammed in

Figure 3–6 are the subject of somewhat more debate, and probably only occur in certain circumstances and in organisms that employ special reproductive and ecological strategies. The first of these, designated parapatric or type II by Bush (Figure 3–6), has also been referred to as stasipatric by M. J. D. White. In this mode of speciation the isolation event is intrinsic rather than extrinsic. An initial isolate, which is very small relative to the original population, is formed by an alteration in behavior of the founding individuals. The isolation event can be in the form of a closed mating system or deme, as in rodents, or the selection of a new host species by a parasitic organism. By virtue of this isolation, gene flow between the founder group and the parent species is cut off. Again, because gene flow is stopped the two new populations may diverge further from one another. Note that in this model, because the initial event is one that is intrinsic and results in reproductive isolation, it is possible to consider the speciation event as nearly instantaneous in a geological time scale. Furthermore, because the initial isolation is intrinsic the parent and founder population do not necessarily have to be separated geographically.

Finally there is sympatric speciation. In this mode the separation and therefore isolation of the new species takes place wholly within the boundaries of the parent species population; that is, there is no geographic isolation to allow or support the establishment of the reproductive isolating mechanisms discussed previously. In order to achieve this mode of speciation, isolation must occur completely and rapidly within the range and habitat of the parent. Therefore this type is found only under very specific conditions and in organisms with definite prerequisites in their reproductive strategy. Interestingly, this mode may be fairly common in plants. Nearly instantaneous reproductive isolation can occur in plants through the formation of tetraploids; that is, two related diploid plants can produce infrequently, by the production of meiotically unreduced (nondisjunct) pollen and ova, a hybrid plant that is tetraploid (4n) relative to the two diploid (2n) parents. As pointed out by de Wet, the formation of tetraploids more commonly occurs as a two-step process by the production of first a triploid (3n) individual from the fusion of a normal haploid and an unreduced 2n gamete. The 3n individual then produces an unreduced 3n gamete, which when fertilized by a normal gamete produces the necessary tetraploid.

The single-step process can be mimicked experimentally, as in a cross performed between the radish *Raphanus sativa* and the cabbage *Brassica oleracea*. Both plants have a 2n chromosome number of 18. Hybrids can be formed, but they are completely sterile. The nine chromosomes from each parent do not pair and disjoin properly. However, on rare occasions it is possible to obtain diploid pollen and ova from these plants; that is, the reductional division of meiosis has failed for all nine chromosomes. If a plant is formed by the fusion of 2n pollen from one species and 2n ova from the other, a plant of 2n = 36 is formed. There are nine pairs of chromosomes from *R. sativa* and nine pairs from *B. oleracea*, and the plant is tetraploid. This tetraploid is fertile and will perpetuate itself by self-fertilization. Note, however, that the tetraploid plant will produce fertile offspring only with itself and not if crossed back to either parent. By virtue of its new chromosomal constitution it is reproductively isolated from the two parental species and constitutes a new, albeit artificial, species. In nature this type of polyploidy can be produced by the fusion of gametes from the same species (autopolyploidy) or from separate species (allopolyploidy). The effect in either case is the same as above: The tetraploid is reproductively isolated from the two diploid parents. Any cross between 2n and 4n individuals will result in triploid or 3n progeny. These individuals are viable, but are sterile because they usually produce grossly unbalanced or aneuploid gametes. Normally in meiosis chromosomes pair; as long as there is an even number of sets of chromosomes this can be accomplished. If, however, there is an odd number, as in the 3n condition, normal pairing and segregation of chromosomes is disrupted. For any given chromosome, some gametes will get two of one homolog and only one of another. When an unbalanced gamete of this sort combines with a normal haploid gamete the chromosomal constitution of the resultant zygote is also unbalanced. In most cases this condition is lethal. Only if the tetraploid can mate with another tetraploid will fertile 4n offspring result. This type of isolation has occurred in the *Gilia transmontana* species group. These plants are small annual herbs found in the Mojave Desert in the western United States. As noted by Day, and separately by Grant, there are five distinct species that are normally self-fertilizing. Two of these species, *G. transmontana* and *G. malior* are tetraploid, whereas *G. minor*, *G. clokeyi* and *G. aliquata* are all diploid. Breeding experiments have shown that crosses between any two of the five species are sterile. Therefore, despite their similar morphologies and the fact that they all grow sympatrically, they do not interbreed. Based on these morphological similarities Day has concluded that *G. transmontana* is the 4n derivative of the two diploids *G. minor* and *G. clokeyi*, whereas *G. malior*, also 4n, is derived from *G. minor* and *G. aliquata*. The present distribution in conjunction with the chromosomal constitution of these plants can be taken as indirect evidence of a possible sympatric origin.

A further point as to the special nature of plants with regard to this mode of speciation can be made. Many plants can be monoecious (both sexes on a single plant or flower) and self-fertilizing. This property makes polyploidy possible as an instantaneous isolating mechanism. Normally the production of both a diploid male and female gamete is a rare event. Their subsequent fusion is even rarer. Therefore the probability of producing two 4n organisms that could then breed to establish a new species is vanishingly small. However, the ability to cross with oneself obviates this difficulty. The dioecious nature of most animals makes this type of isolating mechanism improbable. Moreover, the sex-determining mechanisms of animals, almost always genically or chromosomally determined, makes polyploidy difficult. Any change in chromosomal and gene dosage with respect to the sex-determining genes or chromosomes tends to produce abnormalities in sex determination, and therefore sterility, and so polyploidy is indeed rare in animals.

It should be noted, however, that in certain cases where reproductive difficulties have been overridden, as in those animals that can produce parthenogenetically, polyploidy is found.

That speciation and therefore evolution in plants offers many special cases is perhaps best summarized in a statement by Grant pointing to the importance of allo- and autopolyploidy in higher plants:

The fundamental characteristics of plants have not only affected the nature of plant species, but they have also had a profound effect on plant macroevolution.... plant phylogeny has involved a repeated anastomosing of previously separate lines. If a phylogenetic tree is the extension of the normal pattern of animal speciation, plant speciation has often led to the formation of a phylogenetic web.

This statement should not be taken to indicate that a sympatric mode of speciation is unique to plants and only occurs by the formation of polyploids. This mechanism is simply the most efficient and rapid. A case can also be made for the sympatric mode occurring in animals by considering behavioral isolation. Bush has invoked this method to account for speciation events in certain insect species that are parasitic on plants. A single gene change could influence host selection by this type of animal and, by altering the plant host selected by a single individual, result in isolation from the remainder of the species group.

There are a variety of modes of speciation and sundry mechanisms by which reproductive isolation can be achieved in plants and animals. But there are two other crucial elements. The first is the amount of time the "event" takes, and second, the amount of genetic divergence that occurs in order for isolation to be affected. Because it is accepted that speciation by the allopatric mode can be a time-consuming process we will not seek to document that fact. Instead, we will seek minimal estimates of the two parameters of time and genetic divergence, assuming that any amount greater than that estimated should also be sufficient for isolation and speciation.

Time and Genetic Divergence

Rates and modes of speciation are well exemplified by the large group of species of drosophilid flies endemic to the Hawaiian Islands, which have been so elegantly studied by H. Carson, his collaborators, and students. Our choice is dictated by several outstanding characteristics of both the Hawaiian Archipelago and the drosophilids found there. The islands are well isolated from the mainland and therefore, like Darwin's Galapagos, offer a natural laboratory for the study of evolution. Moreover, the geological ages of the various islands making up the chain are well known from both potassium-argon dating and magnetic declination data. As one moves on a modern map (see Figure 3–7) of the island chain from north to south the islands become progressively younger. This is caused by the movement of the Pacific crustal plate over a "hot spot" below the earth's surface.

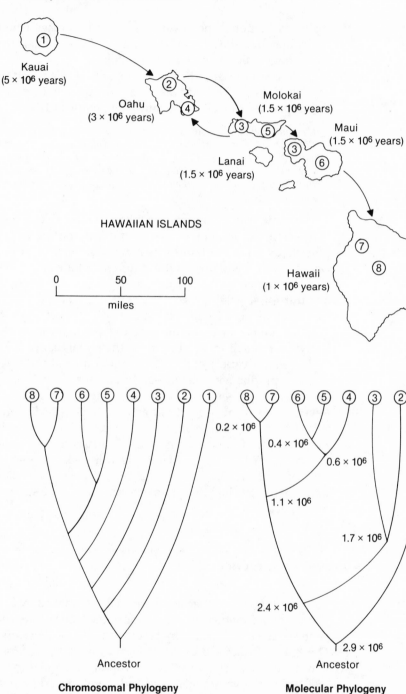

FIGURE 3–7. Map of the major islands of the Hawaiian archipelago. The time given below each named island represents the age of that island based on its time of volcanic origin. The arrows between the islands represent the occurrence and direction of the several migratory and founding events hypothesized in the evolution of the Hawaiian Drosophilidae. Below the map two dendrograms representing the phylogeny of a portion of the planitibia group of the Hawaiian drosophilids are shown. The tree on the left is derived from a phylogeny based on the inversions observed in the polytene chromosomes of these species. The tree on the right is deduced from calculations based on similarities of electrophoretic mobility of enzymes encoded at 10 different loci. (The actual coefficients of similarity are presented in Table 3–2.) The times given at each of the divergence points are based

As the plate has migrated (to the northwest at about 9 cm/year) over this spot, a series of volcanoes have been created. Thus, the island of Kauai was formed in the Pliocene, about 5 million years ago, and Oahu, 3 million. The three islands of Molokai, Maui, and Lanai originally existed as a single large body formed 1.5 million years ago, which subsequently separated into the three smaller islands seen today. Finally, the largest and most recently formed island, Hawaii, still resides over the hot spot, as is evidenced by its still active volcano. The oldest portions of this youngest island were formed in the Pleistocene somewhat less than 1 million years ago. The sequential origin of these islands and their isolation from both the Asian and North American continents have conspired to produce a situation in which colonization of new territory and niches can be closely monitored by examination of the extant species of organisms. A further advantage lies in the fact that Hawaii is tropical and thus has a diverse biota. This fact is amply shown by the drosophilids that we are going to consider. Carson and Kaneshiro estimate that there are more than 500 endemic species of these flies in the relatively small area of these six islands. Therefore the event we wish to study, speciation, has occurred with some regularity in this isolated and temporally well-characterized environment.

A second and equally important aspect of this specific case of analysis is the particular organism under consideration. Not only do these insects offer a large number of species to be analyzed, but by virtue of their genetic properties they yield information about their phylogenetic relationships more easily than most species. This is most clearly seen in their polytene chromosomes. In nearly all of their larval tissues dipteran flies possess cells that undergo DNA replication without cell division.

The replicated interphase chromosomes magnify themselves to the point where they are visible in the light microscope. These chromosomes can be seen to consist of a series of darkly staining bands and less densely stained interbands. Each chromosome from any single species has a very characteristic pattern of thick and thin bands along the length of its arms. When different species are compared for their banding patterns, rearrangements of the genetic material can be detected. This is made simple by the existence of somatic pairing of homologs, as in others of the Diptera. Further, if it is possible to obtain offspring from an interspecific cross, the polytene chromosomes of the resultant progeny can be observed and the presence of genetic rearrangements detected. This tool can be supplemented in these species by also observing the chromosomes from more ordinary cells at metaphase.

---

on a clocklike rate of substitution for the enzymes analyzed. The metric used to set the average rate of substitution is the time of origin of the various islands. The phylogenies of the eight species derived from these two independent means are nearly identical and are both consistent with the notion of interisland migration and the foundation of new species during colonization. The circled numbers on the islands and at the top of the two dendrograms indicate the location and phylogenetic relationship of the following species: (1) D. picticornis, (2) D. substenoptera, (3) D. neopicta, (4) D. hemipeza, (5) D. differens, (6) D. planitibia, (7) D. silvestris, and (8) D. heteroneura. [Derived and adapted from the data of W. E. Johnson et al., Genetic variation in Hawaiian Drosophila, Vol. II. Allozymic differentiation in the D. planitibia subgroup, in Isozymes, Vol. IV: Genetics and Evolution, Academic Press, (C. L. Market, ed.) New York, 1975.]

Carson and F. Clayton have for several years been studying the polytene and metaphase chromosomes of the species of drosophilids endemic to Hawaii. They have found that the great majority of species analyzed thus far have the same basic metaphase karyotype of five rods and one dot, and a diploid chromosome number of 12. The polytene chromosomes, which offer a much greater resolution of chromosome morphology, do not confirm this seeming conservatism but reveal instead a large number of chromosomal inversions. Interestingly, these inversions are not mere random reassemblies of the chromosomes from one species to the next but exist as sets and subsets; that is, if one species is arbitrarily designated as having a standard sequence of chromosome bands, other species can be compared to it by observing how many inversions differ between the two. When this is done with more and more species and the "nonstandard" species are compared, a definite pattern emerges and a phylogeny can be produced by assuming that, for example, if species A differs from standard by a constellation of five or six inversions and species B differs by these same five or six plus three more, it is likely that B is more distantly related to standard than is A. By just such a process Carson has been able to produce a phylogenetic tree for more than 100 of the Hawaiian drosophilids.

A small segment of this chromosomal tree is presented in Figure 3–7 for a portion of what has been called the *planitibia* group of these *Drosophila*. The cladogram presented in Figure 3–7 represents the chromosomal relationships of 8 of the 17 described species of this group. When this cladogram is compared to the island distribution of these eight species, an intriguing pattern emerges and a possible temporal component can be added to the picture. *D. picticornis* is found only on the island of Kauai, the oldest island, and can be shown to be chromosomally more closely related to *D. attigua* and *D. primaeva* than are the other members of the *planitibia* group. These two species are thought to be among the most ancient on the islands and to be closely related to the mainland species believed to have originally colonized the islands and started the entire lineage of Hawaiian drosophilids. On the island of Oahu *D. substenoptera* and *D. hemipeza* are both found. The former is more closely related to *D. picticornis* and is probably derived from a *picticornis*-like ancestor that colonized the island of Oahu subsequent to its formation, approximately 2 million years after the island of Kauai arose. Based on its chromosome morphology *D. hemipeza* is apparently not directly related to either *D. picticornis* or *D. substenoptera*, but derives from a back-migration from the island of Molokai through a *D. neopicta*-like ancestor. This hypothesis is consistent with the fact that *D. hemipeza* contains a *neopicta*-like constellation of inversions as well as two other inversions on the X chromosome, one of which is unique to *D. hemipeza*. The direct derivation of *D. hemipeza* from *D. substenoptera* would therefore violate the rules of parsimony used in the construction of the phylogenetic tree. Moreover, this aspect of the chromosome phylogeny is supported by biochemical evidence produced in an analysis of enzyme polymorphisms in these species, to be presented shortly.

Also found on the island of Molokai is *D. differens*, and on the adjacent and once contiguous island of Maui is the species *D. planitibia*, the species from which the group derives its name. Both of these species have identical polytene chromosome arrangements and yet are reproductively as well as geographically isolated. Since *D. neopicta* is found on both islands it is likely that either *D. differens* or *D. planitibia* arose in a speciation event from *D. neopicta*. Which of the two chromosomally identical species gave rise to the other is less clear, and whether *D. differens* gave rise to *D. planitibia* by a migration and colonization event from Molokai to Maui or the reciprocal cannot be substantiated.

The final pair of species, *D. silvestris* and *D. heteroneura*, are found only on the largest and youngest of the islands, Hawaii. On the basis of their chromosome morphology both of these species, which are chromosomally identical (like *D. differens* and *D. planitibia*), may have arisen subsequent to a colonization event by a *D. planitibia*-like ancestor. The two large island species exist nearly sympatrically and yet interbreed only rarely in nature. In the laboratory, however, matings can be obtained and viable and fertile offspring produced. This would seem to fly in the face of our original definition of species. Nevertheless their distinctive morphologies, to be presented later, as well as their natural isolation can be taken to indicate that their separation has occurred only very recently.

The fact of the natural separation of these two sympatric species brings to light the major, albeit not sole, cause of isolation among these species. The primary mechanism is prezygotic and involves aspects of sexual and reproductive behavior. Males of many of the species of Hawaiian drosophilids are very territorial and will defend against invaders. Females are lured to these leks or reproductive territories and a very rich courtship and coital behavior ensues. Changes in these behaviors, coupled with changes in morphology associated with sex, account for a majority of the isolating factors and therefore speciation events observed in this group of organisms.

The amount and type of variability found in enzymatic proteins has been measured for many of the identified species, most especially for the *plantibia* group. The electrophoretic mobility of a series of different enzymes has been determined and compared for the eight species for which we have presented a chromosome phylogeny. Using 10 of these enzymes Carson and his collaborators have calculated a coefficient of similarity for all pairwise comparisons of the eight species. These values are presented in Table 3–2. A value of 1.00, as seen for a *silvestris* × *silvestris* comparison, indicates genetic identity for these enzymic variations, whereas 0.00 indicates no similarity whatsoever. As can be seen from the table, the two sympatric species from the largest and youngest island, Hawaii, are very closely related. In order to present the reader with a metric for comparison, we cite the work of T. Dobzhansky and his colleagues on the *Drosophila willistoni* complex from central and South America. A similar exercise, comparison, and calculation has been made for this group. Among these flies Dobzhansky and co-workers have identified what they consider to be populations, subspecies,

TABLE 3–2. Coefficient of Similarity for Pairwise Comparisons of Eight Species[a]

	sil	het	pla	dif	hem	neo	sub	pic
silvestris	1.00	0.96	0.74	0.71	0.56	0.39	0.30	0.22
heteroneura		1.00	0.76	0.72	0.58	0.41	0.30	0.23
planitibia			1.00	0.85	0.78	0.49	0.40	0.39
differens				1.00	0.74	0.50	0.40	0.36
hemipeza					1.00	0.40	0.40	0.48
neopicta						1.00	0.59	0.39
substenoptera							1.00	0.29
picticornis								1.00

[a]From W. E. Johnson et al., Genetic variation in Hawaiian *Drosophila*, II. Allozymic differentiation in the *D. planitibia* subgroup, in *Isozymes*, vol. IV: *Genetics and Evolution*, Academic Press, New York, 1975; used with permission.

incipient species, sibling species, and morphologically distinct species, given in order of descending taxonomic similarity. The S-values or coefficients of similarity found for these five catagories are: 0.97, 0.79, 0.79, 0.56, and 0.35, respectively. Comparing these values to those in Table 3–2 we find that *D. silvestris* and *D. heteroneura* are no more allozymically distinct than are two separate populations of the same species of *D. willistoni*. Moreover, we must go to *D. neopicta* before we find an S-value indicating an allozymic difference as great as that between morphologically distinct species of the *D. willistoni* group.

Using these S-values it is possible to create a dendrogram analogous to that for the chromosomal phylogeny. This phylogeny, based on allozymic similarity, is presented in Figure 3–7, adjacent to that for the polytene chromosomes. The similarity is quite striking, and supports and enhances the points made from inversion-derived relationships. Thus, the allozymic data support the hypothesis of a back-migration of *D. hemipeza* to Oahu from the Molokai, Maui complex. This can be seen by virtue of the fact that *D. hemipeza* is more closely related to *D. differens* and *D. planitibia* (S-values 0.74 and 0.78, respectively) than to *D. substenoptera* (S = 0.40).

By utilizing these data Carson has been able to place a time scale on the rate of enzymic or protein change in these species. This is possible based on the accurate values for the ages of the islands on which these species reside and the fact that there could have been no drosophilids present before the island came to be. His best-fit estimate is a 1% accumulation of genetic difference in 20,000 years. Given that these flies breed rather slowly (two generations per year) this divergence occurs in about 40,000 generations. The other point to be noted is that the change would appear to be constant and support other observations of the clocklike evolution of proteins.

Chromosomal and allozymic variations can be used to map phylogenetic relations among these fascinating beasts. However, what about morphology? These flies are classified as separate species on morphological grounds. The most pertinent case is a comparison between the morphologies of *D. silvestris* and *D. heteroneura* on the island of Hawaii. These two sympatric species are identical in karyotype, have a similar basic set of chromosome sequences, and a similar pattern of allozymes. They are, however, morphologically quite distinct. There are striking color differences in the body and wings and a rather odd (for drosophilids) head shape difference between *D. heteroneura* and *D. silvestris* adults (Figure 3–8). In the wild these two species appear to be strongly isolated by a prezygotic mechanism involving an intricate premating ritual specific to each female type. But because this mechanism can be overcome in a laboratory setting and fertile F_1 hybrid flies obtained, it has been possible for F. C. Val to perform reciprocal crosses between these two species and follow the pattern of inheritance of the morphological characters used to differentiate these two species taxonomically. Based on the patterns of inheritance of head shape and size, face color, body pigmentation, and wing spotting, she has been able to make an estimate of from 14 to 19 loci that control the morphological differences between these two species. Because the island of Hawaii is geologically no older than 700,000 years and these two species are endemic to that island their divergence time can be no greater than this value. By using their allozymic difference for 25 loci, which is essentially nil (S = 0.96), they would appear to have diverged even more recently than the earliest origin of this volcano, perhaps as recently as 70,000 years. If this is the case, this particular speciation event offers us evidence for how rapidly a striking morphological change can occur.

FIGURE 3–8. *Dorsal view of the heads of* D. heteroneura *and* D. silvestris. *These two species occur sympatrically on the large island of Hawaii, the youngest of the Hawaiian chain. Despite the striking morphological differences in head shape between the two species, they are interfertile, producing viable fertile offspring in the laboratory. Genetic analysis of these progeny has demonstrated that only a few loci control the head shape difference.* [Redrawn from Val, 1977.]

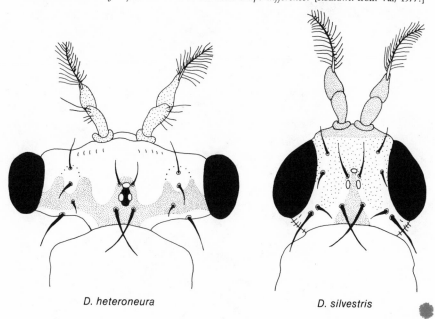

D. heteroneura   *D. silvestris*

In order to demonstrate that this type of change is not peculiar to island biota or the tropics, we can also offer the case of two sympatrically occurring insect species, the North American lace wings *Chrysopa carnea* and *C. downesi*, which have been studied by C. A. and M. J. Tauber. Like the two sympatric *Drosophila* species, *C. carnea* and *C. downesi* are prezygotically isolated in the wild but can be induced to breed and produce fertile offspring in the laboratory. Their isolation in the wild is a consequence of their seasonal and diurnal rhythms. *C. carnea* reproduces to yield three separate generations each summer. In the fall the adults of the third generation enter reproductive diapause and overwinter in that state. On the other hand, *C. downesi* produces only one generation each summer, and reproduces only in the spring. The remainder of the year is spent in diapause. Photoperiod controls the pattern of diapause for each species. A further difference between the two species is color. *C. carnea* is light green during the summer months, but animals in the diapausing population in the fall change to a reddish-brown color. *C. downesi* remain dark green in color all year.

Breeding experiments carried out in the laboratory have shown that the differences between these two species are controlled by three genes. Two of them control the photoperiod-diapause response, whereas a single locus controls the color difference. Although there is no metric for the time of divergence of these two species, it is clear that the amount of genetic change is small and no long period of isolation would be required for such a divergence to occur.

Neither of these cases proves that all speciation events occur in a like manner in a short time frame. However, they do demonstrate two salient points: Speciation and morphological evolution can occur quite rapidly, and they do not require either gross chromosomal rearrangement or quantitatively radical genetic change.

Four

Evolution and the Organization of Eggs and Embryos

Even now multitudes of animals are formed out of the earth with the aid of showers and the sun's genial warmth. So it is not surprising if more and bigger ones took shape and developed in those days, when earth and ether were young...There was a great superfluity of heat and moisture in the soil. So, wherever a suitable spot occurred, there grew up wombs, clinging to the earth by roots. These, when the time was ripe, were burst open by the maturation of the embryos...

Lucretius, On the Nature of the Universe

Spatial Organization and the Initiation of Morphogenesis

The complex changes in size and shape that comprise morphological evolution can occur with substantial rapidity and are achieved primarily not by changes in structural genes, but rather in the regulatory elements that determine the course of development. In later chapters we return to the nature of these regulatory elements and their evolution. At this point it is necessary to consider the processes by which information encoded in the genome is expressed in the increasing morphological complexity of the developing embryo.

The egg is not simply an isotropic reaction vessel containing DNA and the other components required for its transcription and translation: The egg has an internal structure that is as intimately involved as the DNA in determining the course of embryonic morphogenesis. The very initiation of the developmental process and the changes in internal organization of the egg as it begins to cleave in response to its activation by the sperm has been the source of some of the most profound issues in the historical development of embryology as a science.

Early embryologists were faced with the seeming paradox of epigenesis, the complex structure of the animal arising from the apparently structureless egg. Or, as the problem was stated in 1764 by Charles Bonnet, the leading embryological theorist of the eighteenth

101

century, "If organized bodies are not preformed, then they must be formed every day, in virtue of the laws of a special mechanics. Now, I beg you to tell me what mechanics will preside over the formation of a brain, a heart, a lung, and so many other organs?" There was only one logical solution to this dilemma. Bonnet proposed that the egg contains, preformed, a miniature but complete organism that during development grows in size, but not in complexity. This idea rings strange to modern ears, and indeed it led to a great deal of silliness, such as the concept of emboitement, in which the embryo initially present in the egg was visualized as containing an ovary complete with eggs containing yet smaller embryos, and so forth. Attempts were even made to calculate how many embryos lay nestled in Eve's ovaries.

Preformation theory did not require that the tiny precursor embryo in the egg was a homunculus organized identically to the adult it was destined to become, only that the preformed structures were present in some organized entity capable of unfolding to manifest the adult form. However, Bonnet's ideas could not withstand the rise of the science of embryology as exemplified by the work of von Baer, which showed that embryos do in fact increase in morphological complexity as development proceeds. Von Baer was able to demonstrate that all of the differentiated tissues of vertebrates arise from three morphologically simple germ layers: the central nervous system and skin from ectoderm, muscle and skeleton from mesoderm, and digestive organs from endoderm. Later in the nineteenth century Kowalevsky showed that the germ layers themselves arise epigenetically.

A strict preformationalism was replaced by an equally strict theory of epigenesis in which complex structures were envisaged as arising *de novo* from an unstructured egg. This idea in turn was soon to prove unsatisfactory. As observations of the early development of various marine embryos increased in number and detail during the late nineteenth century it became clear that morphogenetic information was present in eggs, and that various regions of the egg were not equivalent with respect to their developmental potentials. This was not preformationalism in the sense of Bonnet. Although eggs contained no miniature organism, neither were they structureless. A hypothesis cast in molecular terms was advanced as early as 1877 by E. R. Lankester:

> All differentiation of cells, the development of one kind of cell from another kind, is dependent on internal movements of the physiological molecules of the protoplasm of such cells....The molecules...are already present before they are made visible to the eye by segregation and accumulation on opposite faces of the differentiating cells. Though the substance of a cell may appear homogeneous under the most powerful microscope, excepting for the fine granular matter suspended in it, it is quite possible, indeed certain, that it may contain, already formed and individualized, various kinds of physiological molecules. The visible process of segregation is only the sequel of a differentiation already established, and not visible.

A century after Lankester's prescient hypothesis we are beginning to gain an understanding of the "physiological molecules" involved in morphogenesis. Newly fertilized eggs contain three information systems that interact during development: the DNA of the nuclear genome, regionalized informational macromolecules in the cytoplasm, and a cytoskeletal matrix that governs the position of localized molecular events in the cytoplasm. The interactions of these informational elements in development, and their changes and roles in the evolution of the major phyla of multicellular animals comprise the theme of this chapter.

As was pointed out by von Baer, development within any group of animals is generally conservative. The mechanistic basis of this observation is obvious. Related organisms are varied expressions of a common body plan based on an inherited common pattern of development. Morphogenesis requires an enormously complex cascading set of interactions in the embryo: Early stages of development are particularly resistant to evolutionary change because any change introduced into early development has such drastic effects on the entire later course of development. New body plans require significantly modified patterns of development; therefore modifications in early development may be expected when organisms have diverged to the extent seen between higher categories, such as classes or phyla. Indeed, in some cases radical transformations of egg and embryo organization have accompanied the evolution of new groups. Remarkably enough, however, early development in some groups of phyla has proved so conservative that while adult morphologies and later stages of development have been profoundly altered, the organization and cleavage of eggs have remained stubbornly similar. This conservativeness has provided one of the principal foundations for arranging the metazoan phylogenetic tree shown in Figure 4–1.

The main trunk of the metazoan phylogenetic tree is divided into two great branches, each containing several phyla. At first glance there would seem to be little logic in grouping together such dissimilar phyla as chordates and echinoderms in the deuterostomes, or platyhelminth flat worms, annelids, mollusks, and arthropods in the protostomes. But when detailed attention is paid to the embryology of these forms the relationships become apparent.

The deuterostomes are so named because the larval mouth arises some distance anteriorly from the blastopore, or site of invagination, at gastrulation, of the cells that will form the primitive gut of the embryo. Deuterstome eggs, once fertilized, begin to undergo cleavages in which the mitotic spindles are alternately oriented in parallel and at right angles to the polar axis of the egg. The result of this radial mode of cleavage is that the blastomeres (or embryonic cells) produced during cleavage come to lie directly above one another, as shown for an echinoderm in Figure 4–2. Cleavage in primitive vertebrates, such as amphibians, follows the same basic pattern.

Like deuterostomes, protostomes are named for the site of origin of the larval mouth, which in protostomes arises from the blastopore or from a nearby site. Cleavage in most of the major protostome phyla follows the spiral mode illustrated in Figure 4–2. The phyla, including nemerteans, platyhelminths, mollusks, and annelids grouped as spira-

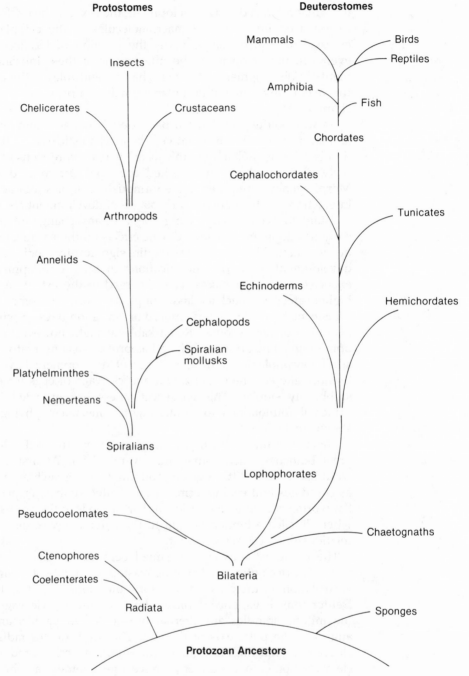

FIGURE 4–1. *A metazoan phylogenetic tree.*

lians in Figure 4–1, all possess a very similar spiral cleavage pattern in their early embryonic stages that, despite the disparate morphology of the adults of these phyla, betrays the closeness of their evolutionary relationships.

Relationships are indicated by another evolutionarily conservative and fundamental aspect of egg organization, the regionalized localiza-

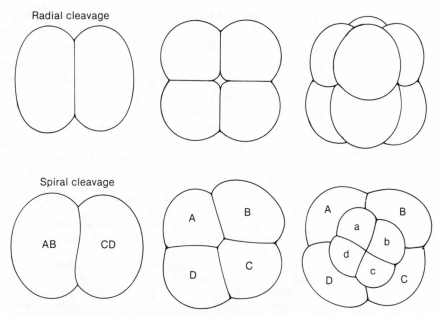

FIGURE 4–2. *Radial and spiral cleavage patterns. The two-, four-, and eight-cell stages of two embryos are shown. Radial cleavage is typical of echinoderms and other deuterostomes, whereas spiral cleavage is typical of many protostomes.*

tion of informational macromolecules in the cytoplasm. Both deuterostome and protostome embryos contain such localized information systems important in the determination of the developmental fates of particular regions of embryos. These localization patterns correlate closely with the patterns of cleavage determined by the timing and placement of the successive mitotic spindles that are reflections of the organization of the cytoskeletal matrix of the egg. The most crucial aspect of the function and evolution of organizational patterns of eggs is the role of such patterns in eliciting, in a regionally precise manner, differential gene expression by the cells of the developing embryos.

Differential Gene Expression in Development

A principal current tenet of modern embryology is that, except in a few specialized cases, all of the cells of an organism, regardless of their particular differentiated states, contain the same DNA genome. Yet gene expression in one type of differentiated cell is demonstrably different from that in another. Differentiated cells evince unique morphologies and maintain distinct patterns of protein synthesis. The messenger RNAs (mRNAs) present are also not identical from cell type to cell type. In accord with these observations the consensus, expressed for example by Davidson in 1976, is that differentiation results from changing patterns of differential gene expression in various cell lineages in the developing embryo.

In bacteria, gene expression is regulated entirely by controls operating at the level of transcription of genes to produce mRNAs. In eucaryotes, regulation of gene action is more complex. Controls operate at the levels of transcription, of processing in the nucleus to yield a messenger from a large and complex RNA transcript, and of transport of mRNA out of the nucleus to the cytoplasm. The translation of mRNA once it reaches the cytoplasm is also regulated by a variety of mechanisms. Our use here of the noncommital term gene expression takes cognizance of the multiplicity of controls that may operate.

The initial determination of blastomeres to follow particular paths of differentiation involves the interaction of the nuclear genome with cytoplasmically localized information. This hypothesis was first clearly stated by T. H. Morgan in 1934 in his book *Embryology and Genetics*:

> *It is known that the protoplasm of different parts of the egg is somewhat different, and that the differences become more conspicuous as the cleavage proceeds, owing to the movements of materials that then take place. From the protoplasm are derived the materials for the growth of the chromatin and for the substances manufactured by the genes. The initial differences in the protoplasmic regions may be supposed to affect the activity of the genes. The genes will then in turn affect the protoplasm, which will start a new series of reciprocal reactions. In this way we can picture to ourselves the gradual elaboration and differentiation of the various regions of the embryo.*

The informational elements of a hypothetical generalized embryo are diagrammed in Figure 4–3, which shows a section of the fertilized egg containing a nucleus and two species of localized cytoplasmic macromolecules indicated by stippling. The cytoskeletal matrix of the egg is represented by the grid. It should be noted that the grid is merely a static representation of a cytoskeletal system which itself appears to change as development proceeds. After the embryo has begun to cleave each cell contains a nucleus equivalent in DNA content to each of the other nuclei, but the nuclei are segregated into different cytoplasmic environments. The arrows in the section represent information flow. Thus the nuclei of the various blastomeres each receives a signal from the particular localized macromolecules. The nuclear response to the signal received depends on the species of macromolecule localized within the blastomeres. This interaction results in the initiation of specific patterns of gene expression by the nuclei (represented by arrows leaving the nuclei). The selective transcription, processing, and translation of

FIGURE 4–3. Regionalized informational systems of eggs and embryos. The cytoskeleton is represented by a grid, the nuclei are black, and regionalized morphogenetic determinants are represented by stippling. Arrows represent information flow. The model is overly static, because in most cases morphogenetic determinants are not prelocalized, but move to their final positions during the first few cleavages. [Modified from Raff, 1977.]

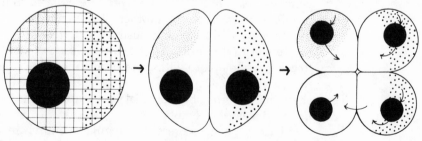

specific portions of the nuclear genome results in the biochemical and morphological differentiation of cells in the embryo. One further important interaction is indicated by the arrow passing from one cell to another. This represents the inductive interactions that arise between groups of cells in the embryo, in which a substance produced by one group of cells induces, at a specific time, the specific differentiation of another group of cells. For example, in chordates the notochord induces the differentiation of the overlying ectoderm into neural tissue.

Convincing cases of localized information that result in regionalized, specific gene expression are rare, but one fine example does exist. Tunicates are primitive relatives of the chordates. As adults they are not very dramatic, being merely sessile, saclike filter feeders. However, the young of most species of tunicates are not only motile, but as tadpole larvae possess a surprising and interesting morphology. These larval tunicates, as shown in Figure 4–4, are organized according to the basic chordate body plan of a dorsal nerve trunk and a notochord, or primitive spinal column. The trunk contains the rudiments of the adult tunicate body, but the only functional embryonic structures of the trunk are the three adhesive papillae at the anterior end, the sensory vesicle, which contains a unicellular otolith, and an ocellus, which contains three lens cells, pigment, and about a dozen retinal cells. These sensory structures allow the larva to orient to gravity and light. The motile tail contains a notochord comprised of 40–42 vacuolated cells. Above the notochord is the neural tube, and on each side a band of striated muscle cells. Each band contains 18 cells. The whole is surrounded by a sheath of epidermal cells. The larva does not feed, and its swimming activity lasts for only a few hours, after which it seeks a suitable substrate, settles, and metamorphoses into a sessile, filter-feeding adult.

In 1973 J. R. Whittaker published a study of the regulation of the appearance of two enzymes that are spatially determined in the early embryo of the tunicate *Ciona*. These are acetylcholinesterase, which appears only in the muscle cells of the tail-bud stage, and tyrosinase, which appears only in two brain pigment cells of the tail-bud stage. By inhibiting cleavage at various times Whittaker produced embryos arrested at various cell division stages, such as 2, 4, 8, 16, or 32 cell (the normal embryo has about 1,000 cells at the time of appearance of the enzymes, which is 9–12 hours after fertilization). Embryos arrested in cleavage remained alive, and produced tyrosinase and acetylcholinesterase at the same time as normal embryos. Most importantly, the production of the enzymes was spatially localized in the cleavage-arrested embryos. In the case of acetylcholinesterase both blastomeres of an embryo arrested at the two-cell stage produced the enzyme, but with arrest at later cleavage stages the synthesis of acetylcholinesterase was found to become progressively restricted to those cells that would normally give rise to the tail muscle. Thus, Whittaker concluded that the capacity for the synthesis of tyrosinase and acetylcholinesterase is localized in the cytoplasm early in development. This interpretation was nicely confirmed by a further study in which Whittaker and his collaborators surgically removed the pair of blastomeres destined to give

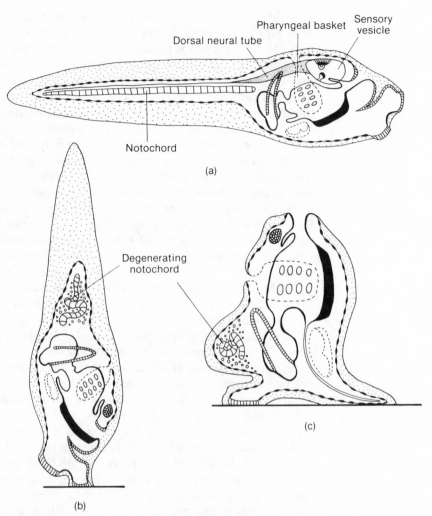

Dorsal neural tube

Pharyngeal basket   Sensory vesicle

Notochord

(a)

Degenerating notochord

(b)

(c)

FIGURE 4–4. *Tadpole larva and metamorphosis of a tunicate.* (a) *Tadpole larva;* (b) *larva attached to substrate and undergoing metamorphosis; and* (c) *completion of metamorphosis with loss of motile and sensory structures.* [Redrawn from E. Korschelt and K. Heider, *Textbook of the Embryology of the Invertebrates,* The Macmillan Publishing Co., New York, 1900.]

rise to muscle from eight-cell *Ciona* embryos. These cells, when subsequently cultured in sea water, produced acetylcholinesterase, whereas the remaining cells of the embryo did not.

Puromycin, an inhibitor of protein synthesis, prevented the appearance of tyrosinase and acetylcholinesterase, indicating that enzyme molecules are actually synthesized at the time of appearance of enzyme activity. Further, treatment with actinomycin D, an inhibitor of RNA synthesis, also prevented appearance of the enzymes if present more than 2 hours prior to the time of appearance of enzyme activity. These results suggest that the mRNAs for the enzymes are not themselves stored, but that they are produced only shortly prior to enzyme synthesis. This regionally specific mRNA synthesis is a result of the action of localized cytoplasmic macromolecules, as modeled in Figure 4–3.

These experiments could perhaps be subjected to the criticism that actinomycin might have inhibited the appearance of the enzyme by poisoning the cells in some nonspecific manner. That this cannot be the case is revealed by another set of similar experiments performed by Whittaker on the appearance of the enzyme alkaline phosphatase, which is confined to the larval endoderm. The localized appearance of this enzyme cannot be suppressed by actinomycin. In this case the mRNA for the enzyme is apparently already present in the egg and is itself progressively localized as development proceeds.

There is an interesting and revealing evolutionary phenomenon associated with the localization of the informational determinants for the appearance of the tunicate larval acetylcholinesterase. Whittaker has also studied some species of the tunicate genus *Mogula* in which larvae fail to develop the expected tadpole form. One of these, *M. arenata*, despite its failure to produce the muscle cells of the tail, and indeed the tail itself, nevertheless does synthesize the localized acetylcholinesterase in the region of the embryo where the tail muscle cells ought to be. Thus, although the capacity for tail morphogenesis has been lost, the localized determinants for acetylcholinesterase have been retained. In some other, presumably older, species of *Mogula*, which also produce tailless larvae, the capacity to synthesize the enzyme has been lost. The importance of the uncoupling of cytodifferentiation, as visualized by the synthesis of specific enzymes or other proteins, from morphogenesis as a mechanism of evolution is discussed in greater detail in Chapter 5.

The studies of Whittaker on tunicates expose two important aspects of the problem: Differential gene expression resulting from the action of localized determinants is crucial to differentiation, and there appears to be more than a single mechanism for storage and expression of localized developmental information.

Embryos actively synthesize proteins using mRNA templates derived from two sources. The first class consists of mRNAs synthesized during oogenesis and stored in the egg until used in development. The second comprises mRNAs synthesized as a result of transcription in the nuclei of the embryo. Both classes contain a large number of sequences, and both are translated during the early stages of development. A sense of the amount and complexity of the oogenetic mRNA may be gained from observing the extent of development possible in sea urchin embryos in which transcription has been blocked. Such embryos develop to the blastula stage, a degree of development that involves a considerable amount of morphogenesis, including cell division, cell shape changes, assembly of cilia, and synthesis of a hatching enzyme. Gastrulation does not occur, however, if embryonic transcription is blocked. Differentiation beyond the blastula stage is largely dependent on gene action in the embryo.

The question of the number of structural genes that must be expressed during the course of development of sea urchin embryos has been investigated by Galau and collaborators. As these workers pointed out, it is not yet known how large a set of genes must be expressed by one cell type to differentiate it from another cell type in the same organism. Nor is it known how many genes provide for the basic "housekeeping" requirements common to all cells. By use of nucleic acid hybridization, Galau et al. determined the number of structural genes represented as active mRNA during various embryonic stages and in various adult tissues. They further determined what proportion of the particular genes represented in gastrula mRNA were also represented in the mRNAs of the other stages and tissues studied. Their finding was that during development very large numbers of structural genes are expressed as mRNA. For example, in the gastrula mRNAs representing between 10,000 and 15,000 genes are in the process of being translated into proteins. Large numbers of structural genes are similarly expressed in other stages and in adult tissues. Some of these are common to all the stages and tissues studied, but a majority are not. Galau et al. concluded that these profound differences in gene expression between developmental stages or tissues underlie their functional differentiation from each other. Thus differentiation during development involves the differential expression as mRNA of thousands of genes, and is associated with changing patterns of mRNA synthesis by the nuclei of cells undergoing differentiation.

If differential gene action is to be elicited by agents localized in the cytoplasm, there should be evidence that cytoplasmic components can indeed direct the course of nuclear function. This evidence is available from experiments in which nuclei from one type of cell are transplanted into a host cell of a different type. A striking example of this experimental approach, as conducted in J. B. Gurdon's laboratory, was the injection of adult frog brain nuclei into three types of recipient frog cells by Graham et al. and by Gurdon. Adult frog brain nuclei do not normally synthesize DNA or undergo mitosis. Nuclei from these cells were injected into (1) oocytes, which synthesize RNA but not DNA; (2) ovulated oocytes undergoing the completion of meiosis and possessing condensed chromosomes on meiotic spindles; and (3) eggs immediately after activation, which are synthesizing DNA but not RNA. In all cases the injected nuclei changed their activities to conform to those characteristic of nuclei of the recipient cell types. This included, for instance, the condensation of chromosomes and their association with spindles in brain nuclei injected into maturing oocytes, or the synthesis of DNA by brain nuclei injected into activated eggs. Because neither of these activities is characteristic of normal brain nuclei, the new activities of the injected nuclei were determined by the host cytoplasms. Similar nuclear transplant experiments have also shown that the transcription of specific genes in the injected nuclei (those for ribosomal RNA) is subject to regulation by host cytoplasm.

The influence of the cytoplasm on the nucleus extends to the eliciting of specific patterns of mRNA synthesis. DeRobertis and Gurdon injected the nuclei of cells of the frog *Xenopus*, grown in tissue culture, into oocytes of the salamander *Pleurodeles*. By high-resolution, two-dimensional gel electrophoresis they were able to distinguish between the protein synthetic pattern characteristic of *Xenopus* from that of *Pleurodeles*, and that of *Xenopus* cultured cells from that of *Xenopus* oocytes. When nuclei from cultured *Xenopus* cells were injected into

Pleurodeles oocytes the *Xenopus* oocyte pattern of proteins was synthesized, not the cultured cell pattern. The shift was prevented by use of the RNA synthesis inhibitor, α-amanitin. Thus, the effect of the oocyte cytoplasmic environment on the *Xenopus* nuclei was to inactivate the expression of one set of genes and to activate the expression of another set characteristic of oocytes.

The most obvious level of nuclear activity at which cytoplasmic control might be exerted is transcription, and there are cases of specific gene transcription in differentiation. One such example is provided by the Balbiani rings of dipteran polytene chromosomes. Flies and other diptera contain in cells of some of their tissues (salivary glands, Malpigian tubules, midgut) giant polytene chromosomes, which when stained for DNA are seen to possess distinctive patterns of bands. Many individual bands have been shown to correspond to the locations of individual genes, and it is possible to correlate the genetic map with the pattern and relative physical position of bands. In certain differentiated cells a limited number of particular bands can be seen to be puffed out from the axis of the chromosome. These Balbiani rings represent genes that are unusually active in transcription. Four characteristics of Balbiani rings are of significance. These are:

1. Different cell types have different Balbiani rings. Thus, the three cell types of the salivary gland of the midge *Acricotopus* each have the same three giant chromosomes, but the chromosomes of each cell type have a characteristic and unique pattern of puffs.
2. Developmental changes in certain cell types are correlated with changes in puffing patterns. The giant footpad cells of certain flies undergo complex developmental changes that are accompanied by an orderly sequential pattern of polytene chromosome puffs.
3. Balbiani rings are the sites of active transcription in polytene chromosomes. Indeed, Daneholt has been able to isolate a unique species of high-molecular-weight RNA transcribed at a particular Balbiani ring of the midge *Chironomus*.
4. Direct correlation exists between the presence of a specific Balbiani ring and the synthesis of a particular protein. Grossbach has studied two closely related midges, *Chironomus tentans* and *C. pallidivittatus*, which contain salivary glands producing large amounts of secretory proteins. Five proteins are synthesized by the salivary glands of *C. tentans*. *C. pallidivittatus* salivary glands produce these same five species plus another. Production of this sixth protein is correlated with the presence of a puff on chromosome 4 of *C. pallidivittatus*. This puff is not present in *C. tentans*. Grossbach crossed the two species to produce hybrids, and was able to show that production of the sixth protein by hybrids is dependent on the presence of the *C. pallidivittatus* chromosome 4 with its unique puff.

Until recently such data were regarded as demonstrating that differential gene expression in development is primarily a matter of differential gene transcription, as is so clearly the case with the Balbiani rings. However, some recent observations have made it necessary to view this conclusion with some caution, since other levels of control of gene expression might be equally important.

The complexity, or number of different single-copy DNA sequences represented as RNA, is generally 5–10 times higher in nuclear RNA as compared to mRNA. The nuclear RNAs, which are the immediate products of transcription, are longer than and include the precursors to mRNAs. A simple differential transcription model would require that two states of development with largely different messenger RNA populations, as those examined by Galau and his co-workers, would exhibit significant differences in their nuclear RNAs. Kleene and Humphreys compared the nuclear RNAs present on two stages of sea urchin development and found the unexpected. The nuclear complexity was very high (about one-third of the single copy genome was transcribed), and the nuclear RNA sequences present in the two stages were identical. This relationship between nuclear RNAs extends to all stages of the life cycle. Wold et al. have observed that while few of the mRNA sequences translated into proteins by sea urchin blastula embryos are also present in the cytoplasm of adult tissues, these sequences are nevertheless present in both embryos and adult tissue nuclei. It appears that the same very large number of structural genes are transcribed in nuclei in all developmental stages, but that only specific subsets of the transcripts are processed to produce the character-istic mRNAs being translated at any particular stage.

The existence of transcriptional as well as posttranscriptional mecha-nisms for differential gene expression by the nucleus will make the task of understanding the controls of gene action ultimately more difficult, but the existence of these mechanisms does not change the basic embryological issue. The hypothesis, diagrammed in Figure 4–3, is that certain macromolecules localized in the cytoplasm and partitioned into some of the embryonic blastomeres come to elicit specific patterns of gene expression by the nuclei of these blastomeres.

Nature and Action of Localized Informational Molecules

It appears likely that the localized cytoplasmic molecules operating to modify gene expression in a regionally specific manner during early development are diverse. Localized determinants occur in many animal phyla, including ctenophores, nemerteans, annelids, mollusks, ar-thropods, echinoderms, tunicates, and chordates. In some instances, the action of localized determinants becomes apparent as early as at the first cleavage division, as for example, in the spectacular polar lobes of a number of spiralians, including the otherwise dowdy mud snail *Ilyanassa*.

The sequence of events associated with the first mitotic division of the fertilized egg of *Ilyanassa* are shown in Figure 4–5. Shortly before cleavage a prominent bulge of cytoplasm termed the polar lobe begins to protrude from the yolk-rich vegetal pole of the egg opposite the animal

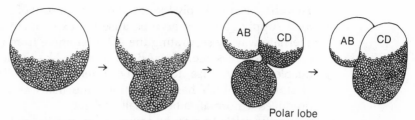

FIGURE 4–5. *Formation and resorption of nonnucleated polar lobe at the time of first cleavage of the embryo of the snail* Ilyanassa. *A nucleus is present in the AB and in the CD blastomere. No nucleus is present in the yolk-rich polar lobe.* [Drawn from life.]

pole at which the cleavage furrow appears. The lobe is oriented perpendicular to the axis of the mitotic spindle. It is important to note that the lobe is nonnucleated and consists of cytoplasm only. When the cleavage furrow begins to deepen, the neck conecting the lobe to the embryo constricts rapidly to a mere thread of cytoplasm. As cleavage is completed, the neck of the polar lobe rapidly increases in diameter, and the lobe is resorbed by one of the blastomeres. The cell into which the lobe is incorporated is designated the CD blastomere, while the other cell is designated the AB blastomere.

The polar lobe is easily removed, and embryos from which the lobe has been deleted continue to develop at the same rate as normal embryos. However, whereas normal embryos give rise to a complex larval form called a veliger, which possesses a variety of structures, delobed embryos give rise only to a ciliated mass of cells (Figure 4–6).

FIGURE 4–6. *Polar lobe function in development.* (a) *Normal veliger larva complete with eyes, foot, shell and internal organs.* (b) *Embryo of same stage from which the polar lobe had been removed at first cleavage. No organized structures are present.* [Photographs courtesy of K. M. Newrock.]

(a)   (b)

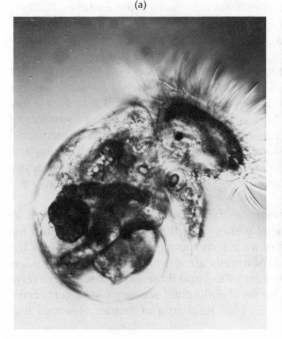

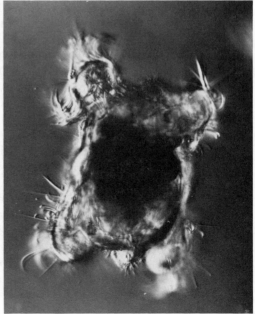

This effect is not simply the result of the removal of a part of the mass or food supply of the embryo because the experiment can be done in an alternate way by separating the AB blastomere from the CD blastomere at the two-cell stage. The AB blastomere, like the lobeless embryo, is unable to give rise to a fully differentiated veliger. But the CD blastomere, which has about the same mass as a lobeless embryo, produces a normal, albeit small, veliger.

Specific mRNAs may be sequestered into the polar lobe of *Ilyanassa*. Lobeless and normal embryos were found by Newrock and Raff to exhibit nonidentical patterns of protein synthesis, even at stages prior to the beginning of morphogenesis. These synthetic differences were also evident between lobeless and normal embryos when both were cultured continuously in the presence of sufficient actinomycin to abolish RNA synthesis. This result was interpreted to indicate that some preformed mRNA species were segregated into the lobe, because the embryos cultured in the presence of actinomycin were translating only those mRNAs already present in the egg cytoplasm at the time of formation of the polar lobe. However, Brandhorst and Newrock, using two-dimensional gel electrophoresis techniques capable of resolving the few hundred most prevalent species, detected no qualitative differences in protein between normal and delobed embryos. They did, however, detect striking quantitative differences that may account for the results of Newrock and Raff. Neither study could detect the potentially very large number of rare mRNA species that could be differentially segregated into the polar lobe.

A comparable problem has been observed in sea urchin embryos. At the fourth cleavage division, three cell types, the mesomeres, macro-meres, and micromeres, which differ considerably from each other in size, are established. These three cell types are, when they appear, committed to separate and distinct fates within the embryo. Rodgers and Gross, as well as Ernst and her co-workers, using nucleic acid hybridization, found that the distribution of RNA of high-sequence complexity is not homogeneous among the three cell types. Similarly, Mizuno et al. and Whiteley et al. have reported that prevalent transcripts of repeated DNA sequences differ between blastomeres. The picture is complicated by the observation of Tufaro and Brandhorst that there are no differences between blastomeres in synthesis patterns of the approximately 1,000 protein species resolvable by two-dimensional gel electrophoresis. The localized sequences detected by Rodgers and Gross and by Ernst et al. may represent sequences too rare to produce sufficient protein to be seen on two-dimensional gels, or they may not represent mRNAs.

The most definitive evidence that localized mRNAs can serve as regulators of morphogenesis comes from the work of Kalthoff and his collaborators on the eggs of an insect, the midge *Smittia*. In insects the anterior and posterior ends of the egg are specified during oogenesis. As normal development proceeds, a head and three thoracic segments form at the anterior end and a series of abdominal segments (and germ cells) form at the posterior end. In 1968 Kalthoff and Sander observed that irradiation of the cytoplasm of the anterior end of the egg with ultraviolet light results in an embryo in which head, thorax, and anterior abdominal segments are replaced by a mirror-image duplication of the posterior end, only lacking germ cells. Normal and double abdomen embryos are shown in Figure 4–7. Two lines of evidence indicate that the anterior determining substance is RNA. The first is the inactivation by ultraviolet light. The ultraviolet action spectrum has peaks of maximum effect at 265 and 285 nm, which is consistent with the involvement of a nucleic acid-protein complex. Interestingly, the effects of ultraviolet irradiation can be reversed by subsequent exposure of the irradiated egg to light with a wavelength of 320–480 nm. Exposure of nucleic acids to ultraviolet light induces the formation of pyrimidine dimers and inactivates the molecule. Photoreversibility results from the action of a light-dependent enzyme that reverses the formation of pyrimidine dimers.

The second line of evidence implicating RNA as the anterior determinant comes from direct exposure of anterior-end cytoplasm to specific enzymes. Kandler-Singer and Kalthoff submerged embryos in a medium containing the enzyme to be tested; the embryos were then punctured in specific regions. Double abdomens were formed as a result of this treatment only if active RNase was admitted to the anterior end. Very few double abdomens resulted from treatment with inactive forms of the enzyme or from experiments in which RNase was allowed to enter regions of the egg other than the anterior end.

There is other, more circumstantial evidence that RNA also functions as a determinant in other organisms. The formation of germ cells at the posterior end of insect (and many other) eggs is determined by the

FIGURE 4–7. *Normal and double-abdomen embryos of the midge* Smittia. *The normal embryo (top) shows a developing head on the left and abdominal segments on the right. The lower embryo has been irradiated with ultraviolet light and has no head, only abdominal segments at both ends.* [From K. Kalthoff, 1969. Photograph courtesy of K. Kalthoff.]

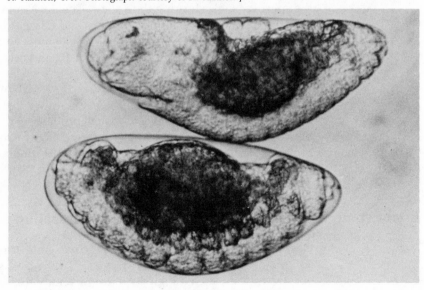

presence of determinants called polar granules, which are readily observed by microscopy. In a series of remarkable experiments Illmensee and Mahowald, using genetically marked strains of *Drosophila*, injected cytoplasm from the posterior pole of eggs of one genetically marked strain into the anterior end of the eggs of a second strain. They discovered that germ cells were induced to form in the anterior region of recipients. Polar granules are, like the anterior determinants, sensitive to ultraviolet irradiation, and appear from cytological staining to contain a large amount of RNA. Structures analogous to polar granules have been found by Dohmen and Verdonk in the polar lobes of some snail embryos. These structures, also found by use of specific stains to be rich in RNA, apparently contain the polar lobe-specific determinants. It would be tempting to seek a relationship to the apparent segregation into the polar lobe of mRNAs observed by Newrock and Raff, but the grounds for doing so are rather tenuous.

Perhaps because current fashion in molecular biology is so strongly intent on nucleic acids there has been less experimental emphasis on proteins as localized determinants of nuclear activity. Yet there is little question that such protein molecules exist. There is a mutation in one amphibian, the axolotl, in which the lack of a specific protein has a drastic effect on development. This mutation, called *o* or *ova deficient*, results in the eggs produced by females homozygous for the *o* mutant allele to stop cleaving and die at about the time when normal embryos are undergoing gastrulation. The *o* mutant is a classic maternal-effect mutant in which only the genotype of the mother determines the developmental fate of the progeny (see Chapter 7). Thus, the eggs of a female homozygous for the *o* allele (*o/o*) will fail to develop, even if fertilized by normal sperm. Conversely, all of the eggs of a heterozygous female (*o/+*), even the half carrying the *o* allele, will develop normally. Since the genotype of the male is unimportant, females of the genotype *o/o* can be obtained by fertilizing the eggs of heterozygous females with the sperm of males bearing the *o* allele. The developmental failure of the eggs of *o/o* females was found by Briggs and Cassens to be preventable if the eggs were injected shortly after fertilization with cytoplasm from normal eggs. The results of Briggs and Justus indicate that the corrective factor missing from the deficient eggs is a protein.

It is clear from a number of studies in which proteins have been injected into eggs that certain proteins readily enter nuclei. That such proteins can influence nuclear behavior has been nicely shown by Benbow and Ford who were able, by exposing isolated frog nuclei to a protein found in the cytoplasm of eggs and embryos, to induce the nuclei to synthesize DNA.

Establishment of Localization and Spatial Organization

Oogenesis is a time of vigorously active gene transcription and of accumulation of mRNAs by oocytes. The mRNAs accumulated in oocytes include such a high-sequence diversity that literally thousands of different protein species can be translated from these templates once these mRNAs become functional in protein synthesis after the start of embryonic development. In the best-studied example, the sea urchin, protein synthesis is very low in the unfertilized egg and rises dramatically a few minutes after fertilization. This initial rise in protein synthesis, and indeed the major part of protein synthesis, at least prior to the blastula stage, is supported by mRNA of oogenetic origin.

Both the *o* corrective protein of the axolotl and the pole plasm of *Drosophila* discussed earlier have been detected in oocytes. Briggs found that *o*-mutant corrective activity is synthesized during oogenesis, and can be detected in an active form quite early in this process. Illmensee and his co-workers similarly assayed *Drosophila* oocytes for active pole plasm. Polar granules could be identified by electron microscopy at the posterior pole of the oocyte in mid-vitellogenesis, or the time of maximal accumulation of yolk, but functional pole plasm could not be detected until late in the maturation of the oocyte. Illmensee et al. therefore proposed that the polar granules that appear during vitellogenesis, while morphologically similar to the polar granules of eggs, may be the matrix upon which functional components are subsequently attached. Finally, Dohmen and Verdonk have found that RNA-rich structures similar to those that are later present in the polar lobes of snails can be demonstrated as early in oogenesis as the beginning of vitellogenesis.

Thus there is little doubt that determinants accumulate in eggs during oogenesis. However, the question of when these materials assume their functional locations still remains open. The model for localization presented in Figure 4–3 makes the simplifying assumption that localization patterns are already established in the egg before cleavage begins. In some cases this is true, but in others localization is progressive and may not be completed until well into cleavage.

The classic example of cytoplasmic localization events triggered by fertilization is that of the movement of pigment granules in the tunicate *Cynthia* (now *Styela*) *partita* described by Conklin in 1905. The sequence of cytoplasmic movements following fertilization is shown in Figure 4–8, which is reproduced from Conklin's 1905 paper. The unfertilized egg is a uniform grayish color, but almost immediately after the entry of a sperm a rapid reorganization of cytoplasm begins. The most dramatic change is the streaming of yellow particles to the vegetal pole of the egg. The yellow material then more gradually spreads from the vegetal pole to cover most of the vegetal hemisphere. Upon movement of the sperm nucleus to one side of the vegetal end of the egg a large part of the yellow material is drawn over with it to form the yellow crescent. The establishment of the localization of the yellow crescent by the movement of the sperm nucleus and its associated aster marks the site of the posterior end of the developing embryo. Other cytoplasmic materials also become localized, so that by first cleavage the egg contains prominent yellow crescent (cr), slate gray yolk (yh), and clear cytoplasmic (cp) regions as well as three other less-prominent regions of colored materials. These localized substances indicate the establishment of determined fates in the regions containing them; thus, the yellow

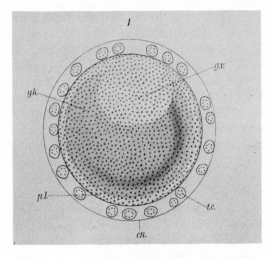

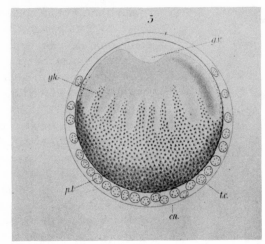

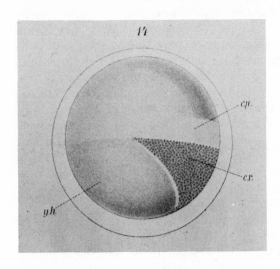

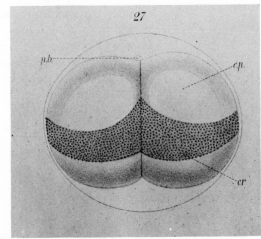

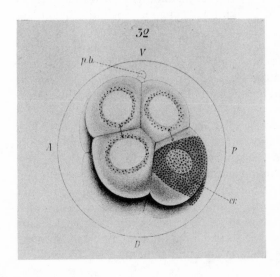

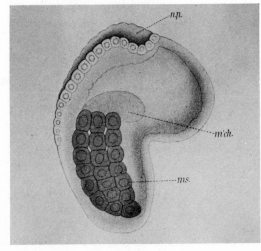

crescent material becomes restricted to the muscle cells of the tadpole, the slate gray material to endoderm, and the transparent cytoplasm to ectoderm. This does not, of course, mean that the colored materials are themselves the determinants, only that they serve as highly visible indicators of cytoplasmic movements involved in the localization of determinants.

In a number of organisms localization patterns are not established until well into cleavage. The investigation of the embryo of the ctenophore *Mnemiopsis* by Freeman has provided one of the most rigorously documented instances of this phenomenon. Ctenophores constitute a small phylum of transparent animals superficially like jelly fish, and commonly known as comb jellies. These animals are biradially symmetrical, and they characteristically bear eight rows of comb plates containing long, fused cilia with which they swim. When disturbed, ctenophores produce rippling flashes of greenish light from specialized light-producing cells, the photocytes, in the meridional canals that lie beneath the comb rows. Both ciliated comb plate cells and photocytes are present in the larval stage, and the differentiation of both results from the action of localized determinants in the embryo.

Development of *Mnemiopsis* is mosaic. If the blastomeres of the embryo are dissociated at the two-cell stage and allowed to develop, each gives rise to a partial embryo containing the structures of a sagittal half of a normal embryo. Blastomeres separated at the four-cell stage give rise to quadrants of the intact embryo that contain both ciliated comb plates and photocytes. Eight-cell embryos consist of two types of blastomeres, four E cells and four M cells. If isolated and allowed to develop, E cells yield partial embryos that contain ciliated comb plates but not photocytes. Isolated M cells, on the other hand, give rise to partial embryos that contain photocytes but lack ciliated comb plates. At the next division, to 16 cells, E blastomeres and M blastomeres divide unequally. Each E cell produces a macromere (E cell) and a micromere (e cell); each M cell likewise produces an M macromere and an m micromere. The pattern of localization of cytoplasmic determinants established in these cells is such that after further development the following pattern of differentiation is observed:

e micromeres $\longrightarrow$ ciliated comb plates
E macromeres $\longrightarrow$ *no* ciliated comb plates
m micromeres $\longrightarrow$ *no* photocytes
M macromeres $\longrightarrow$ photocytes

Freeman performed a set of microsurgical experiments designed to determine when in cleavage localization patterns were established. To

FIGURE 4–8. (Opposite) *Establishment of cytoplasmic localizations in the embryo of a tunicate* Styela partita. (Top left) *Egg with germinal vesicle (gv) still intact;* (upper right) *breakdown of germinal vesicle and streaming of cytoplasm;* (middle panels) *two views of two-cell stage with well-defined yellow crescent (cr);* (lower left) *eight-cell stage;* (lower right) *young tadpole with yellow crescent material confined to tail muscle cells (ms).* [From E. G. Conklin. The organization and cell lineage of the ascidian egg, *J. Acad. Natl. Sci. (Phil.)* **13**:1–119, 1905.]

this end he mapped the regions of the blastomeres of the two- and four-cell embryo destined to produce E and M cells at the eight-cell stage. If the determinants are localized early, as diagrammed in Figure 4–3, then deletion of cytoplasm from the E region or M region of a two-cell embryo blastomere should have the same result as removal of entire E or M blastomeres at the eight-cell stage. That is, deletion of E-region cytoplasm from a two-cell embryo blastomere should result in a loss of the capacity of the cells produced in subsequent development of that blastomere to form ciliated comb plates, whereas deletion of M-region cytoplasm should result in a corresponding loss in capacity to form photocytes. In fact, Freeman found that this prediction was not borne out. Localization of these two capacities was only weakly established in blastomeres of the two-cell embryo, but more nearly completely established in the four-cell embryo, indicating that while the cytoplasmic determinants for comb plate and photocyte differentiation were present at the beginning of cleavage they do not assume their definitive localized positions until the third cleavage.

The timing of the localization of determinants in cleaving embryos is regulated. Studies by Freeman, Guerrier, as well as Dan and Ikeda and others show that the events of localization and early determination of morphogenesis are coupled to the control of the timing of mitotic cleavages, the so-called cleavage clock. The cleavage clock is a poorly understood mechanism by which an embryo keeps track in real time of the number of cleavage cycles experienced and determines the mode of the next cleavage. The existence of this clock has been demonstrated by experiments in which a single early cleavage is suppressed, and then cleavage allowed to resume at the next division. In embryos of the sea urchin, for instance, the fourth cleavage is unequal and produces micromeres as well as larger blastomeres. If an earlier cleavage cycle is suppressed, and then division is allowed to resume, subsequent cleavages occur at the correct time intervals, although they are delayed by one cycle. Accordingly, at the time corresponding to the fourth cleavage in a normal embryo the experimental embryo produces only 8 cells instead of 16. The significant observation, however, is that the experimental embryo, although producing only eight cells, exhibits the unequal *mode* of division characteristic of the normal embryo at this time and forms micromeres. Thus, the timing of the unequal cleavage that yields micromeres is governed by an internal clock tied to absolute time rather than to the number of preceding cleavage cycles actually achieved. The start of the clock appears to be linked to aster formation. In some embryos the clock is set by the completion of meiosis; in others by the initiation of the first mitotic cleavage.

Localization can be demonstrated by cleavage suppression experiments to be coupled to the cleavage clock. The separability of localization and mode of cleavage from the number of cleavage cycles offers considerable evolutionary potential, because if these aspects of early development are under the control of separate genes, then mutations can produce significant changes in the relationship of these events. Indeed, changes of this type have occurred in evolutionary modifications of both spiralian and chordate development. These cases are discussed later in this chapter.

Spatial regulation of localized determinants is affected by the orientation of asters or mitotic spindles that establish embryonic axes. That this orientation is under genic regulation is nicely illustrated by the control of the direction of spiral coiling of the shells of snails. Normally, individuals of the snail *Limnaea peregra* are dextral, that is, the shell and body are coiled in a right-handed spiral. Occasionally, sinistral individuals are found in natural populations. These are complete mirror images of the dextral form, with body and shell coiled in a left-handed spiral. As shown in Figure 4–9, the sinistral or dextral symmetry of snails is detectable from the beginning of cleavage, with the direction of spiral cleavage specified at the second cleavage by the orientation of the mitotic spindles. The orientation of the spindle in a dividing cell determines the location of the cleavage furrow and thus the boundary between the blastomeres resulting from the division.

Symmetry in *Limnaea* is apparently controlled by a pair of alleles of a single gene, and interestingly, it is a maternal-effect gene. The symmetry of progeny is dependent only on the genotype of the mother. The allele for dextral coiling (*L*) is dominant to the allele for sinistral coiling (*l*). But eggs from a mother homozygous for sinistral coiling (*ll*) will develop into sinistral snails even if fertilized with sperm from a parent homozygous for dextral coiling (*LL*). This occurs because the time of effective gene action is during oogenesis when only the sinistral allele is

FIGURE 4–9. *Dextral versus sinistral coiling of the snail* Limnaea peregra *as a consequence of the direction of spiral cleavage in early development.* [Redrawn from T. H. Morgan, *Experimental Embryology*, Columbia University Press, New York, 1927.]

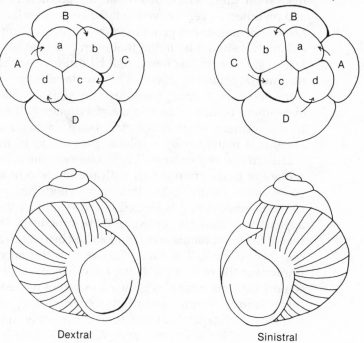

Dextral   Sinistral

present. However, the phenotypically sinistral progeny of this cross are genotypically (*Ll*), and their eggs will all develop into dextral snails because the dominant allele (*L*) will be expressed in the oocyte.

The determination of coiling symmetry in *Limnaea* represents a case in which a rather drastic shift in morphogenesis, the decision to generate a right- or left-handed shell, results from the action of a single gene during the formation of the egg. Moreover, it is apparent that the action of this gene affects components of the cytoskeletal matrix. The cytoskeleton is a dynamic, three-dimensional network of filamentous elements that is responsible for intracellular movements of materials and for changes in cell shape, actions crucial to the establishment of localization patterns in embryos. There are two predominant types of filaments. The first, microtubules, are long, hollow tubules about 25 nm in diameter composed of a group of closely related proteins, the tubulins. The most familiar manifestation of microtubules is the mitotic apparatus, responsible for chromosome movement in mitosis. But microtubules form other cytoplasmic arrays as well. The second major class of filaments, the microfilaments, are solid fibrils about 5 nm in diameter composed of actin, one of the major constituents of muscle. The intricate cytoplasmic networks of microtubules and microfilaments in cultured cells have been depicted in Figure 4–10. Cells are grown on slides, fixed with formaldehyde, and then treated with antibodies specific for actin or tubulin. The attached antibodies are made fluorescent and visualized by ultraviolet dark-field microscopy. Microtubule and microfilament arrays are quite distinct in appearance, location, and orientation.

The arrangement of the cytoskeletal matrix is governed in both time and space. Assembly of microtubules depends on the presence of nucleation sites, which determine the position of microtubule arrays, and on other, as yet not well-defined controls, which determine timing. *Limnaea* demonstrates genetic control of the position of nucleation sites, although it should be noted that even this case has some complicating features. Freeman has found that he can reverse the effect of the allele for sinistral coiling by injection of cytoplasm from dextral eggs into eggs from mothers homozygous for the sinistral allele. The reciprocal experiment results in no change of symmetry. It can be hypothesized that placement of or functional choice among nucleation centers in *Limnaea* is regulated by a soluble component of the egg.

Osborn and Weber have directly observed the microtubule-organizing centers of tissue culture cells with the fluorescent antibody technique. As seen in Figure 4–10, this is a cylindrical, polar structure that contains tubulin, and is located in the vicinity of the nucleus. Each cell appears to contain one or two of these structures. These are most easily seen if the microtubule array is abolished by treatment with colchicine or cold before the cell is fixed. Recovery of the microtubule array after rewarming the cells or removing the colchicine begins at the organizing center, with microtubules growing out from one end of the organizing center and radiating toward the edge of the cell. The fluorescent antibody technique has also been successfuly applied to sea urchin embryos by Harris, Osborn, and Weber, who have been able to visualize

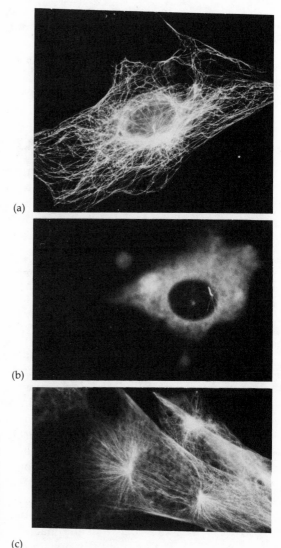

(a)

(b)

(c)

FIGURE 4–10. *Microtubule arrays in mouse cells in culture, visualized by indirect immunofluorescence with an antibody specific to tubulin. (a) Interphase cell, showing cytoplasmic microtubule network. (b) Cell after treatment with colchicine to disrupt microtubules. Note that in both* (a) *and* (b) *the microtubule-organizing structure is visible adjacent to the nucleus. (c) Mitotic cell showing microtubules arranged into asters and spindle.* [From Osborn and Weber, 1976. Photographs courtesy of M. Osborn and K. Weber.]

the microtubules of the spindle apparatus and the asters as well as a curious, transient spiral arrangement of microtubules in the cortex of the zygote.

Microfilaments, like microtubules, have distinct functions in localization events in embryos. Microfilaments are often involved in changing cell shape: Contractility is a basic property of their component protein, actin. For example, microfilaments are the agents of cytokinesis in cleavage. At the end of mitosis a ring of microfilaments forms under the cell membrane in the plane of the metaphase plate. The constriction of the ring pinches the two daughter cells from each other. Similarly, microfilaments have been shown by R. A. Raff and by Conrad and collaborators to cause the constriction of the neck of the polar lobe of mollusks, and thus have a direct function in localization. Positioning of microfilaments in part is dependent on the position of the microtubule arrays. Microtubule-organizing centers have been shown to be present in eggs and embryos, but the recent review by E. C. Raff leaves little doubt that the control of their location remains a central and unsolved problem.

The placement of organizing centers in embryos is at least in part under genic control, but other factors are also significant. For example, in amphibian eggs the point of entry of the sperm determines the plane of the first cleavage, and establishes the dorsal-ventral axis of the embryo. As indicated by the recent studies of Kirschner and his collaborators, the entrainment of localization movements by the events of fertilization appears to be complex. This, along with the previously discussed relationships between localization and cleavage, indicates that the final localization of determinants results from both the organization of the egg in oogenesis and from events set in motion by fertilization.

Evolutionary Changes in the Organization of Spiralian Eggs

Cell fates have been very precisely mapped for a number of embryos with spiral cleavage—particularly mollusks and annelids. Mode of cleavage and size differences between blastomeres have made it possible to follow the developmental fate of particular blastomeres in spiralian embryos. This feature, in combination with the highly mosaic character of the development of these forms, has made spiralian embryos favorite subjects for embryologists. Spiral cleavage is illustrated in Figure 4–2, which shows the first few divisions of a spiralian embryo. The second cleavage gives rise to four blastomeres, A, B, C, and D. At the third, unequal division the first tier of micromeres is formed. These are designated 1a, 1b, 1c, and 1d, and the corresponding macromeres are designated 1A, 1B, 1C, and 1D. The orientation of the mitotic spindles are such that the macromeres and micromeres produced by this and subsequent divisions are arranged in spiral orientation with respect to one another.

At the next division, to 16 cells, the micromeres in the first tier divide equally to produce eight first-tier micromeres. The macromeres divide unequally, producing a second tier of micromeres designated 2a, 2b, 2c, and 2d, and corresponding macromeres. As development proceeds both micromeres and macromeres continue to divide. A system of nomenclature has been devised to designate the complex pattern of cells produced, but for our purposes we need only consider the general arrangement of tiers of micromeres in a spiralian, as presented in Figure 4–11(a). This simplified schematic diagram shows only four cells for each tier. In a real embryo some tiers would of course contain more than four cells because of continued cleavage of micromeres. Thus, this and other similar schematic diagrams used in this section omit a great deal of

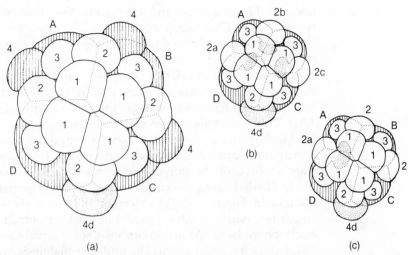

FIGURE 4–11. *Schematic views of cleavage patterns and cell fates in spiralian embryos. (a) A generalized spiralian embryo showing the first three tiers of micromeres (unshaded), which give rise to embryonic ectoderm; the 4d cell (dotted), which gives rise to mesoderm; and the marcromeres (striped), which give rise to endoderm. Schematic views of cleavage patterns and fates for two actual embryos. Crepidula (b) and Unio (c) are also shown. These exhibit similar patterns overall but differ in the sites of origin of embryonic ectomesoderm from second quartet micromeres.* [Redrawn from E. B. Wilson, 1898.]

detail and represent only the general features of the embryos being discussed.

The shading of blastomeres in Figure 4–11 indicates the typical developmental fates of these cells in spiralians. The first three tiers of micromeres (unshaded) give rise to the embryonic ectoderm and some ectodermal structures; the dotted cell (4d) gives rise to mesoderm; macromeres (striped areas) give rise to endoderm, the presumptive midgut. This pattern of cell lineage is remarkably constant for spiralian embryos: The spiral cleavage of polyclad flatworms, annelids, as well as snails and clams are essentially identical. Cell fates in these groups can thus be directly compared.

Despite the constancy of the early cleavage pattern of spiralian embryos, there are nevertheless some important differences that reveal evolutionary modifications in the self-differentiative capacities of blastomeres in mosaic embryos of different groups. These are (1) changes in localization patterns such that the fate of a particular region or blastomere comes to differ from the ancestral condition; (2) changes in relative rates of cell division, resulting in modifications of relative cell sizes and numbers; and (3) changes in the cytoskeletal matrix to produce an alteration in the placement of the mitotic apparatus during cleavage, resulting in shifts in proportion or arrangement of blastomeres in an embryo.

All three of these modifications can be shown to have occurred within the basic spiralian pattern of development. The character of changes in cell fate can be subtle, as in the case of the origin of the ectomesoderm which produces such mesodermal structures as larval muscle. For instance, Figure 4–11(b) and (c) illustrate the origin of ectomesoderm in two mollusks, *Crepidula*, a snail, and *Unio*, a clam. The ectomesoderm of *Crepidula* is derived from three second-quartet micromeres, whereas that of *Unio* is derived from only one.

A striking suite of cell fate changes in annelids is associated with the evolution of oligochaetes from polychaetes. Polychaetes possess a highly differentiated larval stage, the trochophore, shown in Figure 4–12(c). The development of elaborate larval organs requires that many of the blastomeres of the embryo have a temporary larval fate, while the remainder contribute to undifferentiated adult rudiments. The 40-cell-stage embryo of the polychaete worm *Podarke* is shown in Figure 4–12(a). Shaded areas match the areas in the fate map for the *Podarke* blastula in Figure 4–12(b). Most of the areas shown correspond to structures, such as the apical tuft and prototroch specific to the trochophore larva. At metamorphosis some larval tissues, for example, larval muscles and prototroch, undergo histolysis, while other larval tissues, such as the stomodaeum and midgut, transform into equivalent adult structures. Rudimentary adult structures begin to differentiate: Thus, the mesodermal bands give rise to the somites of the trunk.

On the other hand, when larval structures have been suppressed in the course of evolution the majority of the blastomeres of the embryo have a direct adult fate. This has been the case in the evolution of the oligochaetes, which possess no highly differentiated larval stage, and in which development of the segmented adult body begins at gastrulation. Figure 4–12(f) shows the gastrulation of the oligochaete *Tubifex*. No extensive anterior ectoderm with its prototroch is formed. As the midgut invaginates, somite formation leading to the adult body segments begins. This is in contrast to the situation in polychaetes in which, while the stomodaeum and midgut are fully formed in the larva, the presumptive somites are merely present as an undifferentiated rudiment. The resulting modifications of cell fates in the embryo are shown in Figure 4–12(d–f), in which developmental stages of *Tubifex* are drawn for contrast with comparable stages of polychaete development. The cleaving embryo in Figure 4–12(d) shows that in *Tubifex* the pattern of cell division, while still spiralian, has been considerably modified from the polychaete pattern. The micromeres corresponding to those giving rise to the anterior ectoderm of polychaetes are considerably reduced in relative mass in the embryo, whereas the cells yielding presumptive midgut and presumptive mesoderm are greatly increased in relative size.

Cell fates too are modified in the *Tubifex* embryo. Comparison of the fate maps drawn in Figure 4–12(e) with the polychaete fate map in Figure 4–12(b) shows a similar overall relationship of presumptive areas, but many specific changes have occurred. Ectomesoderm, larval ectoderm, apical tuft, and prototroch presumptive regions have all been lost. The blastomeres, which in polychaetes produce these structures, give rise to a different ectodermal structure, the yolk sac, in oligochaetes. Other regions, such as the 4d cell which produces the adult mesoderm in both groups, have the same ultimate fate, but these

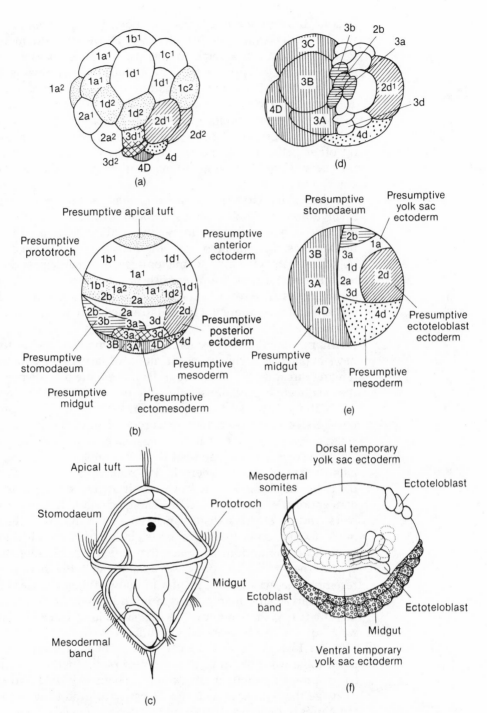

FIGURE 4–12. *Comparison of development of two annelid worms, a polychaete,* Podarke, *and an oligochaete,* Tubifex (a) *The 40-cell embryo of* Podarke; (b) *fate map of* Podarke *blastula;* (c) *trochophore larva;* (d) *cleaving embryo of* Tubifex; (e) *fate map of* Tubifex; (f) *gastrulation leading to direct development of adult segmented body.* [Redrawn from Anderson, 1973, except for Part a, which is redrawn from Treadwell, 1901.]

127

regions have undergone transformations in the developmental pathways by which they reach their fates. However, as demonstrated by Penners, *Tubifex* has retained the rigid mosaic character of annelid development. It appears that the localized determinants have been changed so that they elicit a modified pattern of gene expression in *Tubifex* relative to the ancestral polychaete. Further modification may be seen in the development of the leech *Erpobdella*, which develops in a nutritive cocoon. The 1A, 1B, and 1C cells fail to divide, are overgrown by other cells, and come to function as albuminotrophic cells. The products of the D cell come to make up most of the actual body of the embryo.

Evolutionary changes in the cell fates as drawn for the annelids involve not only modifications in localized inducers of differentiation, but also require changes in relative timing of cleavage by various blastomeres. Macromeres that, for instance, divide several times in polychaetes to give rise to micromeres that participate in such embryonic structures as the prototroch or ectomesoderm do not divide at all in leeches. The role of evolutionary changes in the relative timing of cell division is certainly not limited to annelids; it can also be clearly seen in mollusks.

Among the otherwise generally staid clams, the freshwater unionids have evolved a highly unusual mode of larval life. Unionids live in flowing streams and thus face a distinct problem in dispersal because the adults are essentially sessile, whereas free-swimming larvae would inexorably be carried downstream. Gravid unionid clams produce vast numbers of embryos, which are brooded until the larval glochidium stage is reached. The glochidium resembles a tiny bear trap (Figure 4–13a): To complete development this larva must become attached to the gills or fins of a fish. There it will remain, a small and improbable parasite, until after a few weeks it drops off and assumes the conventional existence of a clam. The glochidium has sensitive sensory hairs, and the slightest disturbance causes its valves with their powerful hooks to snap shut. In some species glochidia are merely released to lie in wait on the bottom in hopes that a fish will blunder by. In other species the female clam possesses a mantle with the margin modified to resemble the eyes and body of a minnow. When the clam is ready to release larvae, the mantle margin is undulated, presumably to attract passing fish to the "minnow" so that suitable hosts will be lured within range of the newly released glochidia.

In his 1898 study of the early development of the freshwater clam *Unio*, Lillie was able to convincingly show that while *Unio* retains the basic spiralian patent, it has achieved changes in rate and patterns of cleavage that correlate with the necessity of producing the specialized structures of the larva. In most mollusks and annelids the micromeres of the first tier form the apical region and prototroch. These structures are lacking in *Unio*, in which the rate of division of the first tier of micromeres is retarded relative to the rate of cleavage of the second tier. This retardation can be seen in a comparison of the distribution of cells present at the 32-cell stage of *Unio* and an "idealized" spiralian in Table 4–1. The second tier of micromeres gives rise to most of the larval

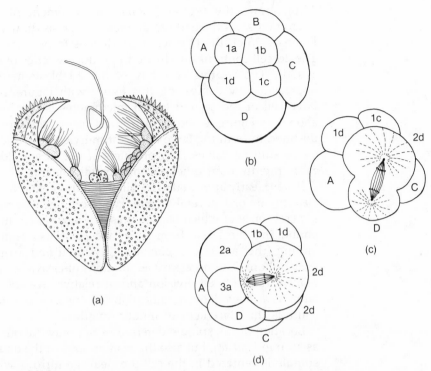

FIGURE 4–13. *Embryos and larva of the clam* Unio. *(a) The glochidium larva with its sensory hairs and "bear trap" valves. (b) Eight-cell embryo, with typical spiralian pattern of cleavage. (c) Production of the relatively large 2a micromere, which produces the larval shell gland. (d) Continued cleavage, with production of the large 2a micromere, which gives rise to the larval adductor muscle.* [Redrawn from Lillie, 1895.]

structures (and mass) of the *Unio* larva. Further, certain embryonic structures in the glochidium are very large relative to other parts of the embryo. One such structure is the shell gland, which produces the relatively massive embryonic shell. This organ is derived from a single second-tier micromere, 2d. Cleavage has been modified so that the 2d micromere is larger than its sister, the 2D macromere, and is in fact the largest cell in the embryo. Figure 4–13(b–d) shows stages in the cleavage of the *Unio* embryo. The eight-cell embryo drawn in Figure 4–

TABLE 4–1. Relative Rates of Cleavage in *Unio* Contrasted with an "Idealized" Spiralian at the 32-cell Stage[a]

	Ideal Spiralian	Unio
First-tier micromeres	16	10
Second-tier micromeres	8	13
Third-tier micromeres	4	4
4d (presumptive mesoderm)		1
Macromeres	4	4
Total numbers of blastomeres	32	32

[a]Modified from Lillie, 1898.

13(b) shows the typical spiralian arrangement of macromeres and micromeres. As the embryo divides further, as drawn in lateral view in Figure 4–13(c), the D macromere divides to produce the large 2d cell. The 2d cell continues to divide to produce a series of smaller cells (Fig. 4–13d). The relative enormity of the 2d blastomere in *Unio* can be appreciated by comparing this figure with Figure 4–12(a). A second large cell of the second tier of micromeres is the 2a blastomere (also shown in Figure 4–13d) which gives rise to the larval ectomesoderm destined to form the large adductor muscle with which the glochidium snaps shut its valves. The products of the 2a and 2d cells also divide more rapidly than other second-tier micromeres.

It was clearly appreciated by Lillie that these modifications of spiralian cleavage involve several factors. While cell fates are not changed in the general sense of which blastomeres produce ectoderm, mesoderm, and endoderm, the larval form is, as we have also seen for oligochaetes among the annelids, considerably modified from the ancestral trochophore. The necessary adaptations of cleavage include changes in relative rates of cell division and in relative sizes of blastomeres. The latter is achieved by modifications in the cytoskeletal matrix, which controls the placement of mitotic spindles.

Location of the spindle determines not only the direction of cleavage as seen in *Limnaea,* but also the relative sizes of the daughter cells. If the spindle is centered in the cell the cleavage furrow, which arises in the plane of the metaphase plate, is equatorially located, and two equal daughter cells result. However, if the spindle is well off-center the resultant cleavage furrow will also be off-center, and one cell will be significantly larger than the other. This is easily seen in the division of the 1D cell illustrated in Figure 4–13(c), which yields two quite disparately sized products.

Control of the rate of cleavage is also localized in the cytoplasm of embryos. In 1904 E. B. Wilson found that at the 16-cell stage of the embryo of the snail *Patella* four cells of the first tier of micromeres were already determined as primary trochoblasts destined to differentiate into the 16 ciliated cells of the prototroch. As cleavage continues these primary trochoblasts are each destined to divide twice more, and then at 10 hours grow cilia arranged in transverse rows. Wilson was able to isolate and culture a single primary trochoblast cell from 16 cell embryos. Isolated trochoblasts divided twice more and no further, and at about the 10th hour produced cilia arranged in the normal pattern. Thus, not only was the differentiative capacity to form specifically ciliated cells segregated in a mosaic fashion, but also the control of rate and number of cell divisions.

Evidence that such control of timing results from gene action during oogenesis comes from cross-specific hybrids between species that differ in rate of development. Hybrids of this type usually develop at the rate characteristic of the maternal species, with paternal characteristics appearing relatively late in development. For example, the frogs *Rana pipiens* and *Rana palustris* differ measurably in rate of cleavage, but hybrids between them develop to adulthood. Cell division during cleavage was shown by Moore to be at the maternal rate. Similarly, hybrids between the sea urchins *Paracentrotus lividus* and *Arbacia lixula* were found by Whiteley and Baltzer to cleave at the rate characteristic of the maternal species. A final, and particularly clear example of control of the timing of development by the egg cytoplasm, comes from the experiments of Minganti, who fertilized enucleated eggs of the tunicate *Ascidia malaca* with sperm of a different tunicate, *Phallusia mamillata*. A significant proportion of the resultant embryos developed as far as the larval stage, and although the only genome present was that of the paternal species, *Phallusia*, the rate of development (which differs in the two species) corresponded to that of *Ascidia*. Thus factors that control the rate of cleavage may be segregated into specific blastomeres in the same manner as factors that initiate specific patterns of differentiation.

Changes in Egg Organization in the Origin of Advanced Protostome Groups

There have been several rather curious major evolutionary trends in early development among spiralians. One of these is the coupling, in several phyla, of an incredibly conservative pattern of spiral cleavage with highly diverse adult body plans. Of the groups we have been considering, annelids, with their highly metameric bodies, bear little similarity to the unsegmented mollusks. Yet both not only possess the same mode of cleavage, but the fates of individual cells are very similar: Thus, the first-quartet micromeres produce larval ectoderm; some second-tier micromeres produce larval ectomesoderm; the macromeres produce endoderm; and the 4d cell ultimately yields the definitive adult mesoderm. In many cases this conservative mode of cleavage may be a consequence of the planktonic larval stages of most spiralians, because a trochophore or similar larva is a basic characteristic of spiralian phyla. Evolutionary modifications in cleavage are most easily correlated with loss or change in larval stages, as we have already seen with *Tubifex* among the annelids and *Unio* among the mollusks.

Another major trend among spiralians has been, in certain groups, a radical shift in early cleavage away from the traditional spiralian pattern. This has occurred in mollusks during the origin of the cephalopods, whose basic body plan is, in spite of advanced nervous, sensory, and locomotor equipment, molluskan. Cephalopods produce very large yolky eggs that develop directly without a distinct larval stage. Cleavage is confined to a thin layer of cytoplasm on the surface of the egg, and as pointed out by Arnold, in no way resembles the spiralian pattern. The changes in cleavage and morphogenesis achieved by cephalopods resemble those of teleost fish and birds, which have also adapted their early developmental processes to the presence of a very large volume of yolk in the egg.

An analogy for the possible course of change in the evolution of the cephalopod egg exists in the eggs of the advanced annelids that possess large yolky eggs and direct development. Cleavage in these forms is modified, and much of the mass of the egg comes to consist of albuminotrophic cells of strictly nutritive function.

Like the cephalopods, the arthropods stem from a spiralian ancestry, but have greatly modified their developmental patterns. Arthropods are generally considered to comprise a single phylum, with close affinity to the annelids. Both annelids and arthropods possess a highly metameric body, serially repeated appendages, a ventral nerve trunk, and a dorsal heart. However, there are a number of characteristics of the three major groups of living arthropods, the crustaceans (shrimp, barnacles, etc.), the chelicerates (horseshoe crabs, spiders, etc.), and the unirames (onychophorans, myriapods, and insects), that have led recent investigators of arthropod phylogeny, such as Manton, D. T. Anderson, and Cisne to consider the arthropods to be an artificial assemblage. With this proposal the Crustacea, Chelicerata, and the Uniramia are classified as separate phyla with independent origins.

Crustaceans are the only arthropods that retain spiral cleavage. Figure 4–14 shows the cleavage of a barnacle. The blastomeres are numbered as suggested by Anderson to correspond to the spiralian convention and to Anderson's views of the possible homologies of these cells with those of spiralian embryos. Comparison of cleavage patterns in this manner appears to be of limited usefulness, because crustacean cleavage is so fundamentally different from classical spiralian cleavage that any attempt to force an analogy is futile. However, fate maps for regions of the surface of the blastula of Crustacea can be compared to those of annelids, as has been done by Anderson. As is seen in Figure 4–15, which presents two such fate maps, the presumptive mesoderm in the annelids lies behind the presumptive midgut, whereas in the crustacean embryo it lies between the presumptive midgut and the presumptive stomodaeum. There has been not only a change in cleavage pattern, but in the basic interrelationships of regions of the egg.

The majority of arthropods other than Crustacea produce very yolky eggs in which the yolk mass remains an undivided syncytium contain-

FIGURE 4–14. *Cleavage pattern of the barnacle* Tetraclita *showing retention of a highly modified spiral cleavage in crustaceans.* [Redrawn from Anderson, 1969.]

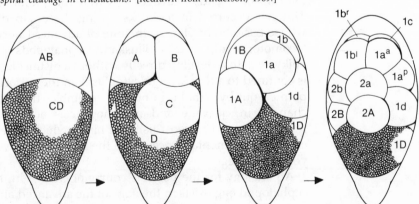

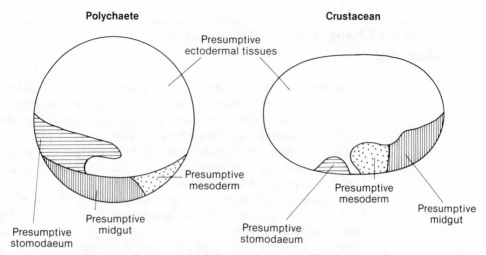

FIGURE 4–15. *Fate maps of polychaete and crustacean blastula-stage embryos.* [Redrawn from Anderson, 1973.]

ing the cleavage nuclei. Subsequent to the cleavage divisions these nuclei migrate to the surface of the egg where, through cellularization, they give rise to the blastoderm. Direct comparison of such a mode of division to spiralian cleavage is of course impossible. Nevertheless, Anderson has compared the fate maps of such arthropods with those of annelids. The onychophorans are the most primitive of the unirames; in fact they are so primitive that they are often put into their own phylum standing between annelids and arthropods. These animals possess some annelid characteristics, including a soft integument, mostly uniform undifferentiated segments, and muscle layers arranged in the annelid pattern. However, some arthropod characteristics, such as jaws, clawed walking legs, and an arthropodlike circulatory system are also present. The possession of trachea suggests an affinity with the insects and myriapods. Comparison of the onychophoran fate map with that of annelids has the interesting result that, unlike the crustacean, the onychophoran embryo has a similar fate map to that of annelid embryos. It therefore appears likely that crustaceans and unirames arose independently. Their origins involved two very different courses of modification of the ancestral spiralian egg. In crustaceans spiral cleavage was retained, but in a completely modified form, and regional fates in the embryo were reorganized. In unirames spiral cleavage was altogether abandoned for a generally syncytial mode of cleavage division prior to blastoderm formation, as an adaptation to very large yolky eggs. But regional fates have maintained their resemblance to those of annelids.

The insects that are the most highly evolved of protostomes are envisaged as having evolved from an annelidlike ancestor in stages resembling the onychophorans, followed by myriapodlike forms. These steps involved the evolution of the unirame leg, organization of the head, and specialization of and reduction in number of segments. The developmental and genetic events underlying these changes are considered in Chapters 7–9.

Evolutionary Changes in the Organization of Chordate Eggs

Spiralians exhibit a pattern of development in which individual blastomeres of the early cleavage embryo are already programmed to follow a specific course of differentiation even when removed from association with the remainder of the embryo. It has been a widely held misapprehension that this mosaic mode of development is typical of protostomes, whereas deuterostomes depend on inductive interactions between cells for determination of cell fates. This idea is incorrect for two reasons. First, some deuterostomes, notably the tunicates, are as highly mosaic as any spiralian. Second, the commitment of particular cells to a defined pathway of development has a temporal element. In what are thought of as typically mosaic embryos this commitment is made very early, but cells become committed sooner or later in all embryos. Thus, all of the cells of the four-cell-stage embryo of the sea urchin (a deuterostome) are equivalent in developmental potential. This is no longer true for 8-cell, and even more obviously, 16-cell embryos in which the blastomeres are clearly committed.

It should be noted that even in spiralian eggs development is not entirely mosaic. Some determinative events do occur quite early, but as development proceeds many inductive interactions increase in importance. One intriguing possibility for the role of mosaic development is that it allows the rapid production of specialized larvae from a limited number of embryonic cells. This mechanism may be particularly advantageous for eggs that develop as plankton, suspended in sea water. The existence of inductive interactions in even the most mosaic of embryos suggests that the relative contributions of self-differentiation and induction might change in an evolutionary lineage, particularly if there is a trend in the lineage toward the loss of specialized larvae. That this process has occurred may be inferred from a consideration of the development of tunicates, amphibians, and mammals, all members of a broad phylogenetic series in the chordate superphylum.

The largely mosaic nature of tunicate development has been demonstrated by two kinds of experiments—experiments in which two embryos are fused early in cleavage and those in which pairs of blastomeres are removed from the embryo and cultured. The former were performed by von Ubisch in 1938. He fused together pairs of embryos at the two-cell stage and found the results were dependent on the relative orientation of the two embryos being fused. Generally, fused embryos developed as double monsters with supernumerary internal organs as, for example, an embryo with a unitary tail containing two notochords, each with an associated neural tube. This result is consistent with the capacity of the blastomeres to differentiate independently of one another. Blastomere deletion experiments, first conducted extensively with tunicates by E. G. Conklin in 1905, have provided more definitive information on the capacities and fates of individual blastomeres. A fate map for the eight-cell tunicate embryo is shown in Figure 4–16. The two anterior animal blastomeres produce head epidermis,

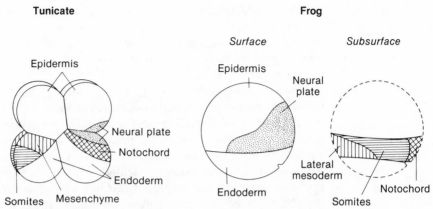

FIGURE 4–16. *Fate maps of tunicate and frog embryos. The amphibian map exhibits similar distribution of fates as the tunicate, but is divided into two layers. Epidermis, endoderm, and neural plate map onto the surface layer, whereas notochord, somites, and mesoderm map to a layer below the surface sheet of cells.* [Modified from Ortolani, 1954; from R. E. Keller, Vital dye mapping of the gastrula and neurula of *Xenopus laevis*, I. Prospective areas and morphogenetic movements of the superficial layer, *Dev. Biol.* **42**:222–241, 1975; and from R. E. Keller, Vital dye mapping of the gastrula and neurula of *Xenopus laevis*, II. Prospective areas and morphogenetic movements of the deep layer, *Dev. Biol.* **51**:118–137, 1976.]

adhesive papillae, and brain; the two posterior animal blastomeres give rise to epidermis only; the anterior vegetal blastomeres produce spinal cord, notochord, and part of the intestine; and the posterior vegetal blastomeres produce intestine, mesenchyme, and muscle.

Conklin also tested the extent to which blastomeres were committed to these fates. By spurting two- or four-cell embryos from a pipette, Conklin was able to kill one or more of the blastomeres to produce half- or quarter-embryos that cleaved normally but remained attached to the nondividing dead blastomeres. These partial embryos could only produce the tissues they were fated to produce in the normal embryo. It can of course be objected that normal development of partial embryos was impaired by the presence of the dead blastomeres, but experiments done by Reverberi with isolated blastomeres have given the same results.

Reverberi and Minganti cultured pairs of blastomeres from eight-cell embryos. All pairs were found to have a limited differentiative capacity in agreement with their positions on Conklin's fate map. One surprising result was that although intestine, notochord, muscles, and mesenchyme were produced by self-differentiation of blastomeres, neural tissue was not. Isolated anterior animal pole blastomeres that should have produced neural tissue produced only epidermis. If eight-cell embryos were dissociated in such a way that contact was retained between the anterior animal pole blastomeres and the anterior vegetal pole blastomeres destined to give rise to notochord, the anterior animal pole cells produced neural tissue. Differentiation of neural tissue thus requires induction by notochord and endoderm cells. This foreshadows the far-greater role of induction observed in vertebrates, but Reverberi has pointed out a significant difference. If in a fractionated embryo contact is maintained between the posterior animal pole blastomeres destined to give rise to epidermis and the anterior vegetal pole blastomeres destined to give rise to notochord, no neural tissue is produced. Posterior ectoderm does not respond to the neural inductive signal from the notochord. This situation stands in contrast to that in amphibians in which inductive relationships have been studied in considerable detail.

Amphibian eggs are easy to obtain, large, and because they can tolerate radical experimental manipulation, have been useful as subjects for nuclear transplantation and grafting experiments. In 1925 Vogt developed a method for determining the fate map of the amphibian embryo. He found that small patches of cells on the surface of the embryo could be permanently stained, without damage to the cells, by applying small pieces of agar soaked in a vital dye. Stained cells could then be followed to their eventual positions in the gastrula. Figure 4–16 presents a fate map that shows the areas of the early amphibian gastrula that will eventually give rise to neural tissue, notochord, mesoderm, and endoderm. These regions occupy the same relative positions as in tunicate embryos, with the modification that the cells that give rise to mesoderm and notochord are not on the surface as in the tunicate, but actually lie beneath a sheet of presumptive endoderm cells. Spemann performed reciprocal transplants in which a piece of presumptive brain tissue was removed from an early gastrula and grafted onto an early gastrula host in a region destined to give rise to epidermis; conversely, a piece of presumptive epidermis was grafted onto the region of host destined to give rise to brain tissue. Grafts healed rapidly and their fates were easily followed because donors and hosts were of two closely related species whose cells differed distinctly in pigmentation. The transplanted cells differentiated in conformity with the host region; thus, their fates were not predetermined.

Spemann and Hilde Mangold later discovered that one region of the amphibian embryo did have self-differentiative abilities. This was the dorsal lip of the blastopore of the early gastrula. The dorsal lip is indicated by the notch on the dorsal side of the amphibian embryo in Figure 4–16. This region is of particular significance because it is the site of the initiation of the invagination movements of gastrulation, and it establishes the principal axis of the embryo. The location of the dorsal lip is itself established soon after fertilization by the location of a gray-colored area called the gray crescent. The location of the gray crescent is opposite the point of entry of the sperm, and its appearance results from a cytoplasmic localization event triggered and spatially defined by fertilization. The dorsal lip region gives rise to a significant proportion of the notochord of the embryo. When Spemann and Mangold grafted a portion of this region onto a host it caused the formation of a secondary embryo containing a notochord and neural fold. The notochord was composed of cells derived from the graft, whereas somites contained both graft and host cells, and the neural fold was almost exclusively composed of host cells. Thus the bulk of cells comprising the secondary embryo was derived from the host, but the grafted dorsal lip induced the differentiation of these host cells into the variety of structures of the secondary embryo.

In an analogous experiment Mangold and Seidel found that fusion of two amphibian embryos at the two-cell stage generally resulted in double embryos. Since the gray crescent, which defines the location of the primary organizer, is already established by the two-cell stage it is clear that fused amphibian embryos will contain two independent organizing centers, and will thus come to form two primary axes. Evidently, early blastomeres of the amphibian lack the extreme self-differentiation capacity seen in tunicates; however, the amphibian does retain the mosaic character of the primary organizer region. The inductive role of the organizer is crucial to the subsequent differentiation of other regions, which depends on a chain of induction events initiated by development of the structures of the primary axis of the embryo.

The developmental strategy of placental mammals is very different from those of both marine invertebrates with small pelagic larvae and vertebrates with large yolky eggs. Most marine invertebrates produce large numbers of eggs containing enough yolk to support rapid development to a stage capable of feeding, whereas vertebrates with yolky eggs, such as the amphibians, produce fewer eggs containing larger provisions of yolk to support a more lengthy development before a feeding stage is attained. All of these eggs also contain stores of ribosomes and mRNA to provide for a rapid beginning of development before enough embryonic nuclei are produced in cleavage to support a high level of protein synthesis with newly transcribed RNA species.

Conversely, placental mammal eggs are small and contain little yolk or other provisions for extended autonomous protein synthesis because development occurs within a container of nutrient medium in the body of the mother. Mammalian eggs have a very slow initial rate of development. A mouse embryo takes three days to complete the first four or five cleavages. By four days it has become a blastocyst of about 100 cells. Implantation occurs at four and a half days. Because there is little stored mRNA present, transcription by embryonic nuclei, which begins as early as before the first cleavage in the mouse, is vital to completion of early development.

Preimplantation development in mammalian embryos results in the formation of the blastocyst (see Fig. 4–17), a hollow blastulalike structure containing two types of cells: trophoblast cells, which form the outside of the embryo, and the inner cell mass, which is located within the cavity surrounded by trophoblast. The trophoblast gives rise to the placenta, whereas the inner cell mass gives rise to the extraembryonic membranes and the embryo itself. Development of the embryo to the blastocyst stage is independent of developmental information gained by interaction with maternal tissue because this stage of development is reached by embryos cultured in a simple nutrient medium containing only pyruvate and salts. Other, more elaborate culture conditions are necessary for early postimplantation development, but the conclusion drawn from these experiments (a good review of this topic is provided by Graham, 1973) is that early postimplantation development is under intrinsic control, with the uterus providing nutrition and support.

Both cell deletion and embryo fusion experiments have been conducted with mammalian embryos to test the degree to which their cells exhibit mosaic development. N. W. Moore and collaborators destroyed

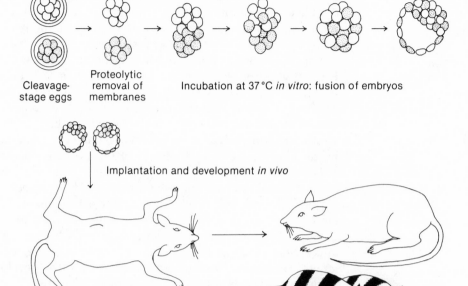

FIGURE 4–17. *Chimeric mice produced by fusion of cleavage-stage embryos of two different genotypes.* [Redrawn from B. Mintz, 1967.]

all but one blastomere of two-, four-, and eight-cell rabbit embryos and transferred the surviving blastomeres into the uteri of foster mothers. Of these, 30% of two-cell, 19% of four-cell, and 11% of eight-cell single blastomeres produced normal baby rabbits. Tarkowski and Wroblewska separated and cultured the individual blastomeres of four- and eight-cell mouse embryos. The fates of each blastomere of only a limited number of individual dissociated embryos could be followed, but some proved very revealing. One four-cell embryo yielded three blastocysts and a trophoblastic vesicle (blastocyst lacking an inner cell mass). Similarly, an eight-cell embryo divided into pairs of blastomeres gave rise to three blastocysts and a morula. None of the mosaic differentiation character-istic of tunicates was apparent. Further, these results contrast with the observations of Ruud in analogous experiments with amphibian em-bryos. Ruud separated each of the four-cell blastomeres and cultured them. The two blastomeres containing portions of the gray crescent region formed small but complete embryos: The other two blastomeres divided but failed to differentiate.

Tarkowski and Wroblewska suggested that predetermined localized regions play no role in differentiation in mammalian embryos. Rather the location of a blastomere in the early blastocyst determines its course of differentiation. Thus, an outside cell becomes trophoblast, and an inside cell becomes inner cell mass. This proposal was tested by Hillman and her co-workers, who transferred marked blastomeres to the inside or outside of unmarked embryos. As predicted by Tarkowski and

Wroblewska, blastomeres differentiated into trophoblast or inner cell mass in accordance with position. Positional control of cell fate and lack of an organizer were also demonstrated in an alternate fashion by Tarkowski and by Mintz, who were able to dissociate morula-stage mouse embryos and aggregate the cells of two embryos bearing different fur-color genes. The resulting hybrid blastocysts were implanted in a foster mother. Normal live young resulted that were chimeras exhibiting patches of both color markers. An experiment of this type is diagrammed in Figure 4–17.

The extent to which the fates of mammalian cells are established by their positions and interactions with other cells in the embryo has been established in a particularly dramatic manner by Mintz and Illmensee. Teratocarcinoma tumors of mice were experimentally produced by implanting an early normal embryo into an extrauterine part of the body cavity. The embryo consequently developed in a disorderly fashion and became a solid tumor containing a rapidly dividing population of stem cells (embryonal carcinoma cells) capable of differentiating into a wide range of tissues. These solid tumors often can be dissociated, and cultured in the peritoneal cavity as an ascites tumor. Ascites tumors consist of embryoid bodies that contain a core of embryonal carcinoma cells surrounded by a sheath of primitive endoderm cells. Mintz and Illmensee injected embryonal carcinoma cells from an ascites tumor line that had retained a euploid chromosome complement into the blastocysts of a genetically marked strain of mice, and obtained healthy progeny that were genetic chimeras composed of normal tissues produced by both host and injected embryonal carcinoma cells. Apparently, the conversion of embryonic tissues to a teratocarcinoma involves disruption in patterns of gene expression and not mutational events, because embryonal carcinoma cells can give rise to normal tissues in response to the specialized environment of the interior of the blastocyst.

In the chordate evolutionary sequence the same basic body plan is retained, but the dependence on mosaic elements in development has apparently become progressively less important until these elements have finally vanished in mammals. Conversely, the role of inductive interactions between regions of the embryo has increased in importance. In tunicates the major induction is that of neural tissue by notochord. This fundamental relationship is retained by more advanced chordates in which the rigid self-determination of other regions of the tunicate embryo has been replaced by a system of determinative events caused by inductive interactions. In cases in which similar tissues or structures result, it seems likely that similar patterns of gene expression occur, although the triggers for gene action may well change in the shift from self-differentiation to dependence on induction.

There is a further and equally fundamental change in chordate development beyond the relaxation of a rigid mosaic pattern of development. In the *Origin of Vertebrates* N. J. Berrill pointed out the significance of changes in the relationship between number of cleavages achieved by an embryo and the onset of the final differentiation of cells. Tunicates typically begin gastrulation between the 64- and 128-cell stage.

According to Conklin, at the 64-cell stage there are 26 presumptive ectoderm cells, 10 presumptive neural plate cells, 4 presumptive notochord cells, 10 mesenchyme cells, 4 presumptive tail-muscle cells, and 10 presumptive endoderm cells. Certain of these cells undergo a limited and discrete number of divisions before final differentiation. Thus, in the tunicate tadpole larva there are 36 tail-muscle cells and 40 notochord cells. The resulting tadpole larva is by necessity small.

Berrill proposed that chordates arose from a tunicate ancestry by neotenous retention of the larval body plan. The strict limitation in cell number and size of the tunicate larva would have severely limited the evolutionary possibilities of any neotenous protochordate. Because there is a practical limit to the size of individual cells, any significant increase in size would have required a greater number of cells of each type. One may or may not accept Berrill's model for a neotenous origin of vertebrates from tunicates, as the fossil record is silent on this point and it is equally probable that adult tunicates represent a specialized terminal addition to the life cycle of animals originally chordatelike as adults. Nevertheless, it is clear from Table 4–2 that there has been a definite shifting of relationships between cleavage cycles and time of differentiation among the chordate phyla. *Oikopleura* is a small pelagic and neotenous tunicate that retains its tail into adulthood. Gastrulation in *Oikopleura* occurs one cleavage cycle earlier than in typical tunicates such as *Styela*. Cell fates are the same, but the number of notochord and tail-muscle cells reflects an earlier time of determination in *Oikopleura*. *Amphioxis*, the most primitive of true chordates, has an egg the same size as that of *Styela*. Since both gastrulation and differentiation are delayed three cleavage cycles, notochord and tail muscle in the *Amiphioxis* larva before feeding and growth begin contain about eight times the number of cells as the *Styela* larva. The vertebrates *Petromyzon*, the lamprey, and *Triturus*, a salamander, have extended this trend even further in the production of large larvae. That there is indeed a genetic basis for a change in the relationship between number of division cycles and differentiation is indicated by the existence of a mutant in *Drosophila* called *giant*, in which homozygotes are double-sized but otherwise

TABLE 4–2. Relationship Between Timing of Cell Determination and Ultimate Larval Cell Numbers in Tunicates and Chordates[a]

| Animal | Gastrulation | | Notochord | Tail Muscle | |
	Cleavage Number	Approx. no. of cells	Approx. no. of cells	Approx. no. of cells	Diameter of Egg (mm)
Oikopleura	5–6	38	20	20	0.09
Styela	6–7	76	40	36	0.13
Amphioxis	9–10	780	330	400	0.12
Petromyzon	11	2,200	500	—	1.0
Triturus	14	16,000	1,200	—	2.6

[a]Modified from Berrill, 1955; used by permission

morphologically normal. This effect results (as discussed in more detail in Chapter 7) from an extra round of cell division during the later stages of larval life.

If there is a moral to the tunicate's tail, perhaps it may serve by providing a useful generalization for the complexities of evolutionary events in early development. In 1933 Joseph Needham pointed out that although developmental processes are closely and elegantly integrated with one another, they are in fact dissociable; that is, it is possible to experimentally separate differentiation from growth or cell division, biochemical differentiation from morphogenesis, and even various aspects of morphogenesis from one another. The importance of this observation to evolution is enormous. Dissociability provides a key to the description of the evolutionary consequences of changes in relative timing in development on morphological evolution, as has been done by Gould in his book *Ontogeny and Phylogeny*, and it places very real constraints on our approaches to the genetic organization of developmental processes.

Five

Interactions Within Embryos

The act of becoming, however, is neither abstract nor instantaneous...
D. A. Sipfle, "On the Intelligibility of the Epochal Theory of Time"

Interaction and Integration

The mounted skeletons of great extinct animals are largely taken for granted by modern museum-goers. But it was not always so. Scientific knowledge of such spectacular forms as mammoths and giant ground sloths dates only to the early nineteenth century when Baron Cuvier astonished the world with his restoration of a host of ancient mammals, thereby revealing a succession of vanished faunas. His ability to reconstruct extinct animals from a jumble of disarticulated bones belonging to many species was a result of his detailed knowledge of comparative anatomy, a discipline largely developed by him. However, Cuvier's sophisticated use of comparative anatomy involved more than an empirical knowledge of anatomy: It was based on his underlying principle of correlation of parts. Thus, a skull bearing long canines and molars adapted for shearing meat, for example, would imply a body and feet adapted to the pursuit and capture of prey—claws rather than hooves. Cuvier realized that not only was the anatomy of extinct animals governed by the same rules as those that operate in living animals, but that all organisms exhibit a fine and profound integration of body organization, a functional and morphological unity. It was this concept that led Cuvier to oppose Lamarck's evolutionary hypothesis, because functional integration requires stability. Minor variability in superficial or functionally peripheral characteristics was permissible in Cuvier's view, but any change in any major part of the body would dissolve the unity of the whole and render the machine an unworkable absurdity. And indeed, this problem of functional integration has continued to plague evolutionary theory. Even if organisms evolve in a punctuational mode in which transitions between stable body plans are rapid, they still cannot avoid the necessity of maintaining a sufficient degree of integration as organisms to remain alive and reproduce throughout the process.

Paradoxically, the basis of both stable integration of structure and

142

change in integration of structure derive from the same cause, the interactions between regions within the embryo during ontogeny.

Regional differentiation in embryos can derive from three kinds of mechanisms. The first is nuclear differentiation, a change in the genetic composition of groups of cells. This mechanism is best known from the work of Theodor Boveri on the development of the notorious parasitic worm *Ascaris*, in which nuclear differentiation is obvious and spectacular. For the first five cleavages of the embryo, each time the cell destined to give rise to the germ line cleaves it produces one germ-line precursor and one presumptive somatic cell. Blastomeres destined to give rise to somatic cells undergo a drastic elimination of portions of their chromosomes and thus become distinctly different from the germ cell precursors, which retain a full chromosome complement. This mechanism is successful in *Ascaris*, but it appears to be something of a biological velocipede: It looks interesting, it works, but there are not many in general use. Although the possibility remains that there are specific and limited genomic rearrangements in some cases, most metazoans (and plants) manage differentiation in other ways.

A second and major mode of regional differentiation depends on the mechanisms of cytoplasmic localization discussed in Chapter 4, in which regionally segregated cytoplasmic "determinants" influence the expression of genes by otherwise genomically identical nuclei. In some cases a mosaic mode of development may govern cell fates, even late in development. It is more common, however, for localization phenomena to be most important in determination of cell fates during very early development, whereas a third kind of control of differentiation becomes dominant later. This final method of triggering and maintaining regionalized differentiation requires signals from outside the differentiating cells. Such signals can derive from the direct influence of the extraembryonic environment or can be generated by other cells within the embryo. The developmental role of signals from both sources is quite clear in mammalian embryos, in which the decision of any blastomere to become embryonic tissue or extraembryonic trophoblast depends directly on the cell's position within the embryo. Blastomeres largely surrounded by other blastomeres give rise to inner cell mass; external cells give rise to trophoblast. As development proceeds, embryonic cells are largely removed from the influence of the external environment, and like passengers growing bored with the scenery, become engaged in conversation with one another. The predominant form of signaling between groups of cells in embryos is induction. Typically, this has been considered as a one-way monologue in which a substance produced by cells of one type causes or facilitates a specific course of differentiation in adjacent cells of another type. There is no doubt that this occurs, but it should be kept in mind that in many inductive systems interactions are reciprocal. One such example is the limb bud of the chick embryo. In experiments pioneered by Saunders and by Zwilling, in which tissues of the limb bud were separated and recombined, it was found that the limb-bud mesoderm induces the overlying ectoderm at the distal tip of the bud to thicken and form an apical ridge. In turn, this apical ridge acts as an inducer of limb structure formation by the mesoderm. Throughout the process the mesoderm continues to produce a factor required for maintenance of the apical ridge.

The regulation of morphogenesis and differentiation by means of interactions between cells of an organism changes during the course of ontogeny. The formation of the neural tube, growth of limb buds, and development of individual organs are classic examples of embryonic induction. All depend on interactions between neighboring groups of cells. In early development individual inductive events are regionally autonomous, and in many cases proceed independently of other inductive events in remote parts of the embryo. But as development proceeds, inductive systems become progressively integrated and their integration plays a large part in ensuring the subsequent canalization of development. The integration of inductive systems results from the formation of cascades and networks. Cascades are characteristic of differentiation because the induction of many structures is dependent on prior induction events. For example, induction of the vertebrate eye lens by the optic cup, which is an outgrowth of the brain, has as an absolute prerequisite the prior induction of the anterior neural system. Networks result from a second characteristic of inductive systems; that is, there may be more than one inducing tissue involved in the induction of a particular structure. In turn, such a structure may act as an inducer to several other tissues.

In the more or less fully developed animal localized interactions continue to play a role, but the problem of maintenance of the organism as a whole becomes critical. Global controls requiring cellular interactions at a distance become necessary and are achieved by humoral agents, hormones. These can be generalized in their developmental functions, such as growth hormone which affects all regions of the body, or thyroxine, which governs the diverse morphological and differentiative processes of amphibian metamorphosis. Other hormones, for example, the thyroid-stimulating hormone or erythropoietin, have more specific functions and target tissues.

Induction and the Appearance of Structure

In superficial examination of the increasing complexity of developing embryos it is possible to observe the appearance of familiar features of the animal: organs, teeth, limbs, and eyes. But embryologists have often found themselves in much the same position as Captain Richard, the protagonist of Ernst Juenger's novel *The Glass Bees*. In the face of the incomprehensible creativity of Zapparoni and his intricate automatons, the glass bees, Richard could only muse: "I had to accept these new creations as they came—I couldn't keep up with the task of interpretation. Much the same thing happens when we watch aquatic animals from a cliff—we see fish and crabs and even recognize jellyfish; but then creatures rise up out of the depth which set us insoluble and disquieting riddles."

One of these riddles is the apparent epigenesis mediated by inductive interactions. The most revealing insights into the nature and role of induction have stemmed from the work of Hans Spemann and his collaborators in the period following World War I. Spemann found that ectodermal cells could be taken from the surface of one amphibian embryo and transplanted onto the surface of a second embryo. Differently pigmented species of the newt *Triturus* were used as donors and recipients, so that transplants could be readily distinguished from host tissues by the presence or absence of pigment granules in patches on the surface of the chimeric embryo. If embryos at the early gastrula stage were used as donors, cells from various regions of the ectoderm would follow the course of differentiation of the region of the host into which they were transplanted. However, if donor embryos had completed gastrulation this was no longer so for most cells. Transplanted cells were found to have already become determined, and so, for instance, transplanted presumptive neural ectoderm cells would differentiate into a patch of neural tissue regardless of the site in the host at which they were placed. These experiments provided a vital piece of information. The course of determination of ectodermal cells in the amphibian is not programmed in these cells: Differentiation of a cell into epithelium or neural tissue depends on its location in the embryo during gastrulation.

The crucial experiment, one of the most spectacular and influential experiments in embryology, was performed by Spemann's student Hilde Mangold (Spemann and Mangold, 1924), and provided a direct demonstration of the induction system responsible for determination of the neural axis. As is so often the case with truly seminal work, Mangold's experiment was to raise some other disquieting riddles.

As shown in Figure 5–1, cells from the surface of the amphibian blastula are moved inward through the blastopore at gastrulation. The result is an embryo with a double-layered structure. This two-layered organization is a necessary prerequisite for the major initial inductive events. But, as recognized by Spemann and Mangold, there is something rather special about the dorsal lip of the blastopore. The dorsal lip is the site of invagination at gastrulation, and although its function is not revealed until gastrulation, the site of the dorsal lip is specified not long after fertilization and is marked by the appearance of the gray crescent.

Mangold's experiment was a simple one: She grafted a section of the dorsal lip of an early gastrula onto the side of a host embryo. However, unlike the earlier experiments with ectoderm grafts the transplanted cells did not develop in conformity with the region into which they had been grafted. Instead, as shown in Figure 5–2, the grafted cells caused the formation of a secondary embryonic axis, a Siamese twin. The secondary embryo lacked (in Mangold's first experiments) a head, but possessed a notochord, paired somites, and a neural tube as well as a gut lumen and renal tubules. Because the graft was made using cells of a different pigmentation than that of the host, it was possible to determine

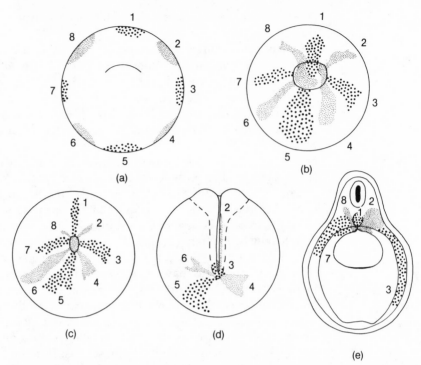

FIGURE 5–1. *Vital dye marking experiments to map cell movements in the gastrulation of the frog* Bombinator. *(a)–(c) Patches of cells marked by staining with vital dye are drawn into the embryo through the blastopore. (d) External view of the resulting neurula. (e) Cross-section of embryo after neural tube closure, showing internal distribution of dye-marked cells. Such experiments allowed the construction of the first amphibian fate maps by Vogt.* [Redrawn from Vogt, 1929.]

which parts of the secondary embryo were derived from transplanted cells and which from host cells. The notochord was found to be composed entirely of transplanted cells, whereas the bulk of the neural tube, somites, and other structures were derived from host embryo cells.

That most of the tissues formed as a result of the graft of the dorsal lip are composed of host cells shows that these cells were induced to follow altered pathways of differentiation by what Spemann came to call the primary organizer. This organizing potential lies in the cells of the dorsal lip that upon invagination gives rise to the archenteron roof, which in turn produces the notochord. The induced axial structures occur in a distinct anterior-to-posterior direction. Experiments in which transplants of regions of the archenteron roof are made show that the ordering of the structures of the central nervous system is determined by inducers produced by the corresponding regions of the archenteron roof. Not only does the underlying notochord mesoderm induce neural structures, but there is an informational specificity in the inducers produced by regions of this mesoderm.

Two generalizations can be made about inducing systems. The inducing tissue must have the capacity to produce whatever inducing substance is necessary, and the target tissue must be competent to

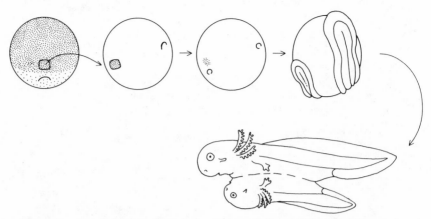

FIGURE 5–2. *Graft of dorsal lip cells from one embryo to the ventral side of a recipient, resulting in the production of a secondary neural axis in the recipient.* [Redrawn from Hadorn, 1974.]

respond. These indispensible properties reflect the dynamics of developmental processes. Production of inducing substances and competence to respond are both transient, and the relative positions of inducing and induced tissues change with the progress of morphogenesis.

The vertebrate eye lens is a specialized structure that functions in the refraction of incident light. During lens development, the presumptive lens tissue invaginates to form a spherical lens rudiment. The cells of the rudiment then elongate to form lens fibers, which are arranged in an orderly array to produce the refractile body of the lens. In the process they lose their nuclei and mitochondria and the lens becomes transparent. Morphological differentiation is accompanied by cytodifferentiation. Lens cells, as reviewed by Bloemendal, become heavily committed to the synthesis of a specialized group of proteins, the lens α-, β-, and γ-crystallins. Crystallin protein synthesis apparently accounts for over 80% of the protein synthetic effort of the lens cells and is supported by long-lived crystallin mRNAs.

The lens arises from the epidermal cells of a restricted region of the head, but in amphibians it has been found that epidermis from almost any region of the embryo can be induced to develop into lens. This can be done either by implanting the optic vesicle in an abnormal position or by replacing epidermis normally destined to become the lens with epidermis from a different part of the body. The optic vesicle, which is an outgrowth of the forebrain that gives rise to the optic nerve and retina, is the main inducer of lens differentiation, but it is not the only inducer.

A particularly informative treatment of the dynamic nature of inductive interactions has been presented by A. G. Jacobson in his analysis of the process of lens induction in amphibian embryos. Morphogenesis in development is accompanied by a great movement of tissues relative to one another. Consequently, the presumptive lens epidermis first, at gastrulation, overlies the endodermal wall of the pharynx-to-be, as diagrammed after Jacobson in Figure 5–3. This endoderm is in fact the first inducer of the lens. As gastrulation proceeds, the heart mesoderm

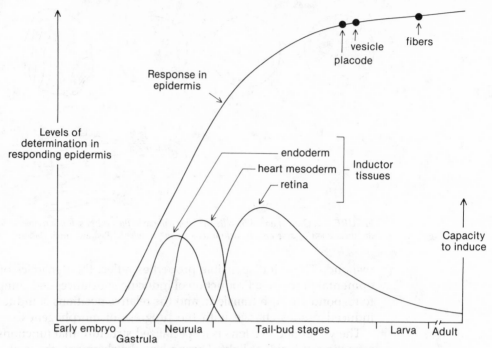

FIGURE 5–3. *Sequential inductive interactions required for lens formation in an amphibian embryo.* [Redrawn from A. G. Jacobson, Inductive processes in embryonic development, *Science* 152:25–34, 1966. Copyright 1966 by the American Association for the Advancement of Science.]

advances until its anterior edge comes to lie beneath the presumptive lens, and it too acts as an inducer. Subsequently, in neurulation the neural plate closes and becomes the neural tube. The optic vesicle begins to evaginate at that time. Closure of the neural tube brings the future lens cells into contact with the presumptive retina, which will from then on serve as the major lens inducer. The degree to which each of the inducing tissues contributes to lens induction was evaluated by Jacobson on the basis of experiments in which inducing tissues were removed and the degree of lens differentiation monitored. In experiments in which the retina was removed, leaving only endoderm and heart mesoderm to act as inducers, 42% of subjects still formed lenses. Jacobson concluded that endoderm and mesoderm together are of equal importance to retina in lens induction. Other organs, such as the nose and ear (as shown in Fig. 5–4), have similar multiple inducers.

Networks of inducers may be important in canalization, ensuring that the normal course of organogenesis will be followed even if one component of the inducing system fails to produce a normally strong signal. And as recognized by Jacobson, multiple inducing tissues may be crucial to the fixing of the precise site at which an organ forms. Jacobson was able to perform a series of experiments in which either the strip of head ectoderm, which would eventually give rise to nose, lens, and ear, or alternately, the underlying neural plate could be rotated 180° at various stages of development. These experiments clearly demonstrated that although the brain provided the strongest inducing activities for the

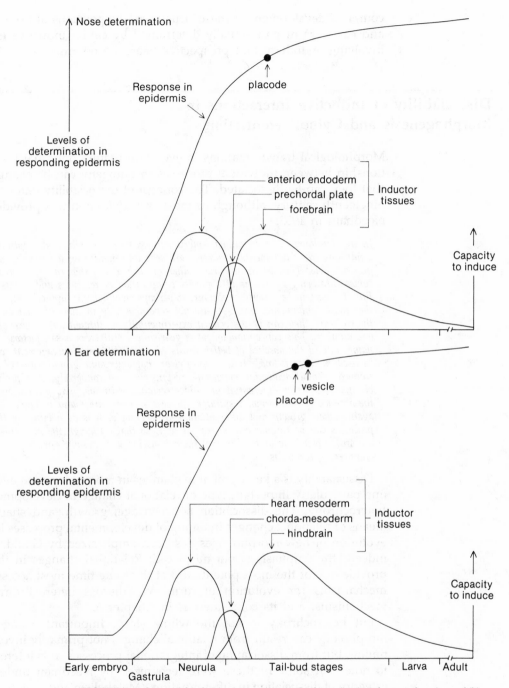

FIGURE 5–4. *Sequential inductive interactions in the determination of nose and ear in amphibians.*
[Redrawn from A. G. Jacobson, Inductive processes in embryonic development, *Science* **152**:25–34, 1966. Copyright 1966 by the American Association for the Advancement of Science.]

149

complete development of these structures, the positions of nose, lens, and ear were in part actually determined by earlier inductive events involving endoderm and prospective heart mesoderm.

Dissociability of Inductive Interactions in Morphogenesis and Cytodifferentiation

Morphological transformations in evolution require only that the relationship between individual processes in ontogeny can be changed—that they can be dissociated. The concept of dissociability dates to the nineteenth century, although its most basic definition was provided by Needham in 1933:

> In the development of an animal embryo, proceeding normally under optimum conditions, the fundamental processes are seen as constituting a perfectly integrated whole. They fit in with each other in such a way that the final product comes into being by means of a precise co-operation of reactions and events. But it seems to be a very important, if perhaps insufficiently appreciated, fact, that these fundamental processes are not separable only in thought; that on the contrary they can be dissociated experimentally or thrown out of gear with one another. This conception of out-of-gearishness still lacks a satisfactory name, but in the absence of better words, dissociability or disengagement will be used in what follows. It is already clear that embryonic growth can be stopped without abolishing embryonic respiration, and conversely, it is probable that growth or differentiation, under certain conditions, may proceed in the absence of the normal respiratory processes. There are many instances, again, where growth and differentiation are separable. It is as if either of these processes can be thrown out of gear at will, so that, although the mechanisms are still intact, one or the other of them is acting as "layshaft" or, in engineering terms, is "idling."

Dissociability is a key element in changes in morphology achieved by one particularly important type of relational change in development—heterochrony. The dissociation between age, growth, and shape by means of relational changes in timing of developmental processes in the evolution of new morphologies has been emphasized by Gould. And indeed, the emphasis is not misplaced: Relational changes in timing provide one of the most potent, and at the same time most accessible, mechanisms for evolution of form. We discuss heterochrony, its mechanisms, and its consequences in Chapter 6.

But heterochrony is not the whole story. Important changes in morphology can result from relational changes not primarily involving timing, but from dissociation of integrational processes. It is interesting to note that some of these were recognized by Needham under his category of dissociation in differentiation. (Metabolism and growth were his other major categories.) Needham listed changes in localization of morphogenetic determinants localized in eggs, cytodifferentiation, and inductive interactions. A good paradigm for the dissociability of such processes is provided by the ease with which morphogenesis and cytodifferentiation can be experimentally uncoupled in the development of a variety of tissues. Several specific examples exist.

In the case of the eye lens discussed earlier, the induction process elicits elongation of lens cells and the production of the lens crystallin proteins. These two processes have been shown by Beebe and Piatigorsky to be dissociable. Beebe and Piatigorsky prepared explants of six-day chick embryo lens epithelium. Explants cultured in the presence of fetal calf serum elongated and synthesized both crystallin mRNA and crystallin protein. Exposure of these explants to the drug colchicine inhibited elongation, but had no effect on crystallin messenger or protein synthesis. A dissociation of a somewhat different sort was found when lens epithelial explants were cultured first in the absence of fetal calf serum, and then after several hours, exposed to fetal calf serum. In these cases cell division was stimulated, but there was no elongation and no stimulation of crystallin synthesis, although crystallin mRNA did increase. These results reveal dissociation between cell division and morphogenesis, and more surprisingly, dissociation even within the pathway of crystallin gene expression.

In vertebrates, a number of internal organs, including lung, liver, pancreas, intestine, and thyroid, arise from endodermal epithelium in combination with mesenchyme cells. These organs have provided very nice experimental systems for examining the inductive interactions responsible for their differentiation because the epithelium and mesenchyme cells destined to form these organs can be dissected out of the embryo and cultured *in vitro*. When normally associated epithelium and mesenchyme are cultured together normal differentiation occurs. Epithelium can also be cultured in combination with mesenchyme derived from other organ primordia. These experiments, which have been reviewed by Wolff and by Deuchar, conclusively show that epithelial differentiation requires inductive signals from the mesenchyme. More interesting, the specific course of differentiation taken by epithelium depends solely on the type of mesenchyme with which the epithelium is combined. Thus, whereas lung endoderm cultured with lung mesenchyme gives rise to the expected bronchial epithelium, lung endoderm cultured with liver mesenchyme gives rise to hepatic cords. Analogous results are found with other epithelial-mesenchymal combinations.

Differentiation in these organ systems comprises both shaping of the organ and cytodifferentiation to yield the appropriate suite of biochemical specializations. The mammalian pancreas offers a particularly good system for experimental dissection of morphogenesis from cytodifferentiation. This complex organ contains two kinds of endocrine cells, the A cells, which produce glucagon; the B cells, which produce insulin; and exocrine cells, which produce and export a suite of enzymes responsible for the hydrolysis of fats, proteins, polysaccharides, and nucleic acids in the digestive tract. The pancreas begins to form in the mouse about halfway through gestation (at about nine days) as a bulge of gut endoderm that projects outward into the surrounding mesenchyme. By about 10.5–11 days of gestation the pancreas rudiment consists of a bulb of epithelium with a constricted base surrounded by a sheath of mesenchyme. In the next few days there is rapid growth and morphogenesis to produce a large number of lobes, the acini, as well as the islets containing the B cells. Cytodifferentiation has been shown by Rutter and his collaborators to begin during this period with the appearance of low levels of synthesis of the characteristic pancreatic enzymes. Between the fifteenth and nineteenth days there ensues a dramatic rise in enzyme synthesis to the fully differentiated levels. Wessells and his co-workers have documented a corresponding intracellular differentiation whose most distinctive element is the appearance in the cytoplasm of masses of zymogen granules that contain the enzymes intended for export from the acinar cells.

Fortunately, from the experimental point of view, Golosow and Grobstein have found that the pancreas organ rudiment undergoes normal differentiation when cultured *in vitro*. Using this well-defined system, Spooner and his co-workers asked if pancreatic cytodifferentiation could be dissociated from morphogenesis. Pancreas rudiments were isolated from 10.5- to 11-day-old mouse embryos and placed in culture. Intact rudiments performed the expected program of growth, morphogenesis of acini, and cytodifferentiation. Epithelia from which mesenchyme had been removed at the time of isolation of the rudiments never developed an acinar morphology, nor did they undergo mitotic growth. Despite the absence of morphological change, the level of the enzyme amylase rose and zymogen granules appeared in the cytoplasm on schedule.

Dissociation of cytodifferentiation from morphogenesis is not limited to laboratory situations. An analogous evolutionary dissociation has occurred in certain tunicates. Most tunicates follow the program of development outlined in Figure 4–4, in which the tadpole larva possesses the chordate pattern of dorsal nerve trunk, notochord, and segmented muscles. The muscle cells of the tail are rich in acetylcholinesterase. There are some species of the family Mogulidae that live on flat sand or mud sea bottoms where there is little need for site selection by the larva, and therefore little necessity for elaborate motile tadpoles. Mogulids, such as *Mogula arenata*, have accordingly abandoned the typical larval tunicate structures, including sensory organs, notochord, and tail muscles. Nevertheless, Whittaker has found that in *M. arenata* embryos the cells that, by homology with the cell lineages of other tunicates possessing tadpole larvae, should have given rise to tail muscles still produce acetylcholinesterase. In another species, *M. pilularis,* both the larval tail and the tail muscle acetylcholinesterase have been abandoned; however, acetylcholinesterase is produced in the muscles and nerves of the adult, suggesting a regulatory change in gene expression because an acetylcholinesterase gene is retained and expressed at a different stage of the life cycle.

The lens, pancreas, and tunicate tail muscle examples are negative in the sense that dissociation is achieved only in conjunction with a loss of structure. It would be more germane to the problem of evolutionary acquisition of novel morphologies or functions to be able to document dissociation events in which cytodifferentiation is achieved in conjunction with a transformation of structure. Examples do exist, with perhaps the best of these provided by the work of Sakakura and his collaborators. In their study, 14-day-old mouse-embryo mammary gland epithelium was recombined either with embryonic mammary mesenchyme or with embryonic salivary gland mesenchyme. The recombined rudiments were cultured within the bodies of syngenic female mice. The morphological results, shown in Figure 5–5, were that mammary epithelium exhibited typical mammary gland morphogenesis when combined with mammary mesenchyme, but underwent salivary gland morphogenesis when combined with salivary gland mesenchyme. Despite their salivary morphology, recombinants of mammary epithelium with salivary mesenchyme behaved biochemically like mammary tissue. Thus, if the host female was lactating, the recombinants exhibited a proliferation of alveoli with enlarged lumens. The typical milk protein, α-lactalbumin, was produced in large amounts, and the alveoli filled with milk. Morphogenesis of mammary epithelium was directed by inductive signals from salivary mesenchyme, but cytodif-

FIGURE 5–5. *Morphogenesis specified by the inducing mesenchyme in recombinants of epithelium with mesenchyme. Panel* (a) shows the typical mammary gland monopodial morphology resulting from mammary epithelium recombined with mammary mesenchyme. Panel (b) *shows branched salivary gland morphology in a recombinant of mammary epithelium with salivary mesenchyme. Despite its salivary morphology the mammary epithelium of this recombinant retains mammary cytodifferentiation.* [From T. Sakakura, Y. Nishizuka, and C. J. Dawe, Mesenchyme-dependent morphogenesis and epithelium-specific cytodifferentiation in mouse mammary gland, *Science* 194:1439–1441, 1976. Copyright 1976 by the American Association for the Advancement of Science. Photographs courtesy of T. Sakakura.]

(a)   (b)

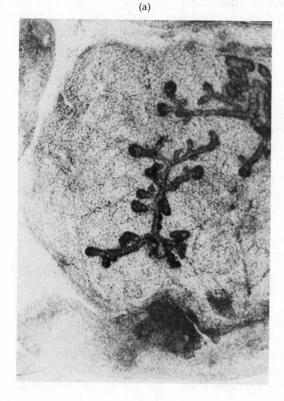
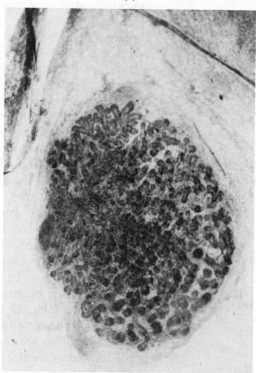

ferentiation was autonomous and had probably already been determined by a separate induction event prior to the time of removal of the
mammary rudiment from the embryo.

Of Hen's Teeth and Lizard's Feathers

If evolutionary changes have indeed arisen as a result of dissociation of
inductive events, it might be expected that remnants of former arrangements would persist—that regulatory genes not elicited in the old way
are still present. The reason for such an expectation is straightforward
and is illustrated in Figure 5–6, which shows O. Mangold's summary
of some of the major induction events within the head and trunk of
amphibian embryos. The diagram emphasizes the generalization already made from the work of Jacobson, that inductive interactions are
complex, with networks and cascades being the rule. The evolutionary
result is that the overall process will resist significant modification, and
when rather major changes occur they will generally do so within the
framework characteristic of the group in question. Drastic transformations involved in the origins of some new groups, such as the
arthropods discussed in Chapter 4, have occurred, but these are rare.

Another consequence of integrated developmental pathways, which
has been widely cited, is that evolutionary changes will be more readily
accepted late in development than early, because changes late in
development simply require fewer adjustments to the affected cascade
of processes. However, this is hardly an exclusive mode of change in
ontogeny. If it were, then something very akin to Haeckelian recapitulation would truly be universal. In fact changes have appeared at all stages
of development, so that although the interactive patterns of amphibians
might be taken as a sort of archetype for vertebrate development, no two
vertebrates develop in precisely the same way—even in very early
development. The early similarity and growing distinctness of embryos
of related organisms described by von Baer's laws should not be read too
literally. Young mammals are never fish or lizard embryos: Their
genomes and developmental patterns are far removed from those of
their ancestors. Yet the basic similarity is there. What has happened is
that multiple inductive interactions, exemplified earlier by the inductions of lens, nose, and ear, have provided for a large degree of
developmental homeostasis, so that changes can be accepted even early
in the overall process without disruption of integration.

An overall conservative pattern of development suggests that embryonic structures may be retained even after they have lost their old
primary function because they still serve as links in a developmental
cascade incidental to the old function. So, too, might regulatory patterns
persist. The test for the retention of ancient regulatory genes is the
resurrection of an abandoned developmental pathway by either mutational or experimental perturbation of the current developmental
pattern. Both have been successfully done. The mutational reincarna-

Induction Processes in the Head

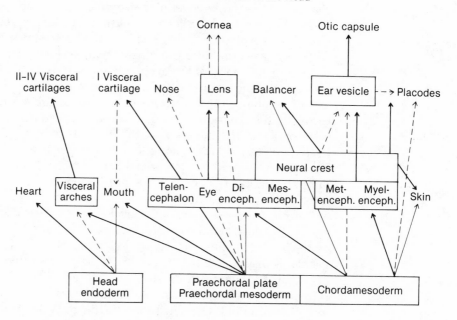

Induction Processes in the Trunk

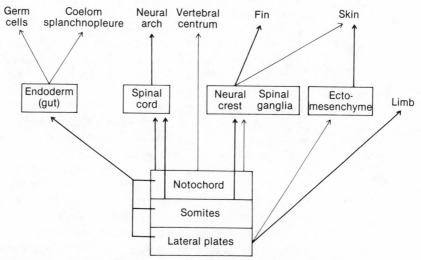

FIGURE 5–6. Inductive interactions and cascades in amphibian development. Top: *Induction processes in the head.* Bottom: *Induction processes in the trunk. Relative strengths of induction are indicated by thickness of the arrows.* [Redrawn from Mangold, 1961.]

tion of old developmental patterns is discussed at length in Chapters 8 and 9. Three evolutionary cases that have been analyzed experimentally have revealed modifications of tissue interaction systems in which new control patterns have been superimposed on the old without the loss of ancestral genetic information. Two of these cases involve modifications that have occurred during the evolution of birds; the third involves changes in the body coverings of reptiles, birds, and mammals.

Archaeopteryx has long been significant as a "missing link" because of its possession of avian traits, such as feathers, in shocking combination with a reptilian, specifically a dinosaurian, skeleton complete with long tail and teeth. Many of the evolutionary changes in birds have led to a greater efficiency of flight through weight reduction accompanied by loss of several of the characteristic features of *Archaeopteryx*, including the teeth, the claws of the hand, the long reptilian tail, and reduction or fusion of some limb bones. Since the last toothed birds date to the Upper Cretaceous, a sufficiently long time has elapsed for avian genomes to have been purged of the genetic information required for tooth morphogenesis. That in fact has not happened, and hen's teeth can actually be conjured forth.

The normal process of tooth development, as reviewed by Thesleff and by Deuchar, requires reciprocal inductive interactions between oral epithelium and the underlying oral mesenchyme. The sequence of events is diagrammed in Figure 5-7. Oral mesenchyme induces the overlying epithelium to differentiate into an enamel organ, which in turn invades the mesenchyme and induces mesenchymal differentiation into the odontoblasts, which secrete dentine. In the absence of an enamel organ, oral mesenchyme gives rise only to spongy bone. There is one further inductive process. The epithelial cells of the inner portion of the enamel organ respond to the presence of the mesenchymal odontoblasts by differentiating into the ameloblasts, which secrete the enamel matrix.

Avian oral mesenchyme and epidermis in their normal interactions produce a beak, not teeth. Hayashi has shown, using heterospecific combinations made between chick and duck oral tissues, that the character of the beak is determined by the specific character of the mesenchyme. Thus, duck embryo mesenchyme in combination with chick embryo epithelium leads to the formation of a beak with the characteristic ridges of a duck's beak. Kollar and Baird have shown in an analogous fashion that mouse oral epithelium will follow the commands of the mesenchyme with which it is recombined.

The revealing and crucial "evolutionary" experiment consists of the recombination of bird oral epithelium with mammalian oral mesenchyme. Kollar and Fisher have performed this experiment by culturing grafts of chick pharyngeal epithelium with mouse molar mesenchyme. The grafts were cultured in the anterior chambers of the eyes of adult nude mice, which provide a suitable, if grotesque, culture milieu. Surprisingly, the chick oral epithelium responded to the mouse oral mesenchyme and produced structures resembling enamel organs. In a few cases, such as that shown in Figure 5-8, recombinants gave rise to complete teeth. Thus, at least one avian genome still retains the genetic information that allows the chick oral epithelium to participate successfully in the sequence of interactions required for tooth morphogenesis and synthesis of the enamel matrix. Evolutionary loss of teeth is therefore the result of a change in the developmental program of avian mesenchyme such that the initial steps of the process fail to occur.

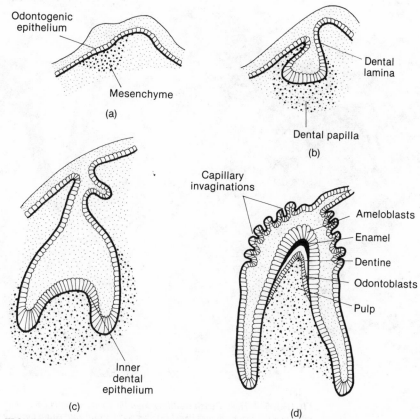

FIGURE 5–7. *Morphogenesis of a mammalian tooth.* (a) *Layer of dental epithelium cells overlying oral mesenchyme.* (b) *Epithelial cells that, as a result of an inductive signal from mesenchyme, invade the mesenchyme.* (c) *Dental epithelium induced to begin formation of enamel organ in turn induces mesenchyme to differentiate into odontoblasts.* (d) *Formation of definitive tooth with secretion of dentine by mesenchymal odontoblasts and enamel by epithelial odontoblasts.* [Redrawn from diagrams provided courtesy of C. E. Smith.]

The reduction of the avian fibula has also resulted from a modification of the developmental program, and not from a loss of the genetic information required for the shaping of a fibula. *Archaeopteryx* possessed a fully developed fibula with articulation surfaces at both ends, whereas as shown in Figure 5–9, the fibula of living birds develops only into a spur of bone lying alongside the tibia. A very thorough investigation of the developmental relationship between the bones of the chick leg has been made by Armand Hampé. Hampé performed a series of experiments in which regions of the developing limb bud were marked by injection of carbon particles. The position of the markers in the fully developed limb allowed Hampé to define the prospective territories of the limb bud (see Fig. 5–9). The major territories correspond to the masses of mesenchymal cells destined to give rise to the femur, tibia, fibula, and tarsal and metatarsal bones. Hampé performed three kinds of experiments that convinced him that the reduced size of the modern bird fibula has resulted from the usurpation of cells in the fibula territory by the tibia territory.

The first of Hampé's experiments involved removal or addition of

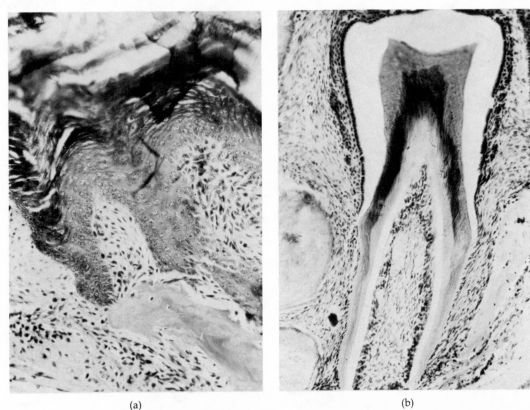

(a)   (b)

FIGURE 5–8. *Hybrid teeth resulting from the combination of embryonic chick oral epithelium and mouse dental mesenchyme. Panel* (a) *shows chick oral epithelium forming a primitive enamel organ and enclosing mouse mesenchyme. Panel* (b) *shows a well-developed hybrid tooth.* [Panel (a) from E. J. Kollar and C. Fisher, Tooth induction in chick epithelium: Expression of quiescent genes for enamel synthesis, *Science* **207**:993–995, 1980. Copyright 1980 by the American Association for the Advancement of Science. Photographs courtesy of E. J. Kollar.]

mesenchyme cells to a limb bud. If he removed cells from both the tibia and fibula territories Hampé found competition to be enhanced, and the fibula failed to appear. When he did the converse experiment of adding mesenchyme cells the tibia was unmodified, but the fibula grew to be as large as the tibia. Competition could be suppressed in a second way. Rotation of the fibula territory by 90° resulted in the tibia and fibula elongating in different directions, thus being unable to compete. Again, a long fibula was formed. These experiments very clearly established the competition between tibia and fibula territories; however, the most spectacular and informative results on the evolutionary changes involved came from a third experiment.

Hampé was able to deprive the tibia territory access to the presumptive fibula cells by the delicate insertion of a small sheet of mica between the two territories in the limb bud. The surprising result was the limb shown in Figure 5–9. Limbs produced in this way possessed not only a "full length" fibula, but one with a distal articulation surface as well. The shape of this fibula and its interaction with two small tarsal bones, the tibiale and the fibulare, closely resembles the homologous structure

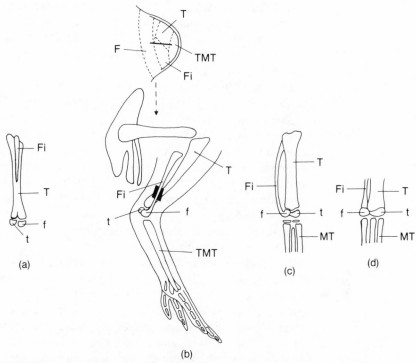

FIGURE 5–9. *Atavism in the development of the chick limb as a result of experimental modification of prospective bone territories in the limb bud.* (a) *Normal lower limb of chick.* (b) *Insertion of mica flake between fibula (Fi) and tibia (T) territories of limb bud and resultant limb with enlarged fibula bearing a distal articulation surface.* (c) *Atavistic chick limb.* (d) *Lower limb of* Archaeopteryx. [Redrawn from A. Hampé, 1959, 1960.]

of the *Archaeopteryx* leg. In the adult chicken the tibiale and fibulare are not distinct, but are fused to the tibia. The separation of fibula and tibia territories by Hampé has resulted not merely in a larger fibula, but in the reactivation of a long-suppressed ancient pattern of development. The genes that shaped the leg of *Archaeopteryx* are still present in the chick, but developmental interactions have been modified in such a way as to remove the opportunity for their expression, at least in the ancient manner.

The examples of regulatory changes in tooth loss and reduction of the fibula in birds indicate ways in which regulatory mechanisms affecting cellular interactions can produce evolutionary changes in morphology. But they suffer in one respect: Both involve structural reduction. If the goal of evolution were a sort of morphological nirvana this would be fine, but morphological evolution also entails modification of structures and sometimes the appearance of novel structures. Unfortunately, evolutionary loss of structures, as in the cases already considered or in others, such as the loss of eyes in the Mexican cave fish studied by Cahn and by Sadoglu, who showed a genetically linked reduction in the ability of the retina to induce lens in these fish, are more readily subjected to experimental analysis than instances in which structure has been gained or modified. The spectacular transformations in body coverings that have occurred in reptiles, birds, and mammals have entailed intricate changes in regulatory gene systems, and the developmental processes that underlie the development of scales, feathers, and hair are amenable to study. Although the control of morphogenesis in these systems is still poorly understood, an outline of the evolutionary changes in regulatory systems have become apparent through the work of P. Sengel and his collaborators, notably D. Dhouailly.

Skin consists of two layers, a superficial epidermis derived from the embryonic ectoderm and an underlying dermis derived from mesoderm. The development of scales or other epidermal structures depends on inductive signals from the dermis. Presumably, feathers and hair had their evolutionary origins in reptilian scales. All of these structures are composed of a family of related proteins, the keratins. The close homology between scales and feathers is indicated by the occasional transformation into feathers of the tips of the scales with which the legs of birds are covered.

The feathers of ducks have a morphology distinctly different from those of chickens. The source of developmental information for feathers was tested by Sengel and his co-workers by preparing cross-species combinations of dermis and epidermis. In these recombinants feather morphogenesis was determined by the dermis. The general architecture, size of feathers, and the number of barbs was characteristic of the species from which the dermis was derived: Only the shape of the barbule cells was determined by the epidermis. Further, the distribution pattern of feathers was also directed by the dermis. This kind of morphogenetic determination by dermis was found to occur generally through further experiments with chick embryos in which dermis and epidermis from the dorsal region, which gives rise to feathers, and the tarsometatarsal region of the leg, which gives rise to characteristic large scales, were recombined. The determination of epidermis to produce feather rudiments or large scales was always coincident with the pattern of morphogenesis characteristic of the site from which dermis was taken. For example, dorsal epidermis, which normally produces feathers, produced large scales when recombined with tarsometatarsal dermis.

Although both scales and feathers are composed of keratins, the spectrum of keratins present differs significantly. In recombinants of chick dorsal and tarsometatarsal dermis and epidermis, the epidermis, regardless of site of origin, exhibits a pattern of keratin gene expression

FIGURE 5–10. (Opposite) *Morphological fates of recombinants between lizard epidermis, which normally produces rows of small scales, with dermis of other vertebrate classes in organ culture. (a) With chick tarsometatarsal dermis, resulting in formation of a scale pattern like that of chick tarsometatarsus; (b) with dorsal chick dermis, resulting not in feathers but in formation of a hexagonal feather pattern; (c) with dorsal mouse dermis, resulting in scales corresponding to the primary hair follicle pattern; (d) with mouse upper lip dermis, resulting in formation of large scales arranged in a typical whisker pattern surrounded by small scales corresponding to the pelage hair follicle pattern.* [From D. Dhouailly and P. Sengel, Interactions morphogènes entre l'épiderme de reptile et de derme d'oiseau ou de mammifère. *C. r. Acad. Sci. Ser.* D **277**:1221–1224, 1973. Photographs courtesy of P. Sengel.]

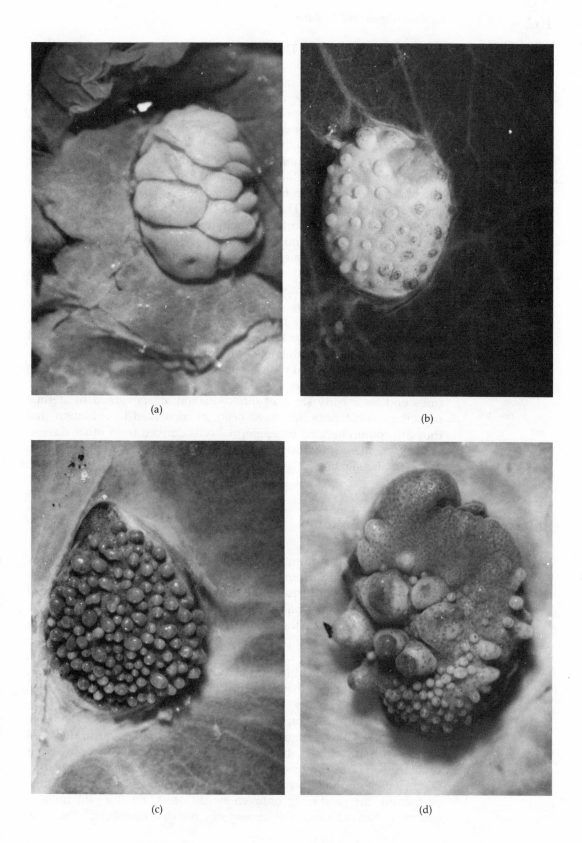

(a)

(b)

(c)

(d)

dictated by the dermis. Thus, the inductive signals to the epidermis from the dermis specify position of epidermal structures, morphological specificity, and keratin gene expression pattern. But the question remains: How have these informational signals been modified during the evolution of higher vertebrates?

Heterospecific recombinants are possible not only within a class, but between classes. Dhouailly has carried out a remarkable series of experiments with recombinants between lizard, chick, and mouse dermis and epidermis which suggest that two induction steps occur, and that these steps differ in specificity. Some of Dhouailly's results are shown in Figure 5–10. The response of lizard epidermis to chick or mouse dermis is particularly instructive. Lizard epidermis responds to chick tarsometatarsal dermis to produce large scales indistinguishable in appearance and arrangement from chick tarsometatarsal scales. Lizard epidermis also responds to chick dorsal dermis, which in chicks determines the arrangement and morphology of feathers. However, lizard epidermis is unable to interpret the specific signals for feather morphogenesis. Instead, the epidermis produces arrested scales arranged in a typical hexagonal feather pattern. In a like manner, lizard epidermis combined with mouse dermis forms not hair but abortive scales arranged in typical mouse hair patterns. Recombinants between chick and mouse produce similar effects. Position of epidermal structures and their early steps of differentiation are governed by dermal regulatory substances that have been so conserved in evolution that they are comprehended by epidermal cells derived from other classes. Full differentiation of scale, feather, or hair buds apparently requires class-specific signals.

Interactions and Macroevolution

The examples drawn from amphibians and birds illustrate evolutionary consequences arising from interactive changes relatively late in development. As a consequence, the changes achieved represent modifications of an established body plan rather than new departures. This is not to minimize the role of innovations, such as those in body coverings, that have had profound effects on physiological and behavioral adaptations of the organisms possessing them. But we are still faced with real difficulties in accounting for relatively large-scale evolutionary changes that have produced novel structures or even radically new body plans. These macroevolutionary events seem necessary to account for the rapid initial radiation of metazoan phyla or the appearance of divergent groups, such as bats, that appear in the fossil record with unsettling abruptness. If major changes in ontogenetic processes occur in early stages of development they can result in large-scale changes in body plan. Of course, such changes also have a high probability for producing ontogenetic disasters, resulting in inviable embryos. Nevertheless, the appeal of "hopeful monsters," the term coined by Richard Goldschmidt to refer to profound evolutionary departures achieved in this way, is great because in the rare cases in which hopeful monsters might be viable they provide the basis for explaining macroevolution.

Not surprisingly, there has been a good deal of vehement opposition to the notion of hopeful monsters both because the genetic mechanisms proposed for it were dubious, but also because this idea is in basic contradiction to the gradualistic idea grounded in the population genetic theory of allele replacement in populations. For example, G. G. Simpson noted in his autobiography that he was greatly stimulated by Dobzhansky's book *Genetics and the Origin of Species,* which was published in 1937, and began work on *Tempo and Mode in Evolution,* in which he intended to integrate paleontology with this approach to evolution. At the same time Simpson was "stimulated to a lesser degree and in quite the opposite way" by the ideas of Schindewolf and Goldschmidt, who looked to large-scale mutations affecting the organization of the genome and altering early development. The ideas of Schindewolf and Goldschmidt were excessive and reaction against them was reasonable. But it is unfortunate that the colorful and loaded term "hopeful monster" has inspired a strongly negative reaction to the importance of Goldschmidt's basic proposal that mutations affecting development are central to evolution.

Macroevolutionary changes in development need not be extreme. We propose that in fact the initial steps for rapid, and ultimately, large evolutionary transitions require only that key regulatory genes be few in number and accessible to nonlethal genetic alterations in their functions. Initial, "easy" genetic changes, which may have significant effects on the organism and become established in a small population, are of necessity viable, and present open avenues for selection of successive genetic changes. Profound change may be rapid in this way without recourse to any instantaneous hopeful monsters. It is already clear from our discussion of evolution in the Hawaiian drosophilids in Chapter 3 that changes in a small number of genes can account for striking morphological changes. Further genetic evidence for the number and operation of regulatory genes is presented in Chapter 8.

There seem to be two principal means for producing changes in tissue interactions within embryos that can lead to macroevolutionary changes. It should be kept in mind that these are not the only means; heterochrony, to be discussed in Chapter 6, provides other equally important mechanisms. Large-scale changes in interactions can result from either resurrection of old integrational patterns in new locations or the appearance of new interactions. We suggest that the first of these mechanisms can be seen in the method of acquisition of an external shell by the pelagic octopus *Argonauta,* and the second in the origin of the pentameral symmetry of echinoderms.

The most ancient cephalopods possessed external shells. As the animal grew it moved forward in its shell and walled off successive gas-filled chambers that provided buoyancy to counterbalance the weight of the shell and viscera. The living species of *Nautilus* are the last remnants of these once prevalent animals, now displaced by the more familiar Coleoid cephalopods, squid and octopus. Coleoids have entirely abandoned the external shell and have compensated for its loss by behavioral adaptations, including increased coordination, better swimming ability, and heightened senses and intelligence. Squids and cuttlefish retain an internal shell or shell rudiment; octopods have lost even the shell rudiment.

It thus presents a curious quandary that there exists a small group of pelagic octopods of the genus *Argonauta* in which the female (shown in Fig. 5–11) produces a shell closely resembling that of an ammonite or nautiloid in external shape. This shell differs from the shell of ancient cephalopods in two significant ways: It has no chambers, and it is secreted not by the mantle but by a pair of specialized arms. The tips of these arms are greatly expanded to form broad membraneous appendages. During secretion of the shell the arms are held over the body and the calcite shell is secreted by glands on the membrane. The two halves of the shell are joined to form a keel ornamented with knobs that apparently mark the location of suckers on the arms. The shell of the argonaut cannot be derived from the ancient cephalopod shell. It is a convergent structure, independently acquired by a member of a group in which the old shell had been lost. It is a novelty.

Since the shell of the argonaut serves to shelter the female's egg mass as well as herself, it seems likely that the shell arose as a simple noncalcareous envelope for the eggs of the argonaut ancestor. This was secreted by glands on the surface of the arms. Production of a calcareous

FIGURE 5–11. *Female of the paper nautilus,* Argonauta argo, *with her shell held by pads on two specialized arms.* [Modified from Young, 1959–1960.]

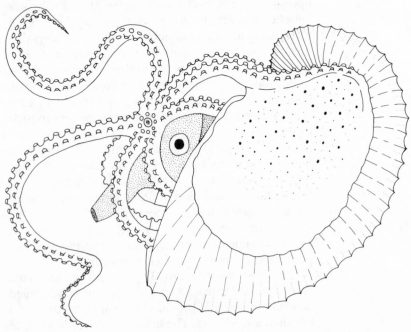

shell became possible by supplementing the glands already present on the skin with glands capable of calcium carbonate secretion. From what source would such shell glands have arisen? They may have arisen *de novo*, or they might, more probably, represent the reactivation of an ancient and long-suppressed developmental program for the differentiation of shell glands. Instead of being developed in the ancestral position in the mantle, the glands were developed on the arms. Suppression of old patterns of expression of genes controlling morphogenesis probably long precedes their loss. This has certainly been the case in the example of hen's teeth, where close to 100 million years have elapsed since the ancestors of modern birds had toothy bills. If the genes entailed are involved in other induction events they may not be lost at all, although they are no longer expressed to form the ancestral structure. Such suppressed developmental genetic systems provide the potential for acquisition of new structures if reactivated in new positions or at a different stage in development.

Changes in position of an induction system early in development can have a profound effect on body organization. Localizational changes include movement relative to other structures, increase or decrease in relative size, and increase or decrease in the number of elements of a repeated structure. It is the last of these that we discuss with respect to the peculiar symmetry of echinoderms. Echinoderms exhibit a number of unique features, but it is the origin of the pentameral symmetry characteristic of all living and most extinct classes that has been a major puzzle to phylogeneticists.

Strathmann has explored the limitations of diversity of forms of ambulacral systems in echinoderms. The ambulacral systems of echinoderms possess structures, such as podia or brachioles, that are involved in respiration, locomotion, and feeding. As size increased in the course of evolution, echinoderms were confronted by the physical dilemma that if shape were to remain constant, the ratio of ambulacral length and area to body volume would decrease. Compensation to increase ambulacral size has been achieved in various echinoderm groups by increased ambulacral length or width, or by branching or torsion of ambulacra. Fundamental change of symmetry might be added to this list, but with the exception of symmetry changes early in the history of the phylum symmetry has remained constant, suggesting that for most echinoderm classes the pentameral pattern has been selectively optimal. The bilateral symmetry of the larva prior to metamorphosis has been taken to indicate that the echinoderm ancestor was bilaterally symmetrical, and that pentameral symmetry was superimposed over the primitive symmetry (see Hyman for a review of these ideas). Ubaghs, on the basis of the fossil record of the earliest echinoderms, has proposed that the most primitive and ancient echinoderms were bilaterally symmetrical or even asymmetrical forms. Haugh and Bell, on the basis of their recent studies of fossilized internal organs of extinct echinoderm "paleoviscera," have pointed out that not only is pentameral symmetry not basic to the phylum but that a water vascular system may have been absent in some extinct classes. Pentameral symmetry may have arisen in an ancient form possessing a small number of ambulacra. Plausible candidates do exist. Helicoplacoids, with their single forked ambulacrum and paracrinoids with a pair of ambulacra, represent nonpentameral stocks. Bell has suggested that trimeral ontogenetic stages of Ordovician edrioasteroids, which are pentameral as adults, may represent a recapitulation of a more primitive symmetry. The enigmatic trimeral Ediacaran animal *Tribrachidium* might represent such an ancestral echinoderm. The developmental genetics of the situation, at least as judged from modern echinoderms, would seem to make the required changes rapid and "easy."

The symmetry, that is, the number of ambulacra present in the adult echinoderm, is determined by a "counting" process during early larval development. This process is illustrated in Figure 5–12, which shows the development of the hydrocoel in a generalized echinoderm embryo. The hydrocoel, and indeed the entire water vascular system, develops from the left-side coelomic cavity, which forms adjacent to the mouth of the larva. The corresponding right-side coelom degenerates. As growth of the hydrocoel proceeds, five diverticula appear. These interact with the overlying epidermis to induce the growth of the five ambulacra. If, as was done by Czihak, the left coelom of a 3–4-day-old sea urchin larva is destroyed by localized irradiation with an ultraviolet microbeam, the hydrocoel and ambulacral rudiments fail to form. Larvae lacking a hydrocoel fail to develop ambulacra. Conversely, Czihak has also observed that in a few cases in which hydrocoels form from both left and right coeloms, a double embryo with two ambulacral systems results. The determination of symmetry thus depends on the control of a meristic trait expressed early in the development of the adult rudiment in the larva.

The genetic basis for the control of the number of ambulacral rudiments in development of the water vascular system is limited by the

FIGURE 5–12. *Development of the hydrocoel in an idealized echinoderm larva. The hydrocoel is stippled. Note the formation of five diverticula.* [Redrawn from G. Ubaghs, General characters of Echinodermata, in *Treatise on Invertebrate Paleontology,* Part S. Echinodermata I, Vol. 1, 1967, R. C. Moore, ed. Courtesy of the Geological Society of America and University of Kansas.]

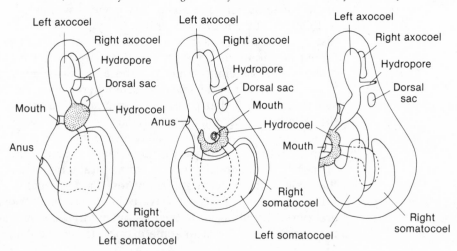

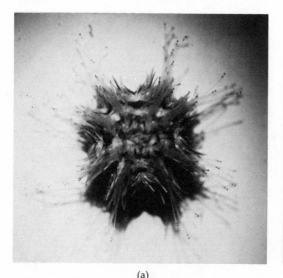 

(a)   (b)

FIGURE 5–13. *Mutant sea urchins exhibiting four-fold rather than five-fold symmetry.* (a) *is a living, laboratory-reared specimen of* Lytechinus pictus *exhibiting the* square *phenotype of four ambulacral tracts.* (b) *is the test of a four-sided specimen of* Strongylocentrotus franciscanus *found in nature.* [Photographs courtesy of R. T. Hinegardner.]

difficulties inherent in using echinoderms for genetic experiments. Nevertheless, Hinegardner has devised methods for breeding sea urchins in the laboratory, and he has obtained some genetic mutants. The most intriguing of these is a symmetry mutant Hinegardner calls *square,* shown in Figure 5–13. Animals expressing this mutation typically have four- rather than five-fold symmetry, and thus have four ambulacra and a four-jawed Aristotle's lantern. The genetic observations described by Hinegardner show that more than one gene is involved in control of this meristic trait. Further, the mutation is not one- to four-fold symmetry *per se:* Rather, it appears to result in loss of control of symmetry. While four-fold symmetry is the most common mutant form, an individual cross may yield two-, three-, four-, five-, and six-sided progeny. Of the abnormal forms only the four-sided progeny are viable because the Aristotle's lantern cannot function in adults of other abnormal symmetries. Houke and Hinegardner have studied the developmental basis for the action of the *square* mutation in living embryos. Whereas normal larvae develop the expected five diverticula from the growing hydrocoel, mutant animals fail to do so. Generally, only four diverticula appear. These abnormal larvae undergo metamorphosis to produce animals with a correspondingly abnormal symmetry.

Echinoids have not exhibited any evolutionary exploitation of mutational changes in this genetic system, but starfishes have. There are a number of species of asteroids that have increased the number of their arms and the associated ambulacra. In at least some of these the basis is a change in the early counting event. A good example is seen in the development of the common West Coast six-armed starfish *Leptasterias hexactis.* Figure 5–14 shows that five hydrocoel diverticula appear in the larva, and shortly thereafter, a sixth. These induce the outgrowth of the

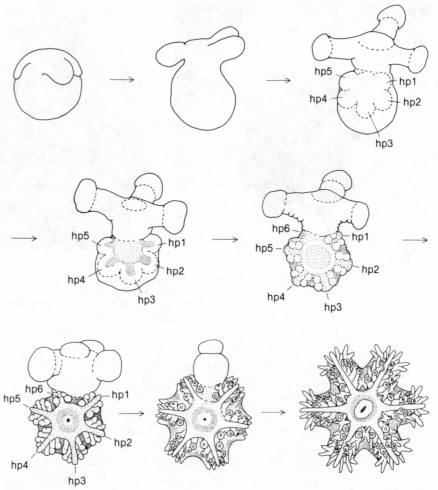

FIGURE 5–14. *Development of the six-armed starfish* Leptasterias hexactis. *The hydrocoel diverticula, or pouches, are labeled hp1–hp6. Note that the appearance of hp6 is later than that of hp1–hp5.* [Redrawn from H. L. Osterud, Preliminary observations on the development of *Leptasterias hexactis*, Publ. Puget Sound Biol. Sta. **2**:1–15, 1918.]

six arms of the juvenile starfish. A similar mode of development (with suitable arithmetic adjustments) occurs in the nine-armed starfish *Solaster endica* described by Gemmill.

The genetics of symmetry in extinct echinoderms appears to have been similar to that of echinoids, and in fact, fossil mutant blastoids bearing three, four, and six ambulacra (as shown in Fig. 5–15) are not uncommon. Studies by Wanner on Permian blastoids and by Macurda on the blastoid *Pentremites* indicate a rather high frequency of symmetry mutants (as high as 1/200) in these populations. Edrioasteroids bearing four, six, and even nine ambulacra have been described by Bell. Unlike the mutant echinoids studied by Hinegardner, mutants of these filter-feeding echinoderm classes did not have to contend with the moving parts of an Aristotle's lantern; thus, individuals bearing highly abnormal symmetries could survive.

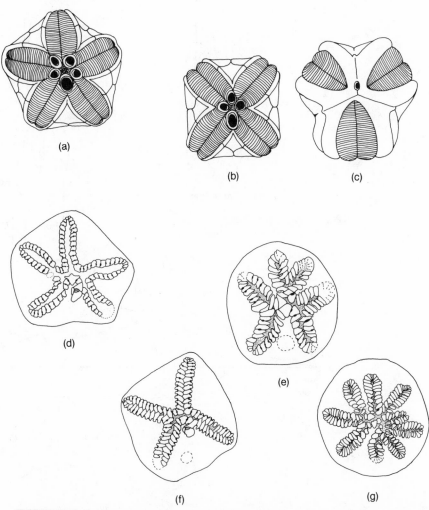

FIGURE 5–15. *Symmetry mutants in extinct echinoderm classes.* (a) *Normal pentameral blastoid.* (b), (c) *Blastoids with deviant symmetries.* (d) *Normal pentameral edrioasteroid.* (e–g) *Edrioasteroids with four, six, and nine ambulacra.* [Blastoids drawn from specimens loaned by J. A. Waters and redrawn from H. H. Beaver, Morphology, in *Treatise on Invertebrate Paleontology*, Part S, Echinodermata I, vol. II. 1967, R. C. Moore, ed. Courtesy of the Geological Society of America and University of Kansas. Edrioasteroids redrawn from Bell, 1976.]

Although the genetic basis for discontinuous changes in numbers of echinoderm ambulacra is difficult to study directly, an analogous situation in the determination of the number of toes developed by guinea pigs has been analyzed very elegantly by Sewall Wright. The forefeet of guinea pigs normally lack a thumb, while the hindfeet lack both a large and small toe. But guinea pigs do occasionally develop a little toe, and inbred lines can be selected in which a perfect small toe including bones, muscles, and the nail are always present. Wright studied several inbred lines with distinct differences in frequency of development of the small toe. Strain D, which had been originally selected by W. E. Castle from a four-toed variant in a wild-type population some 20 years before Wright's study, consistently produced

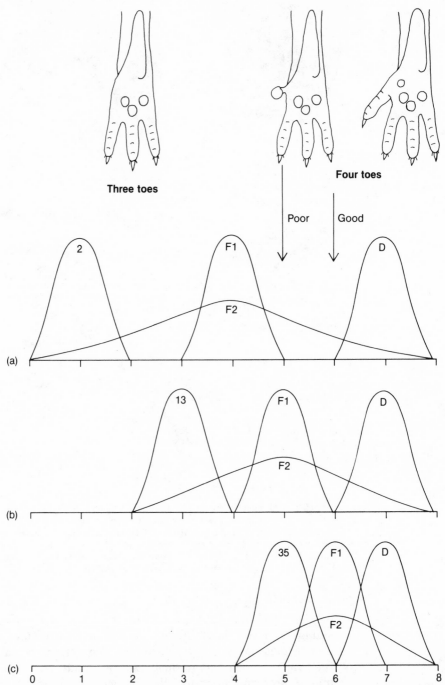

FIGURE 5–16. *Genic control of toe number in guinea pigs. At the top left is a normal three-toed hindfoot of a guinea pig. At the right is a foot with a well-developed little toe. The foot in the center shows imperfect development of the extra toe. Phenotypic distributions as a function of number of alleles for polydactyly are illustrated for crosses between the three inbred strains. The arrows indicate the genic thresholds between three-toed, poor-four-toed, and good four-toed feet. (a) Cross between strain D (four-toed) and strain 2 (three-toed). Strains D and 2 are far apart in genic constitution, and the F1 progeny have insufficient polydactyly genes to produce four toes. (b) Cross between strains D (four-toed) and 13 (three-toed). Strain 13 has more polydactyly genes than strain 2; thus the F1*

TABLE 5–1. Toe Numbers in Crosses Between Strains of Inbred Guinea Pigs[a]

Cross	Percentage		
	Three-toed	Poorly Developed Four-toed	Good Four-toed
F1 progeny:			
2 × D	100	0	0
13 × D	69	31	0
35 × D	6	13	81
F2 progeny:			
2 × D	77	16	7
13 × D	48	19	33
35 × D	25	20	55

[a]Data from Wright, 1934; used by permission.

four-toed progeny. Strain 2 produced strictly three-toed progeny. Two other inbred strains, 13 and 35, were also analyzed by Wright, and as will be seen, differed significantly from each other and from strain 2 in their genetic compositions. Strain 13 was a three-toed line like strain 2, but strain 35 consisted of a population that was approximately 40% four-toed. Wright performed crosses between strain D and the other strains. The characteristics of the F1 generation of each cross are shown in Table 5–1. The cross between strains 2 and D gave only three-toed progeny and the cross between strains 13 and D produced a significant percentage of progeny with a poorly developed fourth toe, whereas the cross between strains 35 and D yielded a majority of four-toed offspring.

Crosses of three-toed F1 animals with each other produced three- and four-toed progeny. The ratios for the F2 progeny from the 2 x D hybrid are consistent with a Mendelian ratio of 3:1, and suggest that the trait is controlled by a pair of alleles of a single gene. However, Wright did a series of back-crosses of the F2 animals to strain D that showed that this simple explanation cannot be true, and in fact that it is more probable that a set of alleles representing four genes separates strains 2 and D. Strains 13 and 35 were separated from strain D by fewer alleles. Wright's conclusions are diagrammed in Figure 5–16, which plots phenotypic distributions as a function of genotype. Because number of toes is a discontinuous trait there is a threshold effect: Development of a four-toed foot requires the presence of at least five alleles for polydactyly.

The discontinuous trait controlled in this manner in guinea pigs occurs late in development and has little overall effect on body structure. However, this kind of genetic control imposed early in development, as

progeny produce imperfect four-toed feet. (c) Cross between strain D (four-toed) and strain 35 (which produces both three- and four-toed progeny). [Redrawn from S. Wright, An analysis of variability in number of digits in an inbred strain of guinea pig, *Genetics* **19**:506–536, 1934, and S. Wright, The results of crosses between inbred strains of guinea pigs, differing in numbers of digits, *Genetics* **19**:537–551, 1934.]

in echinoderms, has a major effect on body structure and may have provided a key basic step in the macroevolution of pentameral echinoderm classes. Hopeful monsters need not be the results of sudden and wholesale remodeling of the genetic program for ontogeny. A minor genetic change can result in an alteration in development that allows a radical change in adaptation: A small key can open a large door.

Six

Timing of Developmental Events: Evolution Through Heterochrony

You'll remember that from his birth Pinocchio had had tiny little ears—so tiny that they were not visible to the naked eye. So you can understand how surprised he felt when he realized that his tiny ears had grown so long during the night that they now resembled two brooms.

Carlo Collodi, Pinocchio

Heterochrony: A Classical Mechanism for Evolutionary Change

The major proposals seeking to account for the mechanistic roots of evolution in ontogeny have centered on heterochrony, changes in relative timing of developmental processes. Haeckel emphasized a relationship between evolution and development such that ancestral adult stages were compressed within the ontogeny of the descendant. There were, of course, intimations that evolution could occur by temporal shifts in ontogeny different from those of Haeckelian recapitulation. The axolotl, in which somatic development is retarded in relation to maturation of the gonads, was studied by Dumeril in the 1860s, while other biologists, including von Baer and Kollmann (who coined the term *neoteny*), documented other cases in which gonadal maturity is achieved in a larval body. In *Ontogeny and Phylogeny* Gould has traced out the history of these ideas in detail, and has demonstrated that in the late nineteenth century none of these apparent exceptions to recapitulation had much effect on the general enthusiasm for the biogenetic law.

Haeckel's recapitulation was grounded in a Lamarckian view that assumed that only the adult stage was of significance in acquiring new traits. Garstang's critical paper in 1922 put an effective end to the acceptance of recapitulation as a universal evolutionary mechanism because Mendelian genetics made it clear that traits could appear and be selected effectively at any stage of development. Further, early studies by Ford and Huxley and by Goldschmidt revealed that genes could

173

function by controlling the rate of biological events. The work of Ford and Huxley, for example, focused on the rate of pigment accumulation in the eyes of the developing crustacean *Gammarus*. *Gammarus* carrying identical pigment gene alleles were found to develop quite different final eye colors, depending on the presence of dominant or recessive alleles of a "rate gene," which controls the rate of pigment accumulation.

The notion of rate genes had an irresistible appeal in the early 1930s to several evolutionary biologists, including de Beer, Huxley, and Haldane. Haldane, for example, discussed genes known to act at various stages of the life cycle—gamete, zygote, larva, immature animal, reproductive organs, and so on—and proposed that heterochrony could be linked to changes in time of action of genes controlling rates of developmental processes. Since genes could act at any stage of development, profound evolutionary changes in morphology, physiological adaptations, or behavior could be achieved by simply changing timing. Huxley, who at this time was concerned with the phenomenon of allometry, could rationalize relative growth as a function of rate genes. Rate genes clearly provided a means of changing relative growth patterns in evolution. Thus,

> it is in general clear that rate-genes may mutate in either a plus or a minus direction, either accelerating or retarding the rates of the processes they affect. In the former case the effect will be in certain respects at least recapitulatory, since a condition which used to occur in the adult is now run through at an earlier stage. In the latter case, the effect will be anti-recapitulatory, since a condition which once characterized an earlier phase of development is now shifted to the adult phase.

This new view of developmental processes as the means to evolutionary change formed the basis for a profusion of nonrecapitulatory phylogenetic speculations in the 1920s and 1930s. A large number of such theories have been summarized by de Beer in *Embryos and Ancestors*. These include such diverse examples as Bolk's conclusion that many human morphological traits represent the retention of ancestral primate fetal characteristics in the adult, Garstang's proposal that vertebrates arose from tunicate tadpole larvae that achieved sexual maturity, and de Beer's suggestion that insects similarly arose from six-legged larval myriapods (an example we shall return to again). De Beer especially stressed the importance of the appearance of new structural features in nonfossilizable larval stages as a mechanism for the origin of major metazoan groups. This was an exciting possibility because "...it is probable that these gaps, that these discontinuities in the phylogenetic series of adults, may to a certain extent also be due to 'clandestine' evolution in the young stages, followed by neoteny and the sudden revelation of these hidden qualitative novelties." In this light, some of the eight modes of heterochrony, which de Beer defined in *Embryos and Ancestors*, had very profound macroevolutionary potential while others, in fact the modes leading to recapitulation, produced only minor modifications in structure.

The most obvious temporal shifts involve the dissociation of rates of development of somatic traits and the rate of maturation of the gonads.

TABLE 6–1. Categories of Heterochrony[a]			
Timing of:			
Appearance of Somatic Feature	*Maturation of Reproductive Organs*	*Name in de Beer's System*	*Morphological Result*
Accelerated	Unchanged	Acceleration	Recapitulation (by acceleration)
Unchanged	Accelerated	Paedogenesis (= progenesis)	Paedomorphosis (by truncation)
Retarded	Unchanged	Neoteny	Paedomorphosis (by retardation)
Unchanged	Retarded	Hypermorphosis	Recapitulation (by prolongation)

[a]Modified slightly from Gould, 1977; used by permission.

In fact, the classic definitions of categories of heterochrony are based on these dissociations. We follow Gould's definitions here of the processes that lead to a recapitulatory result and those that result in paedomorphosis. His four categories of heterochrony and their results are summarized in Table 6–1. If the appearance of a somatic feature is accelerated with respect to gonadal maturation it will result in a formerly adult trait becoming a juvenile trait in the descendant: This is the classic pathway to Haeckelian recapitulation. A second way in which a recapitulatory event can occur is if maturation is so retarded that a formerly adult trait appears at the same stage in development, but that stage is now a preadult rather than an adult stage because of prolongation of development. This phenomenon is called hypermorphosis. Size increase is a common evolutionary trend and it commonly results from hypermorphosis. Acceleration of somatic development also provides a means of size increase and may be of significance in the evolutionary exploitation of positive allometric relationships. The ancestral shape is reached earlier in development and a perhaps more extreme morphology, such as the horns of titanotheres or the antlers of the Irish elk, is gained as a consequence of allometric relationships extended to a larger body size.

Timing shifts can produce a quite opposite evolutionary result to recapitulation, and that is paedomorphosis. In paedomorphosis traits characteristic of the juvenile ancestor are retained by the adult descendant, but there are two quite distinct means by which this can occur. In the most familiar case, neoteny, somatic development is retarded with respect to the course of reproductive maturation. For example, in that workhorse of neoteny, the axolotl (Fig. 6–1), the animal requires the same length of time to mature sexually as related nonneotenous salamanders, but many somatic morphological features of the larval stage are retained so that the sexually mature axolotl remains aquatic and looks like an overgrown larva. There is a second kind of paedomor-

FIGURE 6–1. *Paedomorphosis and recapitulation. The Mexican axolotl (upper left) retains the gills, flattened tail, and skin of the larva although it grows to adult size and sexual maturity. This neotenic paedomorph can be induced to metamorphose into a typical terrestrial adult of its genus (center) by treatment with thyroxine. The clam* Hinnites *exhibits classical recapitulation. As a juvenile* Hinnites *resembles other scallops, but with further maturation attaches to the bottom and continued growth of the shell produces a less-regular oysterlike form.* [Axolotls drawn from photographs provided by G. M. Malacinski. *Hinnites* from life.]

phosis, progenesis, that results from an entirely different process. In progenesis gonadal maturation is accelerated so that sexual maturity is attained in what is essentially a small-sized juvenile body: Somatic development is truncated.

Paedomorphosis by either mode provides a genetically easy evolutionary response to environmental pressures because an already integrated developmental system can be exploited. Neoteny and progenesis both yield paedomorphosis, but are apparently elicited in response to different conditions. Neoteny may, for instance, occur in situations in which the paedomorph can escape a harsh or unstable adult environment and remain in a more equitable and stable larval environment. Amphibian neotenates certainly typify this solution; several groups of salamanders have evolved neotenous adults that remain aquatic.

Neoteny can lead to rather surprising evolutionary possibilities; for instance, half-ton grazing birds. Flightlessness is common among birds on islands that lack mammalian carnivores. Because, as pointed out by Olson and by Feduccia, the flight muscles and bones of the pectoral girdle and sternum account for about 20–25% of the mass of the average bird, their growth and maintenance require a considerable portion of a bird's energy budget. If flight has no immediate advantage, maintenance of an energetically expensive set of structures will not be favored by selection. One rapid mechanism for loss of these features is neoteny, and indeed flightless birds have many features in common with the chicks of flying species. According to Olson and Feduccia, in some groups of birds, such as pigeons and rails, the sternum develops late in ontogeny, after hatching. The contrast with galliform birds is significant, because galliforms have never given rise to flightless species. In a typical galliform (the chicken) the sternum begins to ossify between the 8th and 12th days of incubation instead of after hatching, as in rails. Arrest of part of the morphogenetic program at such an early stage would probably be lethal, but arrest of sternum development in forms in which the processes do not occur until after hatching would probably introduce few complications. Thus, flightless rails have evolved many times because the developmental program of rails supplies a preadaptation for such a mode of change. Once the neotenic change is achieved, large size and other adaptations become attainable. Unfortunately, dodos, moas, elephant birds, and all too many others of what was a splendid array of flightless birds on large and small islands across the world have become extinct as a result of human activity.

Neotenic traits are common among highly social birds and mammals, and it has been proposed that delayed somatic maturation may simplify recognition of social status to avoid conflicts and increase social stability. Human neotenic characteristics reflect an exaggeration of the primate traits of social stability, intense parental care, and long maturation period. Neotenous features of the skull also reflect the enlargement of the brain that has been achieved by prolongation of fetal growth patterns with their attendant allometric trends.

Progenesis, on the other hand, appears to typify a response to environmental circumstances in which high reproductive rate or small size may be particularly advantageous. In some such cases selection may not be for a particular morphology, but for small size *per se*. The result is a reproductively mature organism with larval morphology or with a mixture of larval and adult traits. In the environments open to small organisms some morphological traits will be removed from the selective constraints acting on the ancestral, larger species. Under relaxed or modified selection anatomical characteristics may accumulate changes that make possible transitions between body plans that would otherwise be extremely difficult. This idea may well apply to the origin of insects from paedomorphic myriapods, as suggested by de Beer. Figure 6–2 shows the larva of the millipede *Glomeris* which, at the time of hatching, has only three pairs of legs and a limited number of body segments. More legs and body segments are typically added in myriapod development. If, however, such a larva attained sexual maturity an organism

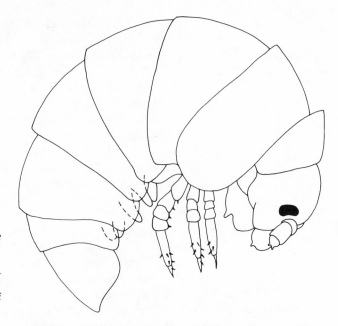

FIGURE 6–2. *Newly hatched larva of the millipede,* Glomeris, *with three pairs of legs.* [Redrawn from L. Juberthie-Jupeau, Action de la température sur le développement embryonnaire de *Glomeris marginata* (Villers). *Symp. Zool. Soc. Lond.* **32**:289–300, 1974, by permission of the Zoological Society of London.]

very similar to a primitive insect would result. The significant point is that although the initial progenetic event has been selected for its immediate survival value, it has also provided a departure point for a spectacular macroevolutionary event. Insect evolution may have been initiated in this manner. However, insects are not merely paedomorphic millipedes: An extensive further set of genetic changes are subsequently required. Not only were wings acquired as a qualitative novelty, but a series of genes were also acquired to regulate the differentiation of all body segments. These genes determine segmental organization to produce an insect with distinctive and very stable differentiation of pro-, meso-, and metathorax and their associated appendages as well as suppression of limb development on abdominal segments. A detailed examination of the role of these genes in insect development and evolution is presented in Chapters 8 and 9.

Many heterochronic events occur in the timing of appearance of one somatic feature in relation to another in an otherwise little altered ontogeny. Such heterochronic changes may be inextricably linked mechanistically to changes in inductive interactions discussed in Chapter 5. Changes in timing of cell division or of movement of tissue layers will directly affect inductive interactions. The consequence may be an enlargement of the induced structure. Conversely, a change in timing might lead to the loss of an induced structure—not because of failure to produce an inducing signal or loss of competence to respond, but because of a failure of the inducing and target tissue to make contact at the proper time. Instances of this kind are known among cave-dwelling vertebrates in which eyes are rudimentary. For example, Schlampp found that invagination of the optic cup in the blind European cave salamander is normal; however, a premature migration of mesoderm inserts itself between the optic cup and the ectoderm, thus blocking induction of the lens at an early stage. Inductive failure also underlies development of rudimentary eyes in the Mexican cave fish. Cahn compared development of the eyes of cave fish with that of the eyes of the normally sighted river species from which the cave form had evolved. In this case there were two heterochronic effects: Appearance of optic buds was delayed, and the rate of mitotic cell division was slower in the retina of the blind cave fish than in the normal retina. The result was a poor induction of the lens and a consequently reduced eye.

Arboreal Salamanders and Frogs Without Tadpoles: Heterochrony and Morphological Adaptation

Why should heterochrony be such a common mode of evolution? The answer to this question appears to lie in a characteristic of evolution pointed out by François Jacob. Evolution proceeds by what Jacob calls "tinkering." Unlike the *de novo* design of machines to make use of optimal engineering principles, evolutionary possibilities are limited by history. History in this sense refers to the particular traits or structures already present and available for modification. The result is seen, for example, in such events as the progressive evolutionary transformation of rudimentary elements of the jaw of the mammal-like reptiles into the specialized bones of the mammalian inner ear. Developmental processes are highly interactive, and interactions result in canalization. As defined by Waddington in 1942, canalization is the buffering or homeostasis of developmental pathways such that these pathways resist distortion by either environmental or genetic perturbations. P. Alberch has suggested that the apparent directionality of some evolutionary lineages may reflect the constraints imposed by the epigenetic interactions that produce canalization. Certain evolutionary changes result in little disruption of existing developmental interactions, whereas other theoretically possible morphological solutions would require such revolutionary changes in developmental processes as to be essentially impossible. Given the stability of developmental pathways, heterochrony provides the course of least resistance for evolutionary change. The timing of processes with respect to one another can often be dissociated without drastic interference with the canalization of individual processes, yet the final morphology can be considerably modified.

Some of the clearest examples of the evolutionary role of heterochronic processes have been documented in amphibians, particularly tropical species of salamanders and frogs that have evolved divergent and unusual reproductive strategies. Most families of salamanders live in the temperate zone. However, Wake has pointed out that the neotropical representatives of one group, the tribe Bolitoglossini, comprise 40% of all living salamander species. One genus of this group, *Bolitoglossa*, is large and diverse, and includes not only typical upland terrestrial forms, but also contains the only salamanders to have successfully established themselves in the lowland tropics of Central and South America.

Many species of *Bolitoglossa* belie Faust's nervous query

> *What are those, in thickets crawling?*
> *Salamanders, belly-sprawling?*

in being arboreal rather than terrestrial, as is more usual. They are fully adapted to arboreal life and exhibit morphological adaptations, including fully prehensile tails and modified feet, that allow them to cling to the undersides of wet stems and leaves. The evolution of these arboreal species has been accompanied by a decrease in adult body size and a characteristic set of changes in foot anatomy. These modifications, as shown in Figure 6–3, are the presence of webbing between digits and

FIGURE 6–3. Feet of terrestrial species of the salamander Bolitoglossa (a *and* b) *and of a progenetic arboreal species* B. occidentalis (c). *Note relatively small size of the* B. occidentalis *foot, its webbing, and the reduction of terminal digits.* [Redrawn from P. Alberch and J. Alberch, 1981.]

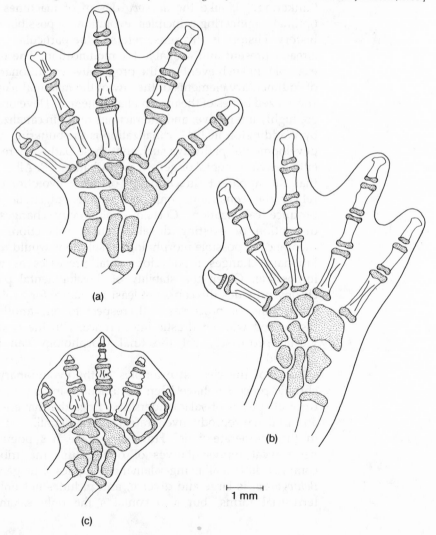

(a)

(b)

1 mm

(c)

reduction or fusion of bones. The terminal phalanges are reduced, the third phalanx of the fourth digit has been lost, and some of the tarsals are fused. Alberch was able to show that the reduced size and peculiar foot structure of the arboreal forms is directly related to their ability to climb on smooth surfaces. The primary means by which salamanders hold on when climbing are surface tension and suction produced by the feet. The peculiarities of foot structure of arboreal species of *Bolitoglossa* result in a better ability of the feet to grip by suction than the longer-toed feet of terrestrial species. Since surface tension is more effective at smaller body size, the generally reduced size of arboreal species is also functionally adaptive.

P. Alberch and J. Alberch compared an arboreal species, *B. occidentalis*, with two typical terrestrial species, *B. rostrata* and *B. subpalmata*. As expected, the arboreal species is smaller than its two congeners, shows the typical foot modifications, and has a modified skull in which the prefrontal bones are absent and ossification is reduced as well. In overall proportions adult *B. occidentalis* resembles juveniles of *B. rostratus* and *B. subpalmata*; it is important to note that the feet of the immature terrestrial species are webbed. These resemblances suggest that in the evolution of the arboreal species selection was for small size and that this was achieved by progenesis, truncation of somatic development.

Patterns of growth and allometry for tail length, weight, foot surface area, and digit length are the same in all three species as would be expected for simple progenesis. *B. occidentalis* simply does not grow as large as the other species. It ceases growth at an earlier stage, and becomes sexually mature. The generation of a paedomorphic shape by truncation is presented in a quantitative way in Figure 6–4, which plots

FIGURE 6–4. *Patterns of foot development in* Bolitoglossa. *In* B. rostrata *and* B. subpalmata *the toes grow out of the pad of the foot, whereas in* B. occidentalis *the growth curve levels off early and the toes never project far out of the pad, with the result that the adult foot is webbed.* [Modified from P. Alberch and J. Alberch, 1981.]

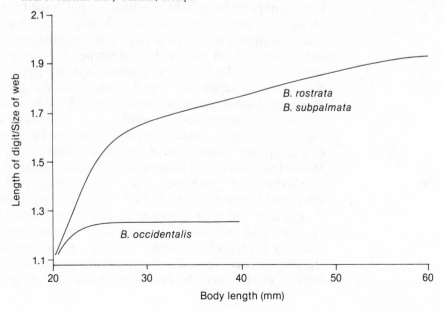

the ratio of toe to web length in the feet of developing *Bolitoglossa*. The initial kinetics of foot growth from a pad are essentially the same in all species; however, toes grow more rapidly than the interdigital webbing. But in *B. occidentalis* differential growth ceases at a size at which the foot is webbed in shape. In the other two species differential growth of digits and interdigital areas of the foot continues, and free digits become pronounced.

By this mode of heterochrony some of the juvenile morphological features of the arboreal species are adaptive, but others may not be. These may be simply the passive consequences of paedomorphosis.

The conclusion that growth of *B. occidentalis* is truncated is further strengthened by the existence of such passive changes, as indicated by the relationship between the sequence of appearance of foot and skull bones in development and the absence of certain of these bones in *B. occidentalis*. In *B. rostrata* and *B. subpalmata* the last skull bone to appear is the prefrontal, which never appears in *B. occidentalis*. Similarly, the third phalanx of the fourth digit of the hindfoot of *B. rostrata* and *B. subpalmata* is the last to appear, and it never appears in *B. occidentalis*. Evidently, development has been truncated at a size just short of the appearance of these particular features.

Although a global change in ontogeny may open a new adaptive zone, complete morphological adaptation will require subsequent changes in genetic systems regulating particular aspects of morphogenesis. Alberch and Alberch have recognized such additional heterochronies in *B. occidentalis* foot morphogenesis. Size and shape of the foot are dissociated in a manner such that shape-change ceases when the animal attains a snout-vent length of 24 mm, but the foot continues to grow until the animal reaches a snout-vent length of 38 mm. There are other dissociations in the foot: Growth rate of the metacarpals is accelerated, whereas the rate of some phalanges is retarded relative to the standard for the terrestrial species. This is an important point because heterochrony is not limited to the classic global modes, but can operate on any individual process in development.

Like salamanders, tropical frogs have entered some adaptive zones very different from those of typical temperate zone species. The degree of predation and competition to which aquatic tadpoles are subjected in the tropics is apparently very intense. This and the availability of a variety of humid terrestrial microhabitats have been conducive to the evolution of reproductive patterns in which eggs are not simply deposited in open water. Table 6–2 contrasts the reproductive modes of a temperate fauna, that of the United States, with two neotropical faunas from sections of Panama and Ecuador. Almost all of the northern species breed in open ponds. A few species representing tropical groups are found in the southern United States. These present a sampling of reproductive behaviors common in the tropical faunas. Modes in which eggs are laid over water or near water in situations such that upon hatching the tadpoles enter ponds or streams reduce the time in which the tadpoles are exposed to the rigors of aquatic life. These adaptations have evolved independently in several families of frogs, but much of the

TABLE 6–2. Diversity in Reproductive Modes of Frogs in Temperate and Tropical Faunas[a]

Reproductive Mode	Percentage of Total Fauna		
	United States	Panama	Ecuador
Eggs and tadpoles in ponds	90	20	37
Eggs suspended over ponds; tadpoles in ponds	—	11	10
Eggs in foam nests; tadpoles in ponds	1	11	8
Eggs and tadpoles in streams	3	11	4
Eggs suspended over streams; tadpoles in streams	—	11	4
Eggs terrestrial; tadpoles carried to water	—	—	7
Eggs terrestrial; direct development	6	30	21
Eggs carried by female; direct development	—	—	3
Unknown	—	—	6
Total number of species in fauna	70	29	78

[a]Adapted from Salthe and Duellman in *Evolutionary Biology of the Anurans,* edited by James L. Vial, by permission of the University of Missouri Press. Copyright 1973 by the Curators of the University of Missouri.

progressive sequence of adaptations freeing tropical frogs from dependence on open ponds can be seen within the single genus *Leptodactylus* discussed by Heyer. There are five species groups in this genus. The Melanotus and Ocellatus species groups exhibit the primitive *Leptodactylus* pattern in which eggs are laid in a foam nest floating on the water. In the somewhat more advanced Pentadactylus group foam nests are placed in potholes in the vicinity of standing water. The nests are destroyed by torrential rains and the tadpoles thus released are washed into the pond. A further step has been taken by the Fuscus group in which the male digs a burrow from which he calls. Eggs are laid in a foam nest in the burrow. Development begins in the burrow, but is completed in ponds into which the tadpoles are swept by rains. Complete independence from aquatic development has been reached by members of the Marmoratus group. In these species eggs are laid in a foam nest in an underground chamber. However, unlike the Fuscus group, the eggs contain sufficiently large stores of yolk for the embryos to complete development and undergo metamorphosis within the nest. Tadpoles are still produced, but they lack the tooth row denticles and spiracles of typical aquatic *Leptodactylus* tadpoles.

The logical end for a series of evolutionary modifications in reproduction such as those seen in species of *Leptodactylus* is direct development with loss of the larval stage. This has been achieved by frogs of several families: *Liopelma* is an amphicoelid from New Zealand; *Arthroleptella* is a ranid from South Africa; *Breviceps* and *Ahydrophrene* are African brevicipitids; and *Eleutherodactylus* is a Central American bufonid. Frogs that lack a free larval stage must be able to provision their eggs with sufficient yolk to support development of a juvenile frog. The key preadaptation, as suggested by Bertha Lutz, appears to have been an increase in the yolk content of eggs of frogs with life histories similar to those of some species of *Leptodactylus*. An increase in yolk appears to be another example of a genetically easy change that makes possible a new evolutionary direction. It is interesting that 50 years ago Noble proposed that evolutionary increases in yolk content of amphibian eggs may occur suddenly because there are species of frogs that produce both large and small eggs.

Direct development has resulted from an elimination of larval structures and an acceleration of development of features of the adult body. The overall process does not correspond to any of the classic definitions of heterochrony because a perfectly conventional frog results from a drastically modified ontogeny. The best example of direct development comes from the definitive work of W. G. Lynn on *Eleutherodactylus nubicola*. Development of this frog is contrasted with that of the North American frog *Rana pipiens* in Figure 6–5. Unlike frogs with a well-defined aquatic larval stage, *Eleutherodactylus* fails to develop lateral line organs, the gut is never coiled, and in *E. nubicola* gills fail to develop. Gills do appear in some other species of *Eleutherodactylus*, such as *E. portoricensis*, observed by Gitlin. Loss of larval structures has not been the result of a global heterochrony because one very prominent larval feature, the tail, is retained in a modified form. Most likely, loss of larval features reflects suppression of inductive systems responsible for the appearance of individual larval structures. There has been a concomitant acceleration in the appearance of some adult structures.

The most pronounced heterochrony affecting external features is in the relationship between the timing of development of the limbs and other structures. A comparison (as in Fig. 6–6) of the time course of limb and tail development in *Eleutherodactylus* with that of *Xenopus*, which exhibits a typical metamorphosis, reveals several differences in timing. In *Eleutherodactylus* the limb buds appear as early as the time of completion of neural tube closure. Limb growth is initiated before differentiation of the major regions of the brain and eye in *Eleutherodac-tylus*, whereas limb development follows those features in *Xenopus* and *Rana*. In *Xenopus* and *Rana* limb buds appear only after rapid growth of the tail has been initiated, in contrast to *Eleutherodactylus*, in which growth of the limbs precedes appearance of the tail. Once both have begun in *Eleutherodactylus*, tail and limb growth are more or less parallel, whereas in *Xenopus* tail growth is more rapid than limb growth until a final spurt of limb growth during metamorphosis. In both species limbs reach their full juvenile length after the start of resorption of the tail.

The acceleration of the development of limb buds is not necessarily reflected in other features. For instance, Lynn compared the order and timing of ossification of the *Eleutherodactylus* skull with that of *Rana temporaria*, which had been studied by Erdmann. The development of the skull of *Eleutherodactylus* differs from that of *Rana* in one very pronounced feature, the total absence in *Eleutherodactylus* of the supra-rostral and infrarostral cartilages that support the mouth of feeding

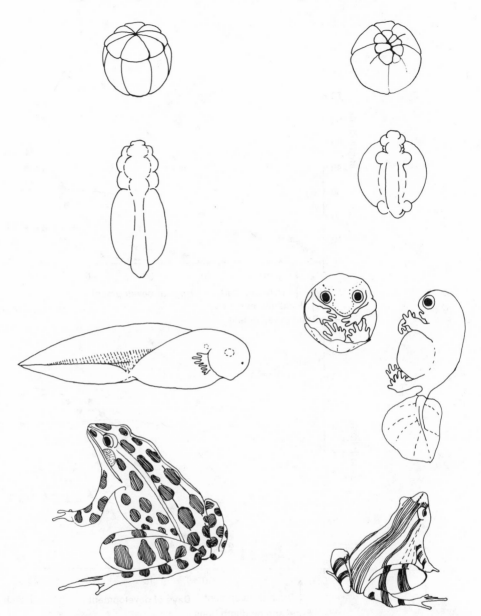

FIGURE 6–5. *Contrast between the developmental pathways of a frog with a tadpole and metamorphosis with a frog exhibiting direct development.* Top left: *Cleavage of* Rana pipiens *egg; below that, in sequence, neurula stage, tadpole with gills and tail, and adult frog.* Top right: *Cleavage of* Eleutherodactylus nubicola; *below that, neurula with precocious limb and tail buds, prehatching froglet with tail modified as respiratory organ, and adult frog.* [Embryonic stages redrawn from V. Hamburger, *A Manual of Experimental Embryology*, 1960. Copyright 1960 University of Chicago; Gitlin, 1944; and W. G. Lynn, The embryology of *Eleutherodactylus nubicola*, an anuran which has no tadpole stage. Carnegie Inst. Wash. Contr. Embryology **30**(190):27–62, 1942. Adults from life.]

tadpoles until metamorphosis. In *Rana* only five bones become ossified before the onset of metamorphosis. During metamorphosis of *Rana* other bones ossify, followed by a final few after completion of metamorphosis. The order of ossification in *Eleutherodactylus* shows

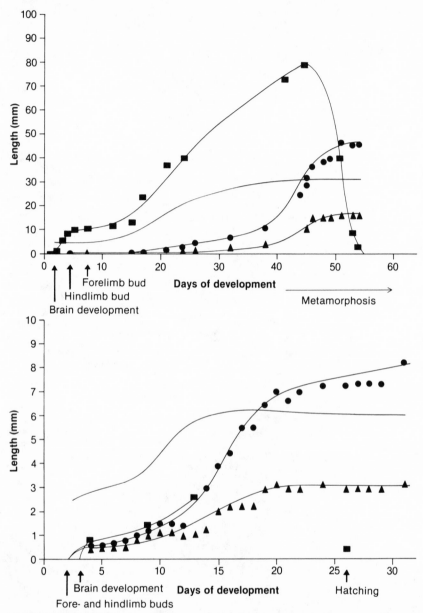

FIGURE 6–6. *Relative timing of developmental events in* Xenopus, *a frog possessing a tadpole stage and metamorphosis, and* Eleutherodactylus, *a frog with direct development. In* Xenopus *(top), the tail appears well before limbs. Brain development is also advanced before limb growth begins. Limbs grow very slowly in the tadpole. This slow limb growth phase is followed by a burst of limb growth during metamorphosis, at which time the tail is rapidly resorbed. In* Eleutherodactylus *(bottom), limbs begin to grow before the tail and before the development of the brain. Except for the resorption of the tail prior to hatching, development of adult features is direct and there is no metamorphosis. Tail (— ▪ —), hindlimbs (— • —), forelimbs (— ▲ —), body length (——).* [Drawn from data of P. D. Nieuwkoop and J. Faber, 1956; and from a figure and data of W. G. Lynn, The embryology of *Eleutherodactylus nubicola*, an anuran which has no tadpole stage, Carnegie Inst. Wash. *Contr. Embryol.* **30**(190):27–62, 1942.]

186

several heterochronic changes when compared with *Rana*. During the period from about the 16th to the 19th day of development the first bones ossify. Two (the angular and the squamosal) are accelerated with respect to their appearance in *Rana*. These bones are followed by a sequence of ossifications such that by four days before hatching the progress of ossification of the *Eleutherodactylus* skull resembles that of *Rana* at metamorphosis. However, some bones are retarded in their appearance. In *Eleutherodactylus* the septomaxilla, pro-otic and men-tomeckelian bones, which appear in *Rana* at metamorphosis, have not yet appeared by 10 days after hatching, nor have the bones that ossify soon after metamorphosis in *Rana*.

If growth of limbs and resorption of the tail are taken as morphological indicators of metamorphosis in *Xenopus* and *Rana*, then the *Eleutherodactylus* embryo has reached a stage corresponding to metamorphosis by a few days prior to hatching. However, not only is ossification of much of the skull not as far advanced as in metamorphosing tadpoles of *Rana*, but some other skeletal elements are also retarded. For instance, Lynn noted that while the ischium, sternum, and episternum ossify in *Rana* during metamorphosis, they have not yet begun to do so at the time of hatching in *Eleutherodactylus*.

The most striking of the accelerated features of *Eleutherodactylus* are the aortic arches. Embryonic vertebrates develop six pairs of aortic arches in association with the pharyngeal, or gill, pouches. In jawed fishes and higher vertebrates no complete first aortic arch is retained by adults; some fishes retain the second arch and most retain arches III–VI. In tetrapods the first and second aortic arches disappear during development as does the fifth, which is retained only by salamanders. The third, fourth, and sixth arches, respectively, give rise to the carotid arch, the systemic arch, and the pulmonary artery. The pattern of development of the aortic arches of frogs is shown in Figures 6–7(a–c), which is redrawn from the work of Millard on *Xenopus laevis*. In that species the first aortic arch appears early in larval development, followed in rapid succession by the third and fourth arches. The second aortic arch appears only as a rudiment. The first and second arches begin to degenerate before the sixth arch appears. The third to sixth arches serve to supply blood to the gills. Later in development the fifth arch disappears and only the third, fourth, and sixth arches remain in the adult.

Figure 6–7 compares the development of the aortic arches of *Eleutherodactylus* with those of *Xenopus*. As in *Xenopus* the aortic arches make their appearance relatively early in development, but the pathway is quite different. The first and second aortic arches fail to appear at all. The third aortic arch appears first, and is followed by the fourth and then the sixth aortic arch. Thus the adult condition is developed directly in an accelerated fashion in an early embryo that never develops gills and has no need to establish the circulatory system associated with gills. It is especially intriguing that the first, second, and fifth aortic arches are not even present as transitory structures in *Eleutherodactylus*. Higher

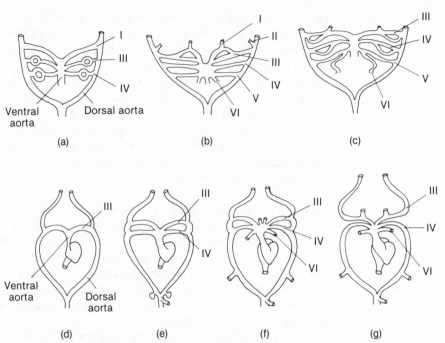

FIGURE 6-7. *The development of the aortic arches in* Xenopus *(a–c) and* Eleutherodactylus *(d–g). All six pairs of aortic arches make their appearance in the* Xenopus *larva, although arches I and II are quite transient. The loops in the arches of* Xenopus *diagrammatically represent the circulation system of the gills. The final form of the heart and aortic arches after metamorphosis of* Xenopus *resembles those shown in (g). In* Eleutherodactylus *embryos arches I and II fail to develop. Instead the final state (g) is directly attained with development of arches III, IV, and VI. There are no gills.* [Modified from N. Millard, 1945; and W. G. Lynn, The embryology of *Eleutherodactylus nubicola*, an anuran which has no tadpole stage, Carnegie Inst. Wash. *Contr. Embryol.* **30**(190):27–62, 1942.]

vertebrates, reptiles, birds, and mammals all exhibit direct development and lack of gills, yet in reptiles and birds all six aortic arches appear in the embryo; in mammals all appear except the fifth. As in *Xenopus*, only the third, fourth, and sixth aortic arches are retained in adults of the higher vertebrates. Retention of transitory aortic arches by reptiles, birds, and mammals suggests a developmental function for these arches, perhaps as elements in an inductive cascade necessary for the appearance of some other structure. Such a role seems to have been obviated in *Eleutherodactylus*.

Direct development by *Eleutherodactylus* illustrates an evolutionary approach to terrestrial life independent of that followed in the Paleozoic Era by the ancestral amniotes. However, it appears likely that a transition from larval to terrestrial reproduction by the ancestors of the reptiles might well have followed a similar course—acceleration of the development of key adult features. Heterochrony would not have resulted in a different adult morphology, but in a different pattern of ontogeny that, in combination with the evolution of the amniotic egg, would have made possible all of the subsequent evolution of terrestrial vertebrate life.

Mechanisms of Amphibian Heterochrony

Heterochronic changes in amphibian development are particularly amenable to experimental study because such changes are common and they are readily defined with respect to the normal metamorphic patterns of related forms. Further, amphibian metamorphosis, for all its complexity, is regulated by a relatively simple sequence of hormonal events. In 1912 Gudernatsch showed that amphibian metamorphosis is primarily under the control of thyroid hormones. The control of thyroid hormone function in amphibians is apparently similar to that outlined for mammals, as recently reviewed by Dodd and Dodd. The mammalian hypothalamus produces a thyrotropin-releasing hormone (TRH), which is transported to the pituitary where it causes the release of thyrotropin, the thyroid-stimulating hormone, TSH. Thyrotropin in turn acts on the thyroid to stimulate the release of thyroxine. In amphibians thyroxine has far-reaching effects on numerous target tissues and induces morphological and biochemical changes, including the resorption of the tail and gills and changes in the character of the skin and the digestive, respiratory, circulatory, excretory, reproductive, and nervous systems. In amphibians, as in mammals, the hypothalamus regulates pituitary and consequently thyroid activity. The exact hormones may differ in some respects because, whereas amphibians respond to mammalian TSH and thyroxine, Taurog and his collaborators have shown that mammalian TRH is not active in amphibians.

The most coherent hypothesis for the hormonal control of amphibian metamorphosis has been advanced by Etkin. Larval development can be divided into three time intervals with respect to hormonal activity. Early in larval life, premetamorphosis, TSH and thyroxine are maintained at a very low level. In the next phase, prometamorphosis, TSH and thyroxine levels rise rapidly; then, after the brief metamorphic climax, hormonal levels decline rapidly. Etkin's model assumes that in premetamorphic larvae the thyroid produces a low level of thyroxine that, through a negative feedback loop, inhibits TSH production by the pituitary. The pituitary also produces a high level of another hormone, prolactin, which acts both as a growth hormone and as an inhibitor of the response of target tissues to thyroxine. Until late in premetamorphosis the hypothalamus apparently exerts no control over the pituitary. However, at the end of premetamorphosis the hypothalamus becomes responsive to the low level of thyroxine, is activated, and consequently stimulates the pituitary to produce high levels of TSH. Thyroxine levels rise and metamorphosis ensues.

Given the hormonal basis of metamorphosis it is hardly surprising that research on the causes of neoteny in salamanders has been largely centered on thyroid function. Dent has classified neotenous salamanders as falling into three categories. These cross taxonomic boundaries because neotenous forms have arisen in several families of salamanders.

The first category includes permanently larval forms that cannot be induced to metamorphose, even if artificially treated with thyroxine. Their tissues are apparently unable to respond to thyroxine. In 1931 Noble considered a number of morphological features, such as the development of limbs and maxillary bones, loss of gills, and reduction of branchial arches, and found that these permanent neotenes differ in the extent to which they correspond to progressive stages along the normal pathway of metamorphosis. *Siren* (whose name belies its appearance) resembles an early larval stage, *Proteus* a late premetamorphic stage, *Cryptobranchus* the beginning of metamorphosis, whereas *Andrias* (*Megalobatrachus*) is equivalent in structure to nearly completed metamorphosis.

A second group of paedomorphs includes species that are consistently neotenous in nature but can be induced to metamorphose in the laboratory. This group includes the axolotl. Studies on the axolotl and similar forms indicate that the tissues of these neotenes exhibit a normal responsiveness to thyroxine and their thyroid and pituitary glands are capable of functioning like those of nonneotenous forms. Neoteny is apparently the result of a failure in the hypothalamic release mechanism. This idea is supported by the observations of Norris and Gern that neotenic *Ambystoma tigrinum* can be induced to metamorphose by injection of a small amount of thyroxine into the hypothalamus, thereby activating the hypothalamic-pituitary-thyroid axis, which results in an elevation of thyroxine production. This rather simple endocrine mechanism is consistent with the genetic observations of R. R. Humphrey (discussed by Tompkins) that neoteny in species of *Ambystoma* is controlled by two alleles of a single gene.

There is a final group of salamander species that generally undergo metamorphosis, but under some environmental conditions are found to be neotenous in the wild. For example, salamanders that metamorphose in warm water ponds may become neotenous in cold water environments, such as mountain lakes. As discussed by Jenkin, this may be caused by developmental retardation of the hypothalamus-pituitary-thyroid axis by cold, or as suggested by Dent, amphibian tissues may not respond to thyroxine in the cold. There may be genetic factors involved in these cases because there is a clear selective advantage for salamanders to be neotenous in cold, mountainous environments.

As is already apparent, heterochrony in amphibians is more varied than neoteny with its very simple endocrine basis. For instance, the role of thyroxine in direct development of the salamander *Plethodon* and the frog *Eleutherodactylus* has been investigated by Lynn and by Lynn and Peadon, who inhibited thyroxine action in these embryos with thiourea. *Plethodon* completed development except for resorption of the gills. Treated *Eleutherodactylus* embryos also developed in a nearly normal fashion with the exception that the pronephros failed to degenerate, the tail was not resorbed, and the egg tooth was not lost. The converse experiment of administering thyroxine to embryos of *Eleutherodactylus* caused premature resorption of pronephros and tail, but had little other effect. Thus, in direct development the retained metamorphic features of development are still under the control of thyroxine, but the majority of the direct development sequence is not.

The potential for evolutionary change in hormonal processes that results in heterochronies is not limited to thyroxine; however, this system provides a good illustration of the possibilities. For example, a change in the time when the hypothalamus is activated by thyroxine would control size at metamorphosis as well as length of time available for development of anatomical features during premetamorphosis. Heterochronic effects can also arise from changes in relative sensitivities of individual tissues to hormones. In fact, change in tissue responsiveness without significant changes in the hypothalamus-pituitary-thyroid axis seems to be the common mode. There may be a change in sensitivity of individual tissues to the growth hormone activity of prolactin, and thus, a change in relative growth rate. There may be a change in tissue sensitivity to the antagonistic effects of prolactin and thyroxine. Finally, individual tissues may change relative to one another in their sensitivity to thyroxine itself. A requirement for a higher level of thyroxine will, for instance, delay a thyroxine-dependent event or even abolish it from the developmental sequence. Action of thyroxine at the cellular level is complex, and changes in thyroxine sensitivity may well involve dissociation of a variety of subtle cellular events. Baxter et al. have discussed the effects of thyroxine on cultured pituitary cells. These cells synthesize prolactin and growth hormone, and respond to thyroxine by alterations in the cell surface and induction of synthesis of the enzyme hyaluronidase and growth hormone. Hyaluronidase is produced in tadpole tails under the stimulation of thyroxine and is associated with the destruction of hyaluronic acid during resorption of the tail. The parallel with cultured cells is striking.

Different tissues acquire cellular receptors for thyroxine and sensitivity to the hormone at different times during development, and they respond in different ways biochemically. Thus, tail tissues degenerate in response to thyroxine (see, for example, the studies of Beckingham-Smith, and Tata), while other tissue types change functions. For instance, skin transforms from that characteristic of larvae to that characteristic of the adult. A part of this change is in the replacement of larval keratins by adult keratins. As shown by Reeves, *Xenopus* larval skin responds to thyroxine by synthesizing adult keratin messenger RNA and translating it to produce the adult protein. Larval skin actually becomes responsive to thyroxine in this respect 20–24 days before the normal rise in thyroxine level triggers the synthesis of adult keratins in metamorphosis.

Dissociation of biochemistry from morphogenetic features of metamorphosis is obvious in some neotenes, and shows that paedomorphs are not necessarily simply giant larvae. They possess a melange of juvenile and adult traits that may provide new evolutionary possibilities. Ducibella has shown that axolotls, while retaining a larval morphology, exhibit the same changes in red blood cell properties, serum proteins, and hemoglobins as observed during metamorphosis of related nonneotenous species. The biochemical situation is more complex in other neotenes, such as *Triturus helveticus*, in which Cardellini and co-workers found that hemoglobins characteristic of both larvae and metamorphosed adults are present. Because there is a low level of thyroxine produced by neotenous axolotls it is probable, as proposed by Ducibella, that the tissues responsible for the biochemical changes probably have a very much lower threshold of sensitivity to thyroxine than other tissues that produce the anatomical changes of metamorphosis.

Morphological and Molecular Recapitulation: A Cracked Mirror

In his discussion of molecular recapitulation, George Wald begins with the statement that "living organisms are the greatly magnified expressions of the molecules that comprise them." This is both a truism and an exaggeration, but it does evoke a general principle applicable to both molecular systems and embryos. In systems in which complex interactions are present, evolutionary change is constrained by the need to maintain functional integration. This is seen on the most basic levels of gene expression in the nearly total conservation of the genetic code from bacterium to mammal with only mitochondria, evolutionary oddities in many respects, showing any variations in the meaning of codons. Complex supramolecular assemblies, such as ribosomes, also show a profound conservation, with the basic structural organization shared by both procaryotes and eucaryotes. Analogous constraints on evolutionary modifications of embryonic processes occur in situations involving inductive interactions between developing structures within an embryo. If, as is often the case, the cascade of inductions is complex it will be maintained through the immensely long history of the major group possessing it. The result is morphological recapitulation.

If morphological recapitulation occurs for a sound mechanistic reason, does the sequence of genes active in development represent an analogous molecular recapitulation? This question is perfectly valid because the switching on and off of genes often accompanies morphological transitions, for example, metamorphosis, and may indicate a linkage between the morphogenetic event and concomitant biochemical transitions. Conversely, if in a developmental sequence showing morphological recapitulation it can be demonstrated that the sequence of molecular events is not recapitulatory, it would indicate that there has been a dissociation between morphological and molecular pathways. As in cases of dissociation discussed in Chapter 5, the dissociation of biochemical events from a conservative morphogenetic sequence opens real possibilities for the evolutionary modification of ontogeny.

The excretion of nitrogenous wastes by vertebrates centers on the disposal of ammonia, the highly toxic product of the deamination reactions responsible for the initial steps of catabolism of amino acids. Aquatic animals, including most fishes, can afford to simply excrete ammonia through their gills into the surrounding water. This option is not open to terrestrial animals because they cannot tolerate high levels of ammonia and cannot afford the loss of water that would be necessary to excrete dilute ammonia directly in their urine. Therefore, terrestrial animals must convert ammonia to a less toxic compound that can be stored in a concentrated form for excretion. The crossopterygian ancestors of the amphibians evolved a pathway, the urea cycle, that made the conversion of ammonia and carbon dioxide to urea, a relatively nontoxic compound, possible. Crossopterygians, which breathed with lungs like their still extant cousins the lungfishes, made their transition to terrestrial life during a time of persistent and widespread periodic drought. When the streams and ponds they inhabited dried up they were able to move clumsily overland in search of other water. The consequences to nitrogen excretion of this sort of activity can still be seen in the existing lungfish *Protopterus* and *Lepidosiren*, which are ammonotelic during their normal aquatic existence. However, during time of seasonal drought these fish aestivate in burrows in the hardened mud of their dried ponds. Aestivation is a different strategy for evading drought than was used by the ancient crossopterygians, but it has the same result in that ammonia cannot be excreted under such conditions. During aestivation ammonia excretion ceases and the lungfish convert ammonia to urea, which accumulates in the body until it can be excreted when the rainy season arrives.

Cohen and Brown have suggested that the crossopterygians were preadapted for ureotelism because the enzymes of the urea cycle already existed and probably served in the synthesis of the amino acid arginine. The ureotelism of the crossopterygians was inherited by the amphibians. Living amphibians repeat in their development the transition of their ancestors from aquatic forms to terrestrial. There is also an interesting, apparently recapitulatory, shift in nitrogen excretion. The tadpole is ammonotelic, but shifts to ureotelism at metamorphosis. Cohen and Brown have documented the enzymic basis for this change. The levels of the individual enzymes of the urea cycle rise sharply as metamorphosis proceeds. Recapitulation is suggested in this case because the transition occurs before the tadpole actually leaves the water.

Not all amphibians become terrestrial upon metamorphosis. *Xenopus laevis,* a wholly aquatic frog, was found by Underhay and Baldwin to be ammonotelic as a tadpole and to begin to switch to ureotelism with metamorphosis. However, upon completion of metamorphosis ammonotelism again resumes prominence, although the enzymes of the urea cycle are present and function in the adult. Is the shift to ureotelism recapitulatory or adaptive? Perhaps it is both. *Xenopus* typically lives in the water and has little need for urea excretion. But during the dry season this frog is capable of aestivation, which would be impossible without the ability to shift from ammonotelism to ureotelism in a manner analogous to that of lungfish. The retention of ammonotelism by *Xenopus* adults is typical of other aquatic amphibians, including paedomorphic salamanders, which never undergo physical metamorphosis.

The concept of recapitulation receives a new twist with the curious second metamorphosis of the common newt *Notophthalmus* (*Triturus*) *viridescens*. This newt starts larval life as a modest olive-green aquatic form possessing gills. As expected, it excretes ammonia. After a few months the newt undergoes metamorphosis to a brightly colored red eft, a wholly terrestrial lung-breathing form that is ureotelic. In two or three years the full-grown eft undergoes a second metamorphosis, resuming an aquatic existence as a sexually mature form. The events of second metamorphosis include the regaining of several larval attributes—a green color, a keeled tail, and functional lateral-line organs. The gills are not regained. Biochemical changes also take place during second metamorphosis. For instance, there is a shift in the major visual pigment in the eye from vitamin A_1 to vitamin A_2. Wald has shown that the presence of vitamin A_1 is typical of terrestrial vertebrates, whereas A_2 is typical of freshwater forms. Thus the newt as a larva utilizes vitamin A_2, switches to A_1 at first metamorphosis, and then again switches visual pigment to vitamin A_2 upon second metamorphosis. Similarly, Nash and Fankhauser found that while the mature newt remains primarily ureotelic, after second metamorphosis ammonia comes to account for about 25% of nitrogen excretion.

As shown in Figure 6–8, the urea cycle has been retained by mammals, but lost in most reptiles and all birds. One of the most widely cited examples of possible molecular recapitulation stems from observations originally made by Needham on nitrogen excretion in chick embryos. However, in this case developmental changes, which at first glance appear to be recapitulatory, are in fact seen to be not so when molecular events are examined in more detail. Needham's data indicated that the chick excretes ammonia during the first few days of development, then urea, and finally uric acid. Uric acid is a purine produced by a pathway entirely unrelated to the urea cycle. Because the adult chicken lacks a urea cycle and is wholly uricotelic, such a pattern of nitrogen excretion bears a striking resemblance to changing patterns of nitrogen excretion in the evolutionary history of birds. However, the details of the sequence of reactions involved fails to bear out any but a most superficial recapitulation. In a critical examination of the problem Fisher and Eakin found that the ammonia content of the egg changes little over the course of development and does not represent active excretion. Urea content does rise, and urea is excreted to the allantois. But no urea cycle is present; urea is generated by the breakdown of yolk-derived arginine by the enzyme arginase. Even though this single terminal enzyme of the urea cycle is active it is not restricted to the liver, as would be expected for a ureotelic organism. Instead, arginase appears to be distributed throughout the body.

There is a similar, potentially recapitulatory set of changes in nitrogen metabolism during snake development. Adult snakes are uricotelic and lack a urea cycle, although they do possess arginase. Snake embryos, however, present a totally different pattern of nitrogen excretion than adults. Clark and his collaborators have found that both the black snake *Coluber constrictor*, which develops in a typical reptilian egg, and the

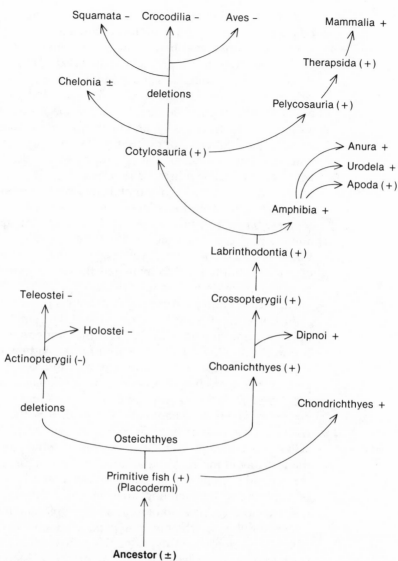

FIGURE 6–8. *The urea cycle and vertebrate phylogeny. Groups with a + have been shown to possess a functioning urea cycle. Groups with a − are known to lack a functioning cycle. Symbols in parentheses indicate postulated presence or absence of the urea cycle. "Deletion" indicates loss of one or more urea cycle enzymes.* [Redrawn from G. W. Brown and P. P. Cohen, Comparative biochemistry of urea synthesis, *Biochem. J.* **75**:82–91, 1960.]

garter snake *Thamnophis sirtalis*, which develops while attached to a placenta within the body of the mother, excrete large amounts of urea. Unfortunately, it is not known if the urea cycle is actually present in these embryos or if, like the chick, they produce urea through the action of arginase alone.

The lack of recapitulation of pathways of nitrogen excretion in the chick embryo is in contrast to the distinctive recapitulation seen in the development of the kidney itself. There are three major types of vertebrate kidneys: pronephric, mesonephric, and metanephric. The vertebrate kidney arises from mesoderm located at each side of the body. The anterior portion of the nephrogenic tissue differentiates into a few tubules organized in a segmental manner, and a duct arises from each tubule to form the pronephric duct that opens to the surface in the posterior of the animal. This pronephric kidney is functional in the larvae of fish and amphibians and is associated with the excretion of ammonia. During development of some species of fish the pronephric kidney is supplemented by, and in all other vertebrates replaced by, the more posterior mesonephric kidney. As the pronephric tubules degenerate the pronephric duct is appropriated by the mesonephros and becomes the mesonephric, or Wolffian, duct. The mesonephros is the adult kidney of fish and amphibians, and in these forms excretes ammonia or urea. Kidney development is somewhat more elaborate in reptiles, birds, and mammals. Nonfunctional pronephric tubules arise and are replaced by a functional mesonephros. The mesonephros is in turn replaced by the later developing metanephros, which becomes the definitive adult kidney. In mammals the metanephros excretes urea; in reptiles and birds it excretes uric acid.

The three kidneys appear sequentially in chick development. The pronephros appears in the second day of development and the tubules never become functional. However, the pronephric duct, which gives rise to the Wolffian duct, is vital to further kidney differentiation. As shown in Figure 6–9, the Wolffian duct induces the development of part of the reproductive system, the Müllerian duct, as well as the tubules of the mesonephros and the metanephros. The mesonephros begins to function at about the fourth day in the chick embryo and reaches a peak of activity by the 14th day. Then it degenerates. Its activities are assumed by the metanephros, which begins to function on the 11th day. If molecular recapitulation were occurring in this system to correspond with morphological recapitulation, the mesonephros would be expected to be excreting urea not uric acid.

A second possible kind of molecular recapitulation has been posed by Zuckerkandl as "whether polypeptide chains that function in the embryo are evolutionarily older than their adult counterparts." Data exist that make it feasible to test for the existence of this type of recapitulation. Zuckerkandl has devised a very simple test for this phenomenon. In terms of the hemoglobins for which sequence data are readily available, Zuckerkandl suggested that one need only compare fetal and adult sequences to their common ancestral sequence. If the fetal globin is recapitulated it should have diverged less from the ancestral sequence than the corresponding adult globin. Since the human β-globins (including the fetal chain γ) share a common ancestral sequence with the α-globins, direct sequence comparisons can be made between human β- versus human α-globin, and human γ- versus human α-globin (see Fig. 6–10). Both differ from the α-sequence by about 55 substitutions. Thus, there have been no differences in evolutionary rates between β- and γ-chains. The "γ"- and β-chains of cattle show the same relationship to α-globin, and the "γ"-chain of cattle is actually more closely related to the β-chain than to the human γ-chain.

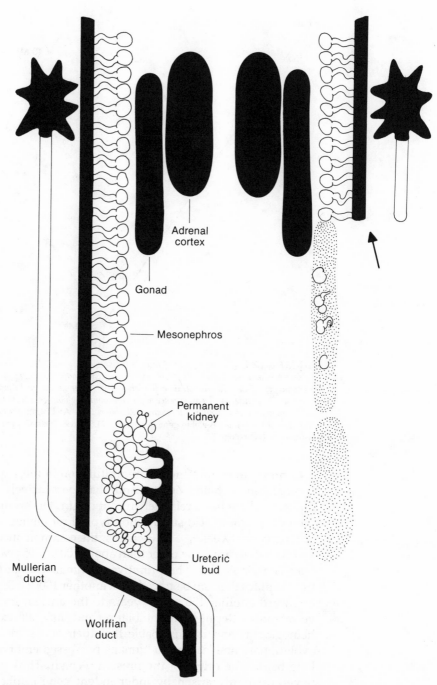

FIGURE 6–9. *Development of the kidney of the chick. The nonfunctional tubules of the pronephros (not shown in this diagram) appear in the second day of development. Kidney development continues with the appearance of the functioning mesonephros at approximately the fourth day, and the permanent kidney (metanephros) at about the eleventh day of development. The Wolffian duct arises as the duct of the pronephros. Its presence is required for induction of development of kidney tubules and for growth of the Müllerian duct. As shown diagrammatically on the right-hand side of the figure, if the Wolffian duct is destroyed these other structures fail to appear.* [Redrawn from P. Gruenwald, Development of the excretory system, *Ann. N.Y. Acad. Sci.* **55**:142–146, 1952.]

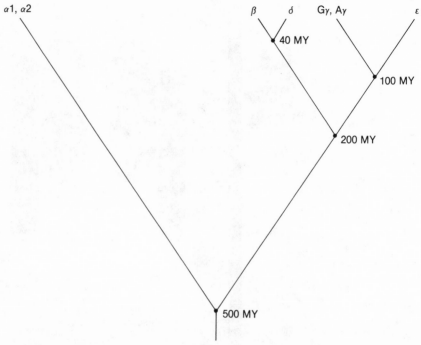

FIGURE 6–10. *Evolutionary tree of human β-like globin genes. The divergence between α- and β-globin genes occurred about 500 × 10⁶ years ago. Efstratiadis et al. suggest a divergence of the adult β-globin gene from the ancestral fetal globin gene early in the evolution of mammals from the mammal-like reptiles, and a divergence of the embryonic ε-globin gene from the fetal chain genes early in the radiation of placental mammals.* [Redrawn from A. Efstratiadis et al., The structure and evolution of the human β-globin gene family, *Cell* **21**:653–668, 1980. Copyright by the Massachusetts Institute of Technology.]

Contrary to confirming any recapitulation in the sequence of the fetal hemoglobin γ-chain, Zuckerkandl has persuasively argued that the human γ-chain has a relatively recent origin. His argument is that given the rate of amino acid substitution typical for globins and the sequence differences between γ- and β-chains, the common ancestor of human γ- and β-chains cannot be older than about 200 × 10⁶ years, or roughly the Late Triassic Period. The primitive mammals of that time had evolved from reptiles that can be traced back another 100 × 10⁶ years, and thence backward another 150 × 10⁶ years to the ancient form ancestral to all vertebrates. Reptiles, amphibians, and fish all express embryonic hemoglobins and it is probable that their ancestors did likewise. The evolutionary line leading to humans possessed embryonic hemoglobins long before the origin of the present γ-chain. Thus fetal non–α-chains have apparently arisen by independent gene duplication events and succeeded each other during hemoglobin evolution.

Seven

Genetic Control of Development

Nature has so much to do in this world, and is engaged in generating such a vast variety of co-existent productions, that she must surely be now and then too flurried and confused to distinguish between the different processes that she is carrying on at the same time.

Wilkie Collins, The Woman in White

Mutations and Changes in Ontogeny

If morphology is the manifestation of a complex set of developmental processes, so too are these processes the manifestation of the actions of a constellation of genes. This has been the essential assumption of this book, and we have assembled at least a *prima facie* demonstration that a portion of the metazoan genome is specifically involved in control of ontogeny and evolves by a mode distinct from that of structural genes. Thus far, however, we have only dealt with isolated specific examples of gene control of morphogenesis and have not attempted to answer the central question of how genes control the course of development.

In a very real sense we return to Roux's program for developmental mechanics, but instead of deleting cells or other embryonic structures to determine their roles in development, as was done by the classical experimental embryologists, the developmental geneticist exploits mutation as a delicate and precise scalpel to delete or modify individual genes.

The genetic paradigm for studying any system theoretically under genetic control is the following: In order to analyze a process, in this case ontogeny, one recovers mutations that alter that process. Once recovered, individuals carrying the mutation are compared in phenotype with normal individuals. This comparison should give a clue as to how the gene in question affects normal development. However, before continuing with the manner in which the comparison is made, it should be noted that there are two basic ways in which mutations can be seen to

199

affect ontogeny. First are disruptive changes, in which the path of normal development is altered to produce morphological abnormalities (e.g., missing structures). In their most cataclysmic form these mutations cause lethality. Second, there are homoeotic changes. This type of mutation causes an alteration in development such that one structure of the organism is replaced by a homologous organ or limb. We will forego any further discussion of this latter type of mutation until the following chapter and concern ourselves here mainly with disruptive changes.

In the analysis of the defects produced by a disruptive mutation it is rarely a simple matter of observing the terminal phenotype of a lethal individual in comparison to the normal, because development is a complex and highly integrated process. The vast majority of events takes place in relation to other events, and indeed, are dependent on one another. This is most clearly seen in the fact that many mutant conditions have pleiotropic effects, in which several morphological alterations are caused by a single genetic defect or change. An example of this is the defect observed in individuals suffering from what is called *Pelger* (*Pg*) anomaly. This trait is inherited as a simple autosomal dominant in humans. *Pg/+* heterozygous individuals are clinically normal but show abnormal nuclear segmentation in one of their white blood cell types (Fig. 7–1). The polymorphonuclear neutrophil leukocytes at maturity normally have four to five lobes on their nuclei. In *Pg/+* individuals there are usually only two, and rarely, three lobes.

FIGURE 7–1. *The* Pelger *anomaly in the rabbit. The polymorphonuclear neutrophil leukocytes of normal (+ / +), heterozygous (Pg/ +), and homozygous (Pg/Pg) rabbits are shown at the top of the figure. Below are the skeletons and adult morphologies of Pg/ + (left) and Pg/Pg (right) rabbits. Note the severe stunting of the limbs in the homozygous individual.* [Adapted and redrawn from Nachtsheim, The Pelger-Anomaly in Man and Rabbit, *J. Hered.* **41**:131–137, 1950.]

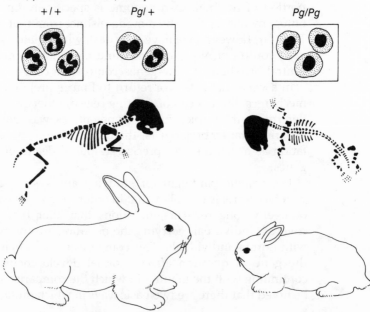

+ / +   Pg/ +   Pg/Pg

This same trait also exists in the rabbit and shows the same pattern of inheritance and blood picture in heterozygotes. By crossing rabbit heterozygotes it is possible to obtain homozygous *Pg/Pg* individuals. The leukocytes of these individuals have no lobes at all and the genotype has low viability. In addition to the leukocyte phenotype the rare survivors demonstrate an extreme dwarfing phenotype in which the limbs and rib cage do not develop fully (Fig. 7–1). The pertinent question to ask at this point is: What is the causative relationship, if any, between these two pleiotropic defects? It is conceivable that both phenotypes are actually the result of a third as yet unknown lesion caused by the *Pg* allele.

A wide range of pleiotropic effects can also be seen in another inherited defect of the blood. This is sickle-cell anemia. It too is inherited as a simple autosomal dominant in humans. The difference between this trait and *Pelger* is that we know the precise biochemical lesion that produces the defect. Individuals suffering from this disease have a single amino acid substitution at position 6 of their β-hemoglobin chains. This single change alters the conformational properties of the resultant hemoglobin tetramer under conditions of low oxygen tension. Red blood cells containing this type of mutant hemoglobin deform to a characteristic "sickle" shape in the peripheral and venous portions of the circulatory system. This shape-change has two direct consequences. The body recognizes the abnormal shape and destroys the sickled red blood cells, resulting in anemia. Also, the sickled cells clump and tend to block capillaries, and thereby local blood supply, which is of course necessary for normal growth and function of the organs. The wide range of defects induced by this single amino acid substitution are summarized in Figure 7–2. What should be apparent is that a naive observer viewing the syndrome from the bottom of the figure faces a very different problem of interpretation than would be the case if the primary defect were understood.

When one is presented with a complex phenotype at the end of a complicated and interactive ontogenetic pathway, it is necessary to consider that a dichotomy exists. The genome can be conceptually divided into two functional parts. One is comprised of those genes involved in what are called "housekeeping" functions and the other consists of those genes directly involved in determination, differentiation, and morphogenesis. The housekeeping functions are those common metabolic and cell-maintenance pathways that, although not directly involved in producing morphology, provide the biochemical *sine qua non* for life. A mutant individual lacking a transfer RNA (tRNA) species or DNA polymerase would indeed suffer severe developmental problems. However, a basic metabolic defect such as these would not necessarily affect a discrete organ, tissue, or time of development. Such mutants must be distinguished from those of real developmental interest.

Two further pieces of information are vital before any definitive conclusions can be drawn as to the nature of the genetic defect. The first is the primary site of action of the gene. That is, is there a specific tissue

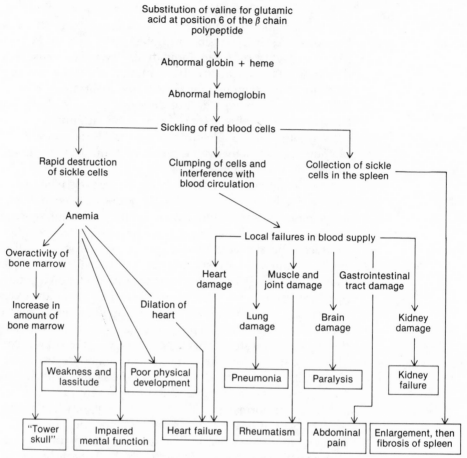

FIGURE 7–2. The pleiotropic effects of the single amino acid substitution in the β-chain of hemoglobin in Homo sapiens *that results in sickle-cell anemia.* [Redrawn from M. W. Strickberger, *Genetics*, 2nd ed., Macmillan Publishing Co., Inc., New York, 1976. Copyright © 1976 by Monroe W. Strickberger.]

or organ in which the gene is active? Moreover, is the gene autonomous in its action? This question relates to the fact that there are actually two types of pleiotropy. In the first, relational pleiotropy, as seen in the example of sickle-cell anemia, there is one primary site of action for the gene (i.e., red blood cells) and all of the other observed defects are related to, or occur, because of that single defect. The second is direct pleiotropy. In this case all of the diverse defects in different tissues and/or organs are caused by the direct action of a single gene. Grünberg has argued, based on his studies of mutations in the mouse, that relational pleiotropy is predominant. However, there are cases of direct pleiotropic effects.

The second piece of information necessary to any meaningful discussion of genetic control of development is the time at which the gene is active. What is the time of onset; is the time of action continuous; does it occur in a single discrete interval; or is gene action required at several discrete intervals? This complex question relates of course to analysis of the nature of the product of the gene being studied.

Analysis of Time and Place of Gene Action

The techniques used to determine the primary site of gene action are similar to and indeed have been borrowed from classical embryology. In their simplest form, an organ or piece of tissue is transplanted from a mutant individual to a normal recipient. The reciprocal graft is also performed. This operation is usually accomplished before any mutant defect is apparent. One can then determine the fate of the developing organ or tissue in its new environment. If the genetic defect of the organ or tissue in question is autonomous to that structure, that is, it is the primary site of action of the gene, it would be expected that the mutant tissue would produce the abnormal phenotype, even in a normal host. Similar kinds of experiments can be performed in tissue or organ cultures in a manner similar to that described in Chapter 5 where mouse, lizard, and chick dermis and epidermis were cocultured. Instead of cross-species associations, mutant and nonmutant tissues or organs are mixed.

The creation of mosaic individuals has been accomplished on an even grander scale by B. Mintz and her collaborators. These workers have been able to fuse entire mouse morula-stage embryos *in vitro*. These "hybrid" embryos are then implanted into pseudopregnant females. These fused or tetraparental mice are comprised of a mixture of cells of two different genotypes, both of which are active. This technique can also be used to analyze autonomy of mutant expression by fusing mutant and normal morulae.

One further technique of this genre should be mentioned: parabiosis. This is the fusion of entire animals rather than just organs or tissues. However, it should be noted that the result in this case is not a true integrated mosaic. All of these techniques require that transplanted tissues, organs, or fused embryos are compatible. In lower vertebrates, such as amphibians, this causes little problem; however, mammals have the added complication of transplant rejection and care must be taken to ensure that the mutant and normal individuals are immunologically compatible.

The remaining two techniques are peculiar to developmental genetics and are used almost exclusively in *Drosophila melanogaster*. These are the production of gynandromorphs and the induction of mosaic individuals by mitotic recombination. Gynandromorphs are adult flies whose bodies are made up of both male and female tissue. By taking advantage of the type of cleavage in the diptera and a special ring-X chromosome, this type of mosaic individual can be produced regularly in laboratory cultures. The normal *Drosophila* X chromosome is a rod with the centromere at one end. A mutant form of this chromosome exists in the form of a closed ring. This ring has the interesting property that it is unstable in the first few cleavage divisions of the embryo. Specifically, this instability can result in the loss of the ring chromosome from one of the two daughter nuclei of the first cleavage division. When loss occurs at this stage, the remaining cleavage divisions yield a population of nuclei in which half have the ring-X and half do not. If the zygote begins its division as a rod-X/ring-X heterozygous female, then after loss half the nuclei will be ring-X/rod-X female and half will be rod-X/null-X male (sex is determined in *Drosophila* by the ratio of X chromosomes to autosomes, not by the Y chromosome, as in mammals). After eight syncytial divisions, the egg is filled by a cloud of nuclei. This cloud, however, is not a random mixture of the XO/XX types. These two classes are present as two spatially separate groups, the position of which is determined by the plane of the first cleavage division. Therefore, when this population of nuclei migrates to the peripheral cytoplasm to form the cellular blastoderm, it does so as two adjacent and contiguous groups of either male or female nuclei. When an adult fly is derived from this gynandromorphic embryo it too will be mosaic. As is shown in Figure 7–3, the amount and position of the adult tissue that is male or female is not constant. This occurs because the plane of the first cleavage division is random with respect to the axes of the egg. Therefore, if the plane of division is perpendicular to an anterior-posterior midsagittal plane in the egg, a bilateral gynandromorph will result. Variation from this simple case will result in a greater or lesser portion of male tissue, depending on how many XO nuclei populate those regions of the blastoderm destined to form adult tissues. If a mutant gene is resident on the X chromosome, its effects can be analyzed in a mosaic simply by

FIGURE 7–3. *The production of gynandromorphs in* D. melanogaster. *The loss of the unstable ring-X chromosome at one of the early cleavage divisions results in the formation of two genotypically distinct populations of cleavage nuclei; XX = dark circles, XO = open circles. If at cellular blastoderm the two populations are separated on a midsaggital plane, the resulting adult fly will be bilaterally mosaic for male and female morphology. If the nonring-X carries recessive marker genes, these will be expressed in the XO male half. This is indicated by the short wing and white eye in the upper diagrammed fly. The two lower figures represent the results of ring-X loss followed by nuclear segregation in either an anterior posterior (left) or an oblique plane (right).* [Adapted and redrawn from M. W. Strickberger, Genetics, 2nd ed., Macmillan Publishing Co. Inc., New York, 1976. Copyright © 1976 by Monroe W. Strickberger.]

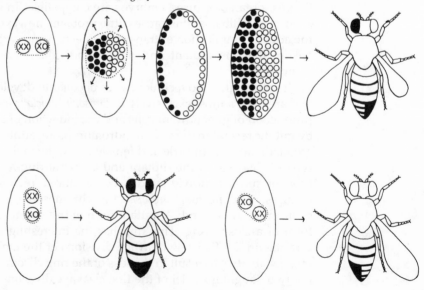

using a rod-X chromosome containing the mutant allele in the above scheme. The resultant gynandromorph will be normal XX female in part and mutant XO male in the remainder of its cells.

The final technique, induced somatic recombination, is not restricted to the X chromosome and therefore has a somewhat wider applicability. In a fly that is heterozygous for a mutant gene and its normal allele, normal mitotic cell division will ensure that every nucleus of that individual is identical and also heterozygous. However, if the developing organism is irradiated with x-rays, recombination can be induced between homologous chromosomes (similar to meiotic exchange). When cell division takes place subsequent to the exchange event, the two cross-over products can segregate from each other (Fig. 7–4). The daughter cells resulting from this process will no longer be heterozygous for the mutant gene, but one will be homozygous for the mutation and the other homozygous normal. The subsequent daughter cells of each of these two reciprocal types will form a clone and, depending on where in the animal and when in ontogeny, will form mosaic spots of differing position and size in the adult. It should be noted that in reality what is produced is an organism of three mosaic cell types because the two homozygous clones will be produced in a background of heterozygous tissue (Fig. 7–4). In this situation it is also possible to determine the survival of the mutant versus normal tissue, or their relative growth

FIGURE 7–4. *A test for cellular autonomy of lethality by x-ray-induced somatic exchange. On the right is the genotypic X chromosome constitution of the heterozygous cell that gives rise to the two homozygous daughter cells after the exchange event.* y = *yellow body color,* l = *lethal, and* sn = *singed bristles. To the left is a dorsal view of the thorax of* D. melanogaster. *The two clones produced by the exchange event are seen as twin spots, one bearing yellow bristles, the other singed bristles. The presence of the "yellow clone" indicates that the lethal is not autonomous in its action.* [Redrawn from E. Hadorn, *Developmental Genetics and Lethal Factors,* Methuen and Company, 1955, English translation, 1961.]

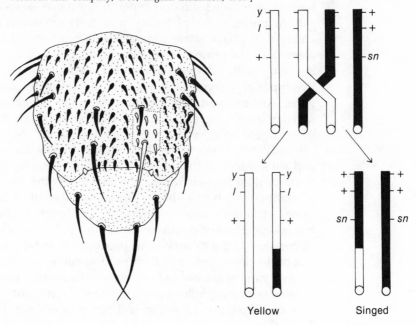

Yellow   Singed

rates, because the single cross-over event should produce twin spots if the markers were appropriately placed in the original heterozygous cell, that is, adjacent patches of the two types of tissue.

The above methods of producing mosaics can yield information as to the site-specificity of gene action for any observed mutationally produced defect. That is, a morphological abnormality produced by a mutant gene can result from the direct action of the gene in the abnormal tissue itself or by the inability of another morphologically unaffected tissue to supply a component necessary to the normal development of the abnormal structure. A case in point is the sex-linked *vermilion* (*v*) gene in *Drosophila*. Flies carrying a *v* mutation have bright red eyes, as compared with the dull red eyes of normal individuals. This defect is produced because the brownish ommochrome pigments normally present in the eye are not formed in *v* individuals. Mosaic analyses can be used to show that the defect in pigment production is not resident in the eye itself. For example, gynandromorphic flies that have a male and therefore *v* mutant head and eye with a female v^+ body have normal eye pigmentation. Recovery of mosaics containing varying amounts of mutant and normal tissue have demonstrated that it is in fact the larval fat body that carries out the synthesis of the missing pigment component step. The product of this step is then apparently transported to the developing eye where it is utilized in eye pigment formation. In this manner it is indeed possible to distinguish between the direct and relational pleiotropic effects of any single genetic lesion.

By coupling these observations with a determination of the earliest observable defect in the mutant organism, it is sometimes possible to deduce probable causative relationships and begin to understand the nature of the genetic defect. However, if the observed lethal phase occurs very early in development before discrete structures or organs are formed, it becomes difficult to determine a precise focus of action of the gene in question. Early lethality can result from two different causes: The embryo may lack some indispensable biochemical function (e.g., an element of the protein synthetic machinery), or some early but discrete morphogenetic activity is aberrant. How then can we distinguish these two types of defects? One way is to determine the time of action of the gene. An indispensable function necessary to all cells should cause lethality at all stages and in all tissue types. A specific morphogenetic gene should be more discrete in its time as well as its place of action.

Ideally, what one would like to do is bypass the early lethal phase, and subsequent to that point, reinstate the genetic defect. This can be accomplished in two ways. The first is by the technique of somatic recombination, but with the added variable of inducing the mutant clones at various times in ontogeny. If the gene under analysis does indeed code for some indispensable metabolic cellular function, mutant clones should fail to survive or produce a normal phenotype, irrespective of the time or place of their production. If, however, the gene has a discrete time of activity only clones produced after this time will survive or be normal morphologically. Analogously, mutant genes that function in a specific tissue or organ will not produce surviving or wild-type clones in these structures. Both discrete time and specific site of action of a single gene are also possible. A mutation in this type of gene would result in the inability of clones to survive or produce normal structures in a specific tissue up to the time of gene action. Subsequent to this time in development, however, the cells should be unaffected.

Finally, although limited in applicability, the most informative method is to recover conditional mutations, specifically, temperature-sensitive lesions. This type of mutation allows one to control the time of onset of the mutant effect to any point in development simply by altering the temperature regimen. (This class of mutation is, of course, not of much use in homeothermic animals.) As an example, let us consider the case of a heat-sensitive lethal in *Drosophila* that is inviable at 29°C, the nonpermissive temperature. If raised at this temperature, mutant individuals succumb in the pupal stage, whereas at 20°C development is normal. Cultures of this mutant are grown first at the high temperature and then shifted down to the low temperature, and the fate of the "shifted" animals is observed. Likewise, reciprocal shift-up experiments are performed. The earliest shift-down experiment at which the lethal syndrome is observed denotes the beginning of what is called the temperature-sensitive period (TSP). In the shift-up one determines the latest time at which the mutant phenotype is no longer expressed. This indicates the end of this sensitive interval. If a discrete TSP is found, this can be confirmed by applying pulses at restrictive temperatures to cultures of mutant individuals. Further, it is possible to do pulse shifts for only portions of the TSP to determine if the mutant phenotype can be partially ameliorated or if certain pleiotropic effects can be obviated. The relationship of the TSP and the actual time of lethality can also be informative, especially if the TSP precedes the lethal phase by a significant time interval. By relating this type of result to early lethals and indispensable functions, it can be seen that if a gene and its product are necessary at all times, a continuous rather than discrete TSP will be realized. If, however, the early function is unique to the early embryo, then once the early period is past, shifts to the nonpermissive temperature should have no detrimental effect. A vital step in the irreversible process of development has been completed: "The Moving Finger writes; and, having writ, moves on...."

Individually, the techniques outlined above can be used to answer specific questions about the nature of developmentally important genetic lesions. Moreover, when several of them are used in consort, truly significant information can be obtained as to how the normal genes are involved in the developmental process. The utility of this approach can be seen in examples, such as the results obtained by D. T. Suzuki and his collaborators in their analysis of two temperature-sensitive lethals in *Drosophila melanogaster*.

The first of these was originally recovered as a simple recessive sex-linked lethal that was also temperature-sensitive. If grown at 29°C, death occurred in the pupal state, whereas if grown at 22°C the flies were normal. Shift studies demonstrated that the TSP immediately preceded the lethal period. During these shift studies, it was noted that some of the surviving adults exhibited a mutant eye color. Subsequent genetic tests show that the TS lethal was actually an allele of a previously described locus called *raspberry* (*ras*), by virtue of its mutant eye color. This particular mutation is known to have pleiotropic effects on pigmentation in that not only is the eye color changed but the pigmentation of the adult testes and larval Malpighian tubules is also altered. Further temperature shift studies on the TS *ras* mutation allowed Grigliatti and Suzuki to determine that the temperature-sensitive period for pigmentation of the Malpighian tubules occurred during the early larval stages, whereas the TSP for both the eye and testis pigmentation took place in the late pupal stage more than four days later in the life cycle. Therefore, the action of this particular gene is necessary at two distinct times in development. However, the question still remains: Is this a relational pleiotropic effect or direct pleiotropy? This question was answered by determining the autonomy of the pigmentation defect. If the lack of pigmentation was caused by the inability of a single tissue to manufacture a common pigment molecule which was then transported to the testes, Malpighian tubules, and eyes, the defects would be relational. However, it was possible to show by using the ring-X method to produce gynandromorphs that the eye and Malpighian defects are specific and autonomous to those two tissues; that is, the tissue must carry the mutant allele in order to express the mutant phenotype. Therefore, it would appear that the same gene, *ras*, is necessary to produce pigment in three different tissues at two discrete periods in the ontogeny of the fly. This result is in contrast to that seen in the relational pleiotropy of the sickle-cell-anemia syndrome and demonstrates that both direct and relational pleiotropy exist.

This point is even more dramatically demonstrated by a second mutation analyzed by Suzuki and his collaborators. This mutation is called *shibire* (meaning paralyzed in Japanese); it was originally isolated as a sex-linked, temperature-sensitive paralytic lesion. Adult male and female flies carrying this mutation have normal mobility at 22°C. If these adults are shifted to 29°C, they immediately fall to the bottom of the culture vessel totally paralyzed. If the flies are subsequently returned to 22°C they begin to move in a few minutes and soon appear normal in all respects. The physiological basis of this adult defect was determined by Ikeda and his co-workers who, by implanting microelectrodes in the flight muscles of normal and *shibire* flies, measured the junctional and action potentials evoked by stimulation of the motor nerve that innervates this muscle. On warming to 29°C, both the junctional and action potentials were lost in *shibire* fibers but not in wild-type flies. However, if the muscle was stimulated directly rather than through the nerve it could be made to contract in *shibire* flies, even at 29°C. Moreover, it was possible to demonstrate transmission of an impulse along a *shibire* mutant nerve fiber at 29°C. Therefore, it would appear that the paralytic defect resides in the neuromuscular junction, which cannot transmit a stimulus at 29°C.

However, this is not the only defect observed in *shibire* mutants. In order to determine if there was any paralysis of larvae and possible developmental defects, Poodry, Hall, and Suzuki performed temperature shifts on developing embryos and larvae. They found that a shift up to 29°C during any stage of development resulted in paralysis and death; therefore, the *shibire* normal function was apparently indispensable to the fly. By performing pulse experiments, it was shown that a shift of > 18 hours to 29°C was sufficient to kill developing larvae and embryos. Heat pulses for shorter periods (two, four, and six hours) revealed unexpected developmental defects. Six-hour heat pulses revealed six critical periods in development when the *shibire*$^+$ gene or its product is needed, or death occurs. One period of high sensitivity occurs at gastrulation, when a two-hour heat pulse is sufficient to kill the animal. This lethality does not occur as in the 18-hour heat pulses from irreversible paralysis, but must result from some other defect.

The heat pulse experiments also revealed temperature-sensitive periods for several visible defects in the adult fly (Fig. 7–5). The most striking of these was the production of a vertical "scar" on the eye facets caused by a disruption of these structures (Fig. 7–6). The scar is produced by a heat pulse of three to six hours administered during a period from about 48 hours before the pupal stage until just after the pupa is formed (Fig. 7–5). By a judicious spacing of these pulses, it is possible to create a fly with a double scar on the eye and demonstrate that the position of the scar moves across the eye in a posterior-to-anterior direction as the temperature-sensitive period proceeds. Interestingly, in this period direction of movement and position in the eye corresponds to a wave of cell divisions that moves across the developing eye in a manner analogous to the scar.

The formation of bristles and hairs on the thorax and head of the fly is also affected by heat pulses, but the sensitive period for these structures is slightly later than that for the eye (Fig. 7–5). Early heat pulses produce duplication of bristles, whereas later pulses result in the deletion of the same bristle types. As with the eye scar, these effects show a posterior-to-anterior movement with progressively later pulses.

The ubiquity of defects produced in *shibire* flies demonstrates that the function of this gene is far less specific than a lesion in the neuromuscular junction. This is perhaps best shown by the fact that gastrulation, an event that takes place long before nerves or muscles are evident or functioning, is extremely sensitive to temperature in *shibire* embryos.

The tentative conclusion from these studies is that the *shibire*$^+$ gene product is a membrane component that is necessary for some types of cell-to-cell interaction or communication. This membrane component would be necessary for different processes that all relate to its primary function in cellular communication, but each different temperature-sensitive period with its corresponding defect would reflect the manner in which a specific function is important at that point in development. Here again we have a case of direct pleiotropy and the demonstration that a single gene can affect several seemingly disparate developmental events. The other point to be made from these results is the usefulness of temperature-sensitive alleles. A nonconditional *shibire* mutation would have produced simply a dead embryo with very little cytodif-

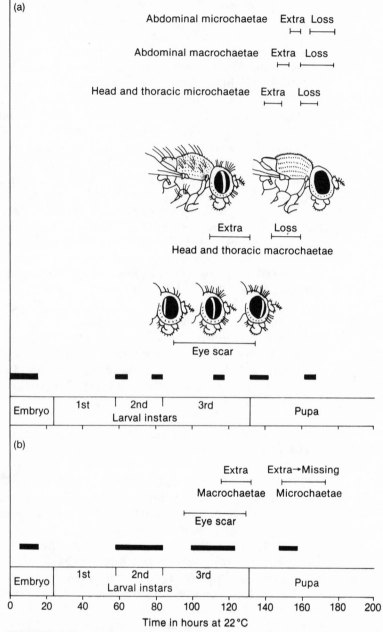

FIGURE 7–5. *Temperature-sensitive periods for the various effects caused by the* shibire[ts] *(a) and* Notch[ts] *(b) mutations. The black bars below the line bars indicate the times at which mutant animals are particularly sensitive to pulses to the nonpermissive tempera-ture. Pulses longer than those necessary to produce the phenotypic anomalies listed result in lethality.* [Part a redrawn from C. A. Poodry, L. Hall, and D. T. Suzuki, Developmental properties of *shibire[ts]*: A pleiotropic mutation affecting larval and adult locomotion and development, *Dev. Biol.* **32**:373–386, 1973. Part b adapted from D. L. Shellenbarger and J. D. Mohler, Temperature-sensitive periods and autonomy of pleiotropic effects of *l(1)N[ts1]*, a conditional *Notch* lethal in *Drosophila, Dev. Biol.* **62**:432–446, 1978.]

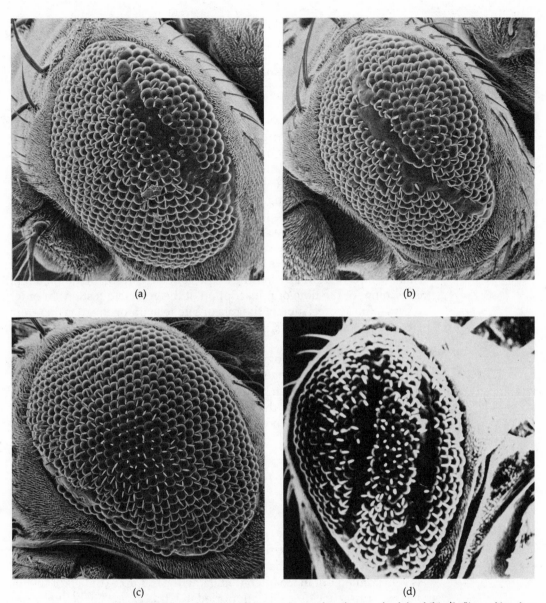

(a)   (b)

(c)   (d)

FIGURE 7–6. *Scanning electron micrographs of eyes of adult* shibire[ts] *flies subjected to temperature pulses during the period that results in eye scars. (a) A pulse early in the period—scar near the posterior margin of the eye. (b) Pulse at the midpoint of the period—scar at the center of the eye (c) Late pulse—scar near the anterior edge of the eye. (d) Double pulse resulting in a double scar.* [Parts a–c from C. A. Poodry, L. Hall, and D. T. Suzuki, Developmental properties of *shibire[ts]*: A pleiotropic mutation affecting larval and adult locomotion and development, *Dev. Biol.* **32**:373–386, 1973. Photographs courtesy of Dr. C. Poodry. Part d taken from D. T. Suzuki, Behavior in *Drosophila melanogaster:* A geneticist's view. *Can. J. Genet. Cytol.* **16**:713–735, 1974. Photograph courtesy of Dr. D. T. Suzuki.]

211

ferentiation, and therefore, no clue as to the nature of the gene product or its ubiquity.

There are, however, certain limitations that must be kept in mind when interpreting the results of an analysis of a temperature-sensitive mutation. The TSP is generally thought to reveal the time at which the gene product is utilized. That is, in the case of an enzyme, the period in which its metabolic function is needed. This, however, is not necessarily always the case. There are also instances in which temperature sensitivity results from an abnormal synthesis of the protein during translation. The protein thus formed is inactive, even at the permissive temperature. Additionally, the protein, once synthesized under permissive conditions, is no longer sensitive to an elevation in temperature. Therefore, the TSP for this lesion reflects not the time of gene action, but rather the time of synthesis of gene product. The relationship of the time of gene transcription, the time of gene product utilization, and the TSP of a temperature-sensitive lesion can only be determined by a more intimate knowledge of the nature of the gene product or an independent assessment of the time of transcription of the gene in question. In order to gain a full appreciation of the role of a gene in ontogeny, it is therefore necessary to adopt several experimental strategies taken from a variety of fields.

Utilizing the methodology of developmental genetics, it is possible to demonstrate the wide range of observable genetic defects with respect to their temporal distribution in ontogeny. In the following discussion, we will produce a picture of the repertoire of genes that is available to the evolutionary process in order for it to effect morphological changes.

Maternal Effect Mutations

In organisms as diverse as sea urchins and frogs, the early cleavage events, and indeed, the majority, if not all, of development prior to gastrulation do not rely on the zygotic genome. The information necessary for performing these initial and crucial steps of ontogeny is specified instead by the maternal genome during production of the egg. As was pointed out in Chapter 4, utilizing the example of shell coiling in *Limnaea*, this conclusion is supported by the existence of what are referred to as maternal effect genes in a wide variety of organisms. Maternal effect genes, when mutated, have a unique pattern of inheritance. If one were to cross two individuals heterozygous for a recessive trait, it would be expected that one-quarter of the progeny would express that trait. However, in the case of *maternal (mat)* mutations the *mat/mat* individuals develop normally. Moreover, males of this genotype are fertile and produce normal offspring when mated to normal females. The homozygous females, on the other hand, produce abnormal offspring. This production of defective progeny results from the fact that these females produce abnormal oocytes that do not complete normal development (Fig. 7–7). The *mat/mat* female survives because she was produced by a heterozygous (*mat/+*) mother that is capable of producing normal eggs. What one would like to conclude

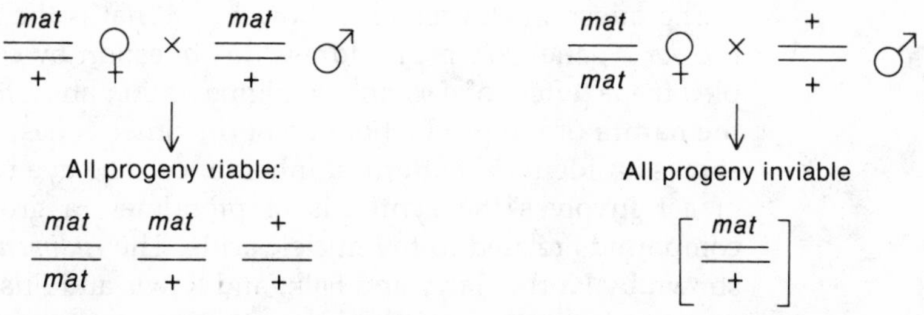

$$\frac{mat}{+} \, ♀ \, × \, \frac{mat}{+} \, ♂ \qquad\qquad \frac{mat}{mat} \, ♀ \, × \, \frac{+}{+} \, ♂$$

All progeny viable: All progeny inviable

$$\frac{mat}{mat} \quad \frac{mat}{+} \quad \frac{+}{+} \qquad\qquad\qquad \left[\frac{mat}{+} \right]$$

FIGURE 7–7. Pattern of inheritance of a maternal effect lethal mutation. The homozygous mat/mat condition is not itself lethal as long as the female parent is heterozygous (mat/ +). However, if a homozygous female is mated all of her progeny die, irrespective of their genotype.

about this type of mutation is that these genes are producing necessary "morphogens" that are placed in the developing oocyte as "instruction" for early development. However, it is also possible that the egg is unable to develop simply because it suffers from some general metabolic defect. A case in point is presented by a group of five different maternal effect lesions on the X chromosome of *Drosophila melanogaster: cinnamon (cin), deep orange (dor), almondex (amx), fused (fu),* and *rudimentary (r)*. These all produce, in addition to their maternal effects, visible adult morphological aberrations, from which their colorful names are derived. Hemizygous males carrying any one of these mutations are viable and fertile, as are heterozygous females. By crossing mutant males to heterozygous females, homozygous females can be obtained, which, when mated to mutant males, are completely sterile. For example, *dor/dor* females produce eggs that arrest development at gastrulation. The remaining four mutations also result in embryonic lethality, the lethal phase varying somewhat from that of *dor*. All five of these mutations have an added perturbation in their pattern of inheritance. If homozygous mutant females are mated to normal males, some progeny are produced. These consist entirely of heterozygous females, that is, eggs fertilized by X-bearing sperm. No males survive. Apparently, the presence of the wild-type allele of the deficient gene can ameliorate the egg's defect, even if it is supplied by the sperm. This of course implies that at least a portion of the zygotic genome is active during gastrulation.

It has been shown by Gehring that injection of wild-type egg cytoplasm into *dor* embryos will rescue the deficient animals. In a further analysis Kuroda developed an *in vitro* test for the *dor*[+] substance. Using cultures of embryonic cells isolated from deficient embryos, he noted that the mutant cells demonstrated abnormalities, such as failure of muscle cells to fuse, in contrast to cells from normal embryos. If the cultured mutant cells were exposed to extracts of normal eggs, the defects were prevented. Moreover, by using cytoplasm from rescued embryos of various stages, Kuroda was able to show that the *dor*[+] substance was produced at and after gastrulation, but not before, by *dor*[−] embryos rescued by *dor*[+] sperm. With this assay, he was also able to show that the rescuing substance is heat labile, and therefore, may be a protein.

The question still remains, however: What is the normal function of the *dor*$^+$ gene? We may address this question by considering first the pleiotropic effect of *dor*, an eye pigmentation abnormality, and second, the nature of the product of one of the other genes, *rudimentary*, which shows an identical pattern of inheritance. The eye pigmentation defect of *dor* involves the synthesis of pteridines, a group of heterocyclic compounds related to the nucleic acids. The *rudimentary* gene has been shown by Norby, Jarry and Falk, and Rawls and Fristrom to encode the first three enzymes in the biosynthetic pathway for pyrimidines. An involvement of *dor* in nucleic acid metabolism is also indicated by the fact that the sterility of *dor/dor* females can be partially alleviated by supplementing the growth media of these flies with extra purines. Therefore, it would appear that the *dor* mutation, and possibly other similar mutations, are producing their defects by causing a deficiency in nucleic acid synthesis or degradation. Such mutations illustrate the problem of distinguishing between nonspecific genetic defects and those specifically affecting mechanisms unique to morphogenesis.

A more likely candidate for a morphogen deficiency is the *oocyte-deficient* (*o*) mutation in the axolotl discussed in Chapter 4. By transplant-ing ovaries from *o*-mutant-bearing individuals to normal hosts, Humphrey demonstrated the defect to be autonomous to the mutant ovary. This result is not expected if the defect is caused by the deficiency of some low-molecular-weight diffusible metabolite. However, despite the extensive work of Briggs and his co-workers, which resulted in the identification of a proteinaceous rescuing substance in normal embryos and oocytes, the precise nature of the defect is not known. One intriguing observation, however, is that the rescuing protein is localized to the germinal vesicle (the oocyte nucleus), implying that the protein in some way interacts with the genetic material. In light of this possibility, it is of interest to consider a newly discovered mutation in *Drosophila*. Mortin and Lefevre have described a sex-linked (X chromosome) dominant mutation that they have called *Ultrabithoraxlike* (*Ubl*). In females heterozygous for the mutation (*Ubl/* +), the halteres or balancer organs common to all dipteran insects are enlarged relative to normal specimens. These halteres are found on the third segment of the thorax and are, in evolutionary terms, vestigial wings that act as stabilizing organs during flight. The *Ubl* mutation mimics another dominant mutation on the third chromosome, called *Ultrabithorax* (*Ubx*), which is homoeotic in nature. A more complete description and discussion of this and other homoeotic loci follows in Chapter 8.

The enlargement of the haltere is indicative of a transformation of that organ toward a fully differentiated wing. This becomes clear in the phenotype of female flies heterozygous for both *Ubl* on the X chromo-some (*Ubl/* +), and *Ubx* on the third chromosome (*Ubx/* +). These flies possess a haltere that has the marginal hairs and bristles characteristic of a wing as well as wing veins, two structures not normally found on the small bulbous haltere. A further discovery made by Mortin makes *Ubl* germane to the discussion at hand; that is, it shows a maternal effect (*Ubx* does not). *Ubl* is lethal in homozygous females and hemizygous males; however, it is possible, by supplying the fly with a duplication containing the normal allele of *Ubl*, to obtain adult female flies of the genotype *Ubl/Ubl/+*. If these females are mated to normal males, they are fertile and produce heterozygous *Ubl/+* offspring. If, however, they are mated to *Ubl/Y/+* males (viable because of the same duplication), they are virtually sterile and unable to produce even *Ubl/+* progeny. What makes this result intriguing and different from the similar rescuable sterility demonstrated by *dor* females is the nature of the gene product of the *Ubl* locus. Greenleaf and co-workers have recovered an additional mutation at this locus that confers resistance to α-amanitin, which is known specifically to inhibit RNA polymerase II from eucaryotic organisms. This polymerase is the enzyme that transcribes those genes that code for RNA (mRNA), that is, structural genes. Moreover, like the *Ubl* lesion at the locus, the α-amanitin-resistant allele interacts with *Ubx* to produce a winglike haltere.

As is examined in Chapter 8, Garcia-Bellido and his collaborators have shown that activity of the *Ubx* locus is necessary as early as the blastoderm stage, where it is involved in the initial determinative events controlling segmentation. This fact, coupled with the demonstration of the maternal effect of *Ubl*, indicates the possibility that a maternally supplied polymerase is necessary to these initial determinative events. It remains to be shown whether the interaction of the altered polymerase with the homoeotic genes is specific to these loci, or if the seeming specificity of the polymerase defect is more apparent than real. Alterations in this subunit of polymerase II may affect binding of the holoenzyme to the promoters of many or all genes; thus, the interaction may be revealing differential promoter strength, with *Ubx* having a weak promoter extremely sensitive to enzyme alterations. Nonetheless, what has been shown is that the function of a gene, *Ubx*, whose correct function in early embryogenesis is crucial to normal morphogenesis, may be dependent in large part on a maternally supplied enzyme important to gene transcription.

Although a direct and unique role of the previously described genes in morphogenesis remains problematical, Rice and Garen have isolated, on the third chromosome of *D. melanogaster*, a group of maternal effect mutations with a seemingly specific morphological involvement. Three of these produce very characteristic and specific defects at the blastoderm stage in embryos derived from homozygous mothers. The first, *mat(3)1*, does not form a normal cellular blastoderm after the initial syncytial cleavage divisions. However, the pole cells do form at the posterior end of the embryo. In the normal embryo, this group of cells is destined to form the gametes, and thus forms the germ line. Therefore, it would appear that germ-line cells and few somatic cells develop in this mutant. Interestingly, despite the lack of a normal blastoderm, the pole cells attempt to invaginate as they would normally do during gastrulation, indicating that at least some of the early cell movements of the embryo do not require cell-cell interactions or adhesions.

A second mutation, *mat(3)6*, also forms only a partial blastoderm. The syncytial nuclei migrate into the cortical cytoplasm as in normal embryos; however, cellularization takes place only in the anterior and posterior ends of the embryo. The distributions of cells in these two

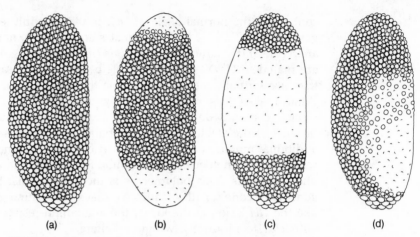

(a)   (b)   (c)   (d)

FIGURE 7–8. The pattern of cellularization in the blastoderm of progeny of three maternal effect mutation-bearing females of Drosophila. *The anterior end of the embryo is at the top, dorsal to the right. The pole cells at the posterior end of the embryo are larger than the somatic cells.* (a) *Normal,* (b) *grandchildless,* (c) *mat(3)6,* (d) *mat(3)1.* [Adapted and redrawn from T. B. Rice and A. Garen, Localized defects of blastoderm formation in maternal effect mutants of Drosophila, *Dev. Biol.* **43:**277–286, 1975; and A. P. Mahowald, J. H. Caulton, and W. J. Gehring, Ultrastructural studies of oocytes and embryos derived from female flies carrying the *grandchildless* mutation in Drosophila subobscura, *Dev. Biol.* **69:**118–132, 1979.]

mutants at the blastoderm stage is shown in Figure 7–8. Again, in *mat(3)6*, like *mat(3)1*, the pole cells attempt to invaginate and the embryo goes through an abortive attempt at gastrulation. In further work on these embryos, Rice and Garen assayed the potentialities of those cells formed by the *mat(3)6* embryos. Previously, Gehring had shown that if a blastoderm-stage embryo is cut in half and these half embryos are injected into the abdomens of adult female flies, cell cultures can be established. These cell cultures can then be assayed for their ability to produce adult structures by a subsequent transplantation from the female's abdomen into a metamorphosing larva. As the larva changes into an adult, so too will any injected cells capable of forming adult or imaginal tissues. By this technique, Chan and Gehring were able to demonstrate that, at the blastoderm stage, the anterior end was already committed to produce only anterior adult structures, and the posterior end, only posterior adult structures. Rice and Garen applied this technique to their mutant embryos and found that only the most anterior head and posterior abdominal structures were formed. They found no thoracic elements. It would appear, therefore, that only very specific portions of the cellular blastoderm are formed in these mutants and that maternal gene products are involved in the cellularization of certain portions of the early embryo.

A third maternal mutation of this genre was studied by Mahowald and his collaborators in another species, *D. subobscura*. This mutation, *grandchildless (gs)*, is inherited as a maternal effect lesion but with a slight difference from those of Rice and Garen. Homozygous recessive *gs/gs* females are fertile and produce viable progeny. However, their male and female progeny are sterile. Thus, the original *gs* female can have no grandchildren. The basis for this phenomenon is that the embryos produced by *gs/gs* mothers do not form pole cells and thereby result in agametic adults. The precise reason for this failure is that the nuclei that normally migrate to the most posterior tip of the cleavage-stage embryo do not do so at the normal time. It has been shown by Illmensee and Mahowald that this region of the embryo contains the "determinants" instrumental in specifying the fate of these pole cells destined to become the germ line. The *gs* mutation apparently prevents this nuclear-cytoplasmic interaction from taking place. Interestingly, at the time of blastoderm formation in *gs* embryos, not only is there a failure of cellularization at the posterior tip of the embryo but at the anterior end as well. This is shown in Figure 7–8, where a comparison of the three blastoderm defect mutations can be made. An intriguing fact is that the spatial distribution of cellularization in *gs* is reciprocal to that in *mat(3)6*.

The final, and perhaps the most striking, of the maternal effect lesions is the *bicaudal* (*bic*) mutation of *Drosophila melanogaster*. This lesion, an autosomal recessive, was first found by Bull and has been more recently analyzed by Nüsslein-Volhard. Homozygous *bic/bic* females produce embryos which, at gastrulation, are seen to have two posterior ends and no anterior. Amazingly, whereas progeny of bicaudal animals fail to hatch from the egg case, these two-tailed monsters complete embryonic development and reach the larval stage before dying. As the photograph in Figure 7–9 shows, all of the normal larval cuticular structures for the

FIGURE 7–9. *Dark-field photomicrographs of a lateral view of the cuticle of larvae produced by normal (top) and homozygous bicaudal mutant females (bottom). Anterior is to the left and dorsal to the top of each animal. The mutant larva has only heavy abdominal denticles on its ventral surface. The pattern of these denticles is one of mirror symmetry around the anterior-posterior midline (mp = mouth parts; ps = posterior spiracle).* [Taken from C. Nüsslein-Volhard, Genetic analysis of pattern-formation in the embryo of *Drosophila melanogaster*, *Roux Arch.* **183**:249–268, 1977. Photographs courtesy of Dr. C. Nüsslein-Volhard.]

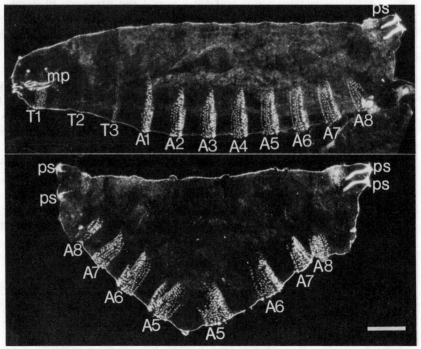

posterior are formed at both ends of the embryo. This phenotype is strikingly similar to the one produced experimentally by Kalthoff in the lower dipteran *Smittia,* which was discussed in Chapter 4. It is therefore reasonable to conclude that *bic* may be the locus responsible for the synthesis of a product whose localized activity is similar to that of the morphogen that was destroyed experimentally by Kalthoff.

The preceding examples demonstrate the involvement of the maternal genome and its products in many of the early events of development. The existence of genes like *dor* and *r* show the influence of maternally supplied enzymes to the metabolism of the early embryo, whereas the defects observed in the *mat* and *bic* studies indicate that positional and organizational cues are also supplied during oogenesis. Finally, the effects of the *Ubl* mutation in *Drosophila* and the *o* mutation in the axolotl reveal the role of maternal information necessary for the proper activation of the zygotic genome.

Mutations Affecting Organogenesis

The early events of development have been shown to be highly dependent on maternally supplied information. However, at about the time of gastrulation, the embryo's own genetic information becomes important to further development and the organism gains control of its own destiny. Developmental events subsequent to the formation of blastoderm require the synthesis of RNA and its translation into protein. We can also demonstrate the necessity of embryonic genetic information by virtue of the large number of mutations that affect events after gastrulation and thus reveal the presence of genes that control these events. These mutations show no pattern of maternal inheritance.

A case in point is the *Notch* locus of *Drosophila melanogaster. Notch* (*N*) is a sex-linked dominant mutation that is also recessive lethal. Homozygous females (*N/N*) and hemizygous males (*N/Y*) die in the embryo at about six hours after fertilization. This time corresponds to a point just subsequent to gastrulation about one-quarter of the way through embryogenesis. Histological and morphological observations by Poulson on these lethal embryos showed that the ventral and lateral ectoderm, which normally gives rise to both epidermis and neural cells, produces exclusively neuroblastlike cells and no epidermis. Therefore, it would appear that the *Notch* locus is necessary for the differentiation of neural versus epidermal tissue from embryonic ectoderm. This, however, turns out to be a somewhat simplistic interpretation. Shellenbarger and co-workers isolated and characterized a temperature-sensitive allele of the *Notch* locus. Flies carrying this lesion grown at 22°C are normal, whereas at 29°C the embryonic lethality described previously is expressed. In shift experiments similar to those on *shibire* presented earlier, it was found that this early embryonic period was not the only developmental stage sensitive to a deficiency of the *Notch* locus and its product. The results of these shift studies are presented in Figure 7–5.

Pulse shifts revealed three additional periods in which normal activity of the *Notch* locus was vital to the mutant organisms. Shifts to the nonpermissive temperature during the second or third larval instars or the pupal stage produced lethality. Moreover, shorter pulses during specific periods of the third larval instar and pupal periods produced the same eye scarring and bristle defects seen in the *shibire* mutants (Fig. 7–5). Therefore, like *shibire*, *Notch* is much more pervasive in its effects than might have been expected from observations on its primary phenotype. All of the structures seen to be affected are ectodermal, and in further experiments utilizing gynandromorphs Shellenbarger was able to show that the observed morphological defects were autonomous to this cell type. Therefore it would seem that again, like *shibire*, *Notch* makes a product that is common to, and necessary for, the functioning of ectodermal cells at several discrete periods throughout development. What should also be noted is that two distinct genes, $Notch^+$ and $shibire^+$, are necessary for the completion of the same set of ontogenic events, and that a deficiency in either gene results in a strikingly similar set of defects.

There exists in the mouse (*Mus musculis*) a complex gene, the *T* locus, that exhibits many analogies to the *Notch* system just described. The first allele of this locus was discovered as an autosomal dominant referred to as *Brachyury* (*T*). In the heterozygous condition, $T/+$, mice have short tails. The homozygotes, T/T, are lethal and die as embryos in the uterus of their mother. Shortly after the discovery of this dominant mutation, it was found that crosses of heterozygotes ($T/+$) to wild caught mice often produced progeny with no tails at all. These tailless mice were shown to be the result of recessive alleles of the *T* locus that were quite common in natural populations of mice. The genotype of these tailless mice was therefore T/t. Crosses of these heterozygous T/t mice produced true-breeding tailless mice. This was subsequently shown to be caused by a "balanced lethal system." That is, not only were the T/T offspring expected from the cross lethal, but the t/t were as well. Therefore, only the heterozygous T/t survived to produce the next generation. This intriguing situation has been exploited both genetically and developmentally by L. C. Dunn and his students, D. Bennett and S. Gluechsohn-Waelsch, in a series of elegant studies.

In genetic studies of the newly recovered balanced lethal strains, it was found that crosses between tailless mice whose recessive *t* allele came from different populations often produced normal progeny. Specifically, the cross $T/t^a \times T/t^b$ produced progeny that were tailless and normal in a 2:1 ratio. The normal-tailed progeny could be shown to be t^a/t^b in genotype. Therefore, not only was this genotype nonlethal, but it was normal morphologically. A further attribute of this locus was discovered in the balanced lethal crosses themselves. Normally, T/t^a crossed to a like heterozygote produces only tailless offspring. However, normal-tailed mice are produced at a low frequency (1/500–1/1000). These normal mice are nearly always produced as rare genetic recombinants in the chromosomal region at or adjacent to the *T* locus in chromosome 17. These recombinant offspring can be shown to be tailed by virtue of the fact that concomitant with the recombination event the original t^a allele has changed to a new t^x allele that complements t^a in the same manner that certain wild-derived t alleles complement each other. The difference, of course, is that t^x is in this case directly related by descent to t^a. This transition of one t allele to other complementing types has been shown to occur with most of the isolated recessives. Some derived alleles, for example, t^x, can in turn, by the same mechanism, produce another complementing allele, for example, t^y. This transition of one recessive t to another forms a graded series, resulting finally in the production of what are called t *viable* or t^v alleles. These are all nonlethal and only express a tailless phenotype in T/t^v heterozygotes; t^v/t^v individuals have normal tails. A further class of recessive t mutations is that of the semilethals, which range in viability from 2 to 51% of normal. Like the completely lethal alleles, they transform to t^v types. Complementation crosses ($T/t^a \times T/t^b$) of all of the wild- and genetically derived recessive alleles have revealed that the 111 extant mutations fall into eight separate groups, all of which fail to complement T. Each group has a different number of members, from a low of one for the t^{w73} group to a high of 66 for the t^v alleles. Of the five dominant T mutations one was induced by x-rays, whereas all of the other mutations known at the locus are apparently spontaneous in origin.

These are a few of the genetic attributes of this complex series of genetic lesions, all of which are related by either complementation and genetic position or by derivation one from the other. But what are the developmental attributes of this complex locus?

Each one of the eight complementation groups produces a different pattern of defects from early to late in development of the embryo. The morphologies of these lesions are summarized in Figures 7–10 and 7–11. The earliest observable defect is produced in homozygous t^{12} embryos. Fertilization and the cleavage divisions of the zygote result in the formation of a ball of cells called a morula. The first evidence of differentiation of cell types in the mouse is the transition of this morula into a second stage, the blastocyst, which consists of an external trophectoderm and an inner cell mass. Homozygous t^{12} embryos do not perform this step and the nondifferentiated "morulae" die before implantation in the uterine wall, an event which in normal embryos takes place at about four days after fertilization. Moreover, t^{12} cells appear to be autonomous in their lethality. Chimeras, which contain both t^{12} embryonic cells and normal embryo cells, still express lethality, and never develop beyond the point when t^{12} embryos normally expire. Therefore t^{12} apparently affects a locus necessary for the first differentiative step in the mouse embryo, the formation of trophectoderm, which will eventually form the chorion and other portions of the zygotically derived extraembryonic elements common to placental mammals.

The next earliest acting allele is t^{w73}. Homozygotes of this type produce a blastocyst. However, the trophectoderm of these mutant embryos does not form the necessary associations with the uterine wall, and the improperly implanted embryos die shortly thereafter.

After successful implantation of embryos in the uterine wall, the inner

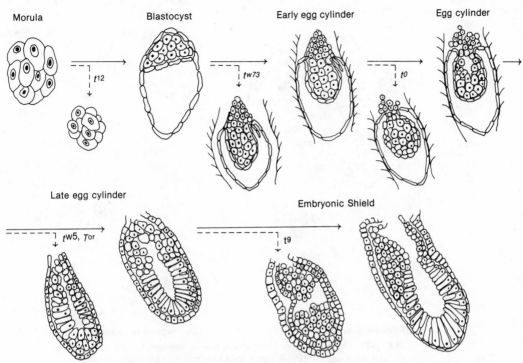

FIGURE 7–10. *Diagrammatic representation of the events and stages in the early embryo of the mouse. The solid arrows show the flow of normal embryogenesis. The dashed arrows show the point of departure from normal development seen in mutant individuals carrying the various* t-*alleles indicated in the figure. For a more complete description of the events and the mutant effects refer to the text.* [Adapted from D. Bennett, The t-locus of the mouse, *Cell* **6**:441–454, 1975. Copyright by the Massachusetts Institute of Technology.]

cell mass begins to grow and further differentiation takes place. One of these differentiative events is the formation of extraembryonic ectoderm and the embryonic ectoderm. The former eventually forms the placenta and portions of the extraembryonic membranes. The latter will form the embryo proper. The mutation t^0 fails to form extraembryonic ectoderm and dies in the early egg cylinder stage.

In normal embryos, further growth of the inner cell mass proceeds and produces an elongated ectodermal mass covered by endoderm, which are together called the egg cylinder. The extraembryonic cells also continue to proliferate and differentiate. Homozygous t^{w5} embryos produce an egg cylinderlike stage, and then the embryonic ectoderm cells become pycnotic and die. The extraembryonic cells, however, do not seem to be affected and the dead embryo persists in the midst of apparently normal extraembryonic development for several days until the extraembryonic cells die as well.

At this point in the development of the mouse, 6.5–7 days postfertilization, differentiation of the embryo proper begins. This is evinced by the development of the primitive streak on the egg cylinder and the formation of a mesodermal layer of cells between the pre-existing embryonic ectoderm and endoderm. Homozygous t^9 embryos do not

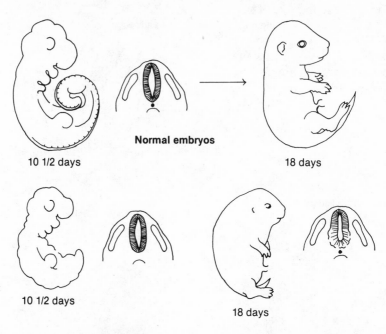

Normal embryos

10 1/2 days   18 days

10 1/2 days

18 days

Homozygous *T* embryo   **Homozygous *t*w1 embryo**

FIGURE 7–11. Drawings of mouse embryos late in development illustrating the affects of two late acting t-alleles. Adjacent to each embryo is a cross-section through the dorsal region showing the nerve cord (cross-hatched tube), the notochord (black dot below the nerve cord), and the somites (ovoid structures lateral to the nerve cord). For a complete description of the mutant phenotypes, refer to the text. [Redrawn from D. Bennett, The t-locus of the mouse, *Cell* **6**:441–454, 1975. Copyright by the Massachusetts Institute of Technology.]

form a normal mesoderm, and therefore, none of the elements derived from this germ layer form. Since these mesodermal cells are derived from the embryonic ectoderm at the primitive streak, it is probable that mutant ectodermal cells are incapable of making this transition. Mutant t^9 embryos transplanted to the peritoneal cavities of normal adults result in malignant tumors consisting entirely of ectodermal tissues.

The primitive streak results in the formation of the three germ layers and the development of the primary axis of the embryo.

Along the majority of the length of this primary axis, the somites are formed, and medial to the paired somites, the chordomesoderm. This chordomesoderm induces the neural ectoderm overlying it to form the neural tube, the structure that will eventually form the spinal chord, and anteriorly, the brain. All of these events take place in t^{w1} embryos. However, once formed, the ventral portion of the neural tube and the brain degenerate. The remaining dorsal cells spatially replace the dead ventral cells, but apparently cannot replace them functionally because these embryos always show a variety of defects and succumb before birth.

The final lethal condition of the T locus is found in individuals homozygous for dominant T. These individuals have a similar lethal phase to t^{w1} individuals, that is, later than the majority of the other recessives. T/T individuals do not extend their primitive streak fully to the posterior end of the embryo. Therefore, none of the structures dependent on the formation of mesoderm in this region ever develop. Most notably, the allantoic placental stalk does not form, meaning that the embryo does not attain its normal placental connections necessary for life. Additionally, no structures posterior to the anterior limb buds of the embryo itself develop. Moreover, despite the fact that the primitive streak is apparently normal anteriorly and somites and notochord do form, the latter structure fails to be maintained and disappears. This results in the anterior end of the animal being severely abnormal.

Despite the seemingly wide range of defects observed in this gallery of monsters produced by mutations in the T locus complex, there is a commonality. As has been pointed out by Bennett, all lethal T alleles result in defects in ectoderm. These defects take the form of either the ability to differentiate, for example, t^{12} and t^o, or to function normally, for example, t^{w73} and t^{w1}. A summary of the described lesions is presented in Figure 7–12. What is apparent from this figure is that the various mutations at the T locus can be envisioned as being involved in a series of binary decisions involving the ectoderm and all of its derivatives. Initially, the morula is made up of undifferentiated cells. The mutation, t^{12}, prevents the first decision, that between trophectoderm and the inner cell mass. Next, t^{w73} prevents the proper functioning of the trophectoderm. The t^o mutation prevents the formation of the extraembryonic ectoderm, analogous perhaps to the effect of t^{12}. In a parallel step, t^{w5} kills the ectoderm of the embryo proper. Finally, t^9, t^{w1}, and T all affect, in different ways, the ability of the embryonic ectoderm either to

FIGURE 7–12. *Hypothetical flow of decisions made during the differentiation of the ectoderm and its derivatives in the mouse. Superimposed on this flow are the various t-allele mutations that block the indicated differentiative events.*

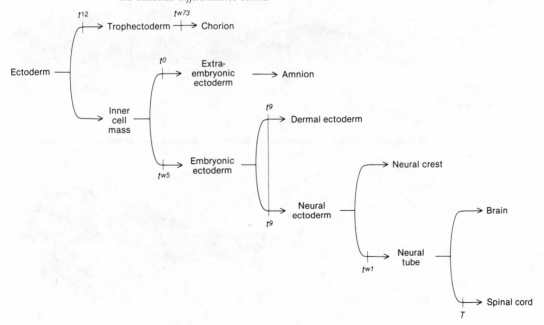

differentiate into mesoderm or the ability of the mesoderm, once formed, to subsequently induce or maintain neural tissue.

The primary defect that is the cause of these far-reaching effects of a single locus is currently a topic of debate and experimentation. However, whatever the proximate cause(s) of the disruptions, it is still self-evident that this locus has a primary involvement in the developmental attributes of one of the three germ layers of the mouse embryo. Because of this fact, like the *Notch* locus of *Drosophila*, it plays an integral role in the development of the organism.

Mutations Affecting Specific Organs

While the *T* locus seems to have quite wide-ranging effects on the entirety of ectodermal development, there are also mutations that are more specific in their defects. An example of such a mutation is *cardiac lethal (c)* in the Mexican axolotl, *Ambystoma mexicanum*. This mutation was originally discovered and analyzed by Humphrey. He found that the gene is inherited as a simple autosomal recessive such that if two heterozygous (*c*/+) individuals are mated, 25% of their offspring die in the early larval stages shortly after hatching. These mutant individuals show normal swimming behavior but are swollen with fluid and have a poorly developed digestive system and gills (Fig. 7–13). The primary

FIGURE 7–13. *Photomicrographs of normal (top) and homozygous* cardiac *mutant (bottom) axolotls* Ambystoma mexicanum. *The bloated appearance of the mutant individual is caused by the accumulation of excess fluid.* [From Kulikowski and Manasek, 1978. Photograph courtesy of F. J. Manasek.]

cause of these defects is the failure of the heart to develop and beat. Thus the animals have no circulation of blood, and respiration probably occurs by diffusion through the skin, thereby allowing mutants to survive for only a limited time period. The autonomy of this heart defect was shown by Humphrey through the production of parabiotic twins of normal and *cardiac* individuals. This process is diagrammed in Figure 7–14. Mutant and normal embryos were selected before heart formation, and a block of tissue was removed from the side of each individual. The two embryos were then conjoined at the wound site and allowed to heal together. When such fused individuals were allowed to complete development, it was found that the normal partner ameliorated the swelling and other defects and allowed survival of the *c/c* individual. However, the heart of the *c/c* partner never developed beyond a simple nonbeating tube, and circulation in the mutant partner was wholly supported by the normal partner.

It has been shown that many vertebrate organs, including the heart, develop as the result of certain inductive interactions during development. Specifically, Jacobson and Duncan have shown that the heart of the salamander is induced from heart-forming mesoderm by the anterior endoderm. The reason that *c/c* individuals fail to form a heart could be caused by a failure of anterior endoderm to induce the process, or from a failure of cardiac mesoderm to respond to the inducer. In order to determine which was the case, Humphrey transplanted normal heart-forming mesoderm into *c/c* hosts and *c/c* mesoderm into normal hosts. He found that the *c/c* mesoderm was capable of forming a beating heart under the influence of the normal anterior endoderm, whereas the mutant hosts did not yield normal heart formation. This result can be

FIGURE 7–14. *The technique of parabiosis. Two animals of different genotype, indicated by presence and absence of stippling, are fused during early embryogenesis and the ability of the normal individual to rescue the mutant member of the pair is ascertained.*

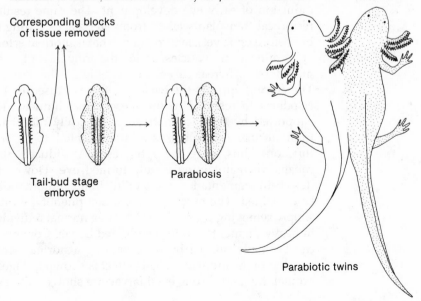

Corresponding blocks of tissue removed

Tail-bud stage embryos

Parabiosis

Parabiotic twins

interpreted as indicating a defect in the inductive potential of the *c/c* anterior endoderm. However, it is also possible that the mutant individuals are actively inhibiting heart formation. This latter possibility was made less tenable by Lemanski and his co-workers. Mutant heart mesoderm, as well as normal, was cultured *in vitro*. Under the conditions used, normal cardiac mesoderm will beat vigorously while the mutant heart mesoderm does not. If heartbeat in the mutant tissue was being suppressed *in situ*, the *in vitro* culture should have removed that influence. Furthermore, when the *c/c* mesoderm was cocultured with normal anterior endoderm, the mutant tissue could be seen to contract, showing that the mutant mesoderm was capable of a normal response if supplied with the proper inductive influence. Thus, the *"cardiac lethal"* appears to be expressed by the failure of anterior endoderm to provide the inductive signal that triggers the differentiation of the heart from cardiac mesoderm.

The results of the studies on another autosomal recessive axolotl mutation, *eyeless* (*e*), stand in contrast to those seen with *cardiac*. As the name implies, homozygous individuals (*e/e*) have no eyes. They also exhibit two other pleiotropic defects: dark pigmentation and sterility. The primary defect occurs quite early in the formation of the eye, which is blocked at or before the formation of the early optic vesicle. Like the heart, the eye is formed as the result of an inductive interaction. The eye is induced from anterior neural ectoderm in the presumptive forebrain (anterior medulary plate) by the chordomesoderm, which comes to lie under it during gastrulation. The induced neural tissue grows laterally as a part of the brain and finally evaginates to form the optic vesicles. Van Deusen investigated the nature of the *e* defect by transplanting prechordal mesoderm from the dorsal lip of the blastopore of mutant and normal embryos. Transplants of *e/e* mesoderm to normal blastulae resulted in the induction of optic vesicles. The reciprocal transplant (normal mesoderm to *e/e* blastulae), however, did not result in the induction of early eye development. The same result was obtained in reciprocal transplants taken from later stages during early gastrulation but before optic vesicle formation. That is, normal ectoderm was capable of forming optic vesicles under the influence of either normal or *e/e* mesoderm, whereas *e/e* ectoderm could not be induced by either type. Therefore it appears the eyeless defect is caused by the inability of the ectoderm to respond to the mesodermal inducer. This conclusion is supported by the fact that morphologically differentiated eyes taken from normal individuals and transplanted to *e/e* hosts survive and function. Thus, once the optic vesicle is induced, the *e/e* genotype is capable of maintaining a fully formed eye. However, the pleiotropic defects in pigmentation and sterility in the mutant individuals remain to be explained. The excess pigmentation phenotype can be mimicked by simply removing the optic vesicles of a normal individual, which results in heavy pigmentation in the blinded larvae. Conversely, pigmentation can be made normal by transplanting a normal eye rudiment into a developing *e/e* individual. This effect is shown in Figure 7–15, in which normal, *e/e*, and *e/e*-one eyed larvae are shown. The extra pigmentation

FIGURE 7–15. *Photomicrographs of an eyeless mutant (left) and a normal individual (right) of the axolotl. These two individuals flank a genotypically eyeless animal onto which a normal eye has been transplanted. This results in a reduction in pigmentation from the eyeless condition (left) to a more normal phenotype (right).* [From Epp, 1978. Photograph courtesy of L. G. Epp.]

is a result of the simple presence or absence of the eye, and therefore results from a relational pleiotropy.

Some of the individuals used in transplantation and eye vesicle operations were grown to sexual maturity. The experimentally blinded individuals could be shown to be fertile, whereas the *e/e* recipients of normal eyes remained sterile. Therefore, the sterility was not caused directly by eyelessness. Moreover, van Deusen was able to show that genotypically *e/e* ovaries transplanted into normal hosts were capable of oogenesis. The sterility was shown by further transplantation experiments to result from a defect in the hypothalamus and the inability of the mutant organ to induce gonadotropins from the anterior pituitary. The hypothalamic primordium resides immediately adjacent to the eye primordium in the anterior neural ectoderm. Thus, the nonfunctioning of this ectodermal organ may be caused by the same inability to respond to induction as the eye, and may be the result of a direct pleiotropic activity of the eyeless gene in both the ectodermal cells that give rise to the eye and the hypothalamus.

In both of the previous cases, the existence of the inducer and its effect on the responding tissue is inferred by the developmental attributes of each system. Unfortunately, little concrete evidence exists as to the nature of the inducer or its mode of action. However, in the case of a mutant gene that affects the development of mammalian secondary sexual characteristics, we do possess more definitive information. This is the *Testicular feminization* locus (*Tfm*). This gene is inherited as a sex-linked character in humans, mice, and rats. Examples also may be present in dogs and cattle. Females heterozygous for the mutant gene, *Tfm/+*, are ostensibly normal, but one-half of their genotypically male progeny (*Tfm/Y*) are phenotypically female, although sterile. In order to understand the mechanism of action of this gene it should be realized that all mammalian embryos begin development indeterminant as to sex. Before development of the gonads, XX and XY embryos have both a Wolffian (male) and Müllerian (female) duct system as well as an undifferentiated urogenital sinus. If the embryo is XX in genotype, the Wolffian ducts degenerate, whereas the Müllerian ducts develop into fallopian tubes, uterus, and vagina, and the urogenital sinus into the female external genitalia. This developmental program is the "ground state;" that is, it will take place even in the absence of ovary, for example, in a castrated male. In XY embryos, on the other hand, the gonadal primordium develops rapidly into testes that begin early on to synthesize and secrete testosterone. This hormone then actively promotes secondary male sexual development, inducing the Wolffian ducts to form vas deferens, seminal vesicles, and ejaculatory ducts, and the urogenital sinus to form the external male genitalia. Moreover, the sertoli cells of the testes secrete an anti–Müllerian factor, which causes the regression of the female Müllerian duct.

Developmental studies on *Tfm* "male" mice by Ohno, Lyon, and co-workers have shown that this mutation affects the ability of all tissues in a male to respond to androgens, and thus, by default, female development ensues. *Tfm/Y* mice can be shown to have reasonable levels of circulating testosterone and do indeed produce the anti–Müllerian hormone from their testes. Therefore, these individuals are not simply mimics of castrated males. Moreover, high levels of injected exogenous testosterone do not ameliorate the *Tfm* defect. Ohno and Lyon were also able to demonstrate that the kidney of normal and castrated normal male mice can be induced to produce high levels of the enzyme alcohol dehydrogenase by administered testosterone, whereas the kidneys of *Tfm* "males" could not. Therefore, the lack of androgen response is not only restricted to sex organs but to nondimorphic organs as well. This fact is further substantiated by the fact that *Tfm* human "males" do not form axillary or pubic hair at puberty, which is a normal response to rising hormone levels at this point in human development.

The basis for this defect has been shown by Ohno as well as Meyer and co-workers to reside in the absence of a specific testosterone receptor protein that is apparently ubiquitous in its distribution in male and female tissues. This receptor is not produced in *Tfm/Y* individuals. It is also the case that *Tfm/+* females only produce the receptor in half of their cells because of the Lyon X-inactivation effect. What is also apparent is that although females produce the androgen receptor, it is not necessary to their normal sexual development. Lyon was able to create tetraparental male mice by fusing *+/Y* and *Tfm/Y* blastocysts by the method illustrated in Figure 4–17. Some of the resulting chimeric males proved fertile and transmitted the *Tfm* mutation-bearing X chromosome. Because of this, Lyon was able to create *Tfm/Tfm* homozygous female mice. These females are entirely normal and fertile, thereby demonstrating the lack of necessity of the *Tfm* gene product to normal female sexual development.

The cases of *Testicular feminization* in mammals and of *eyeless* and *cardiac lethal* in axolotls demonstrate the integral involvement of genes in the development of specific organs and organ systems both at early and relatively late periods of development. That this type of gene is important to evolution was illustrated in Chapter 6 by the example of the blind cave fish, a situation analogous to the eyeless condition of the axolotl. That blindness in the cave fish and eyelessness in the salamander do not occur by homologous genetic alterations is illustrative of a further point. The development of any organ or organ system is dependent not on a single gene but on a group of genes, a conclusion also derived from the case of the *Notch* and *shibire* mutations in *Drosophila*, which produce nearly identical defects at similar points in ontogeny. Further, genes of this sort that cause large effects can be modified in their expression by a variety of genetic alterations, including changes in the genetic background and the inclusion of other mutant genes in the developing system. Thus, the production of the organism and its component organs is the result of coordinated sets of gene actions and interactions. Perturbations in these groups at a variety of points are available to the evolutionary process, all of which can lead to morphological change. It should also be pointed out that although the sum of all these sets of gene actions produces a harmonic whole, the sets are not inexorably linked. Mutations can also change one ontogenic process independently of the others, yielding the dissociability that is so necessary to morphological evolution.

Cell Death in Normal Development

The degeneration of the Wolffian ducts in female development and the Müllerian ducts in the male are examples of the initial elaboration of structure and its subsequent necrosis. These processes are, like all the other ontogenetic events we have discussed, under genetic control. A specific example of this fact is demonstrated in the development of tetrapod limbs. In the chicken, limb buds first appear as lateral thickenings of the somatopleure at about 55 hours of development. These buds grow out from the body as protuberances covered by ectoderm and filled with mesodermal tissue. As growth proceeds, the contours of the limb, characteristic of either wing or leg, appear. The process of contour formation is accomplished by a series of cellular necroses of the mesodermal regions of the limb. Early in limb-bud formation the necrotic regions can be seen by staining with certain vital dyes and are found at both the anterior and posterior margins of the bud, where it joins the body wall, and in the center of the bud. These three regions are called the anterior and posterior necrotic zones and the opaque patch, respectively (Fig. 7–16). The anterior and posterior zones are responsible for the contouring of the more proximal regions of the limb and the opaque patch for the separation of the tibia and fibula in the leg, and the radius and ulna in the wing. Later in limb development the separation of the digits is accompanied by interdigital necrosis (Fig. 7–16).

Normal chick

Chick mutant *talpid*³

Normal duck

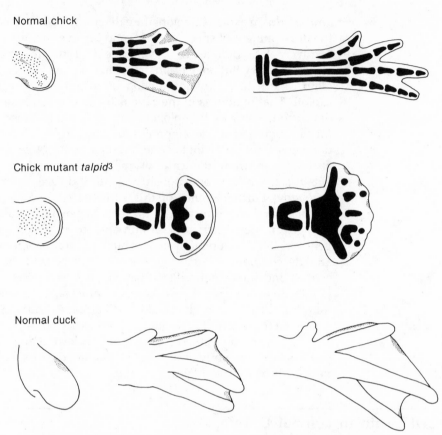

FIGURE 7–16. *The growth of the chick and duck hindlimb. The stippled regions indicate the position of cell death at the posterior and anterior margins of the limb of both the chick and the duck, as well as the intradigital necrosis observed in the chick but not the duck. The* talpid *mutation apparently eliminates the majority of necrosis, resulting in the broad syndactylous paddlelike appendage diagrammed in the center of the figure. The dark areas in the normal and mutant chick limbs represent the areas of formation of cartilage.* [Adapted and redrawn from J. W. Saunders, Jr. and J. F. Fallon, Cell death in morphogenesis, in *Major Problems in Developmental Biology*, M. Locke, ed., Academic Press, London, 1966; and J. R. Hinchliffe and P. V. Thorogood, Genetic inhibition of mesenchymal cell death and the development of form and skeletal pattern in the limbs of *talpid*³ (*ta*³) mutant chick embryos, *J. Embryol. Exp. Morphol.* **31:**747–760, 1974.]

Studies on the posterior necrotic zone (PNZ) of the chick wing bud by Saunders and Fallon have revealed several interesting properties of this group of cells. At 96 hours the PNZ is at its maximal extent, with about 1,500–2,000 cells dying. These cells are phagocytized by a population of nearly 150 macrophages. If the mesoderm of the prospective PNZ is taken from a wing bud 40 hours before necrosis is evident and transplanted to the flank of a host, death still ensues and does so on schedule. This schedule is autonomous to the PNZ cells and is not dependent on the age of the host. However, the mesodermal cells remaining in the bud adjacent to the explanted PNZ do not die. Therefore, it would appear that the cells of the PNZ possess what has been referred to as an internal death clock. They are triggered very early in limb development to die, and once triggered, proceed to their fate.

The clock, however, is not entirely irreversible. If PNZ mesoderm is transplanted to dorsal wing, it does not die. The reason for this was investigated by culturing PNZ cells in combination with a variety of other mesodermal fragments. Mesoderm from the PNZ cultured with somatic mesoderm suffered the same fate as the *in vivo* trunk transplants: The PNZ cells died on schedule. If, however, the PNZ was cultured with dorsal wing mesoderm it did not die. It would appear that the mesoderm from a nonnecrotic region of the wing was affording protection against death by stopping the clock. That this was the case was shown by first transplanting the PNZ cells to dorsal wing, and then removing them from the protective influence of the dorsal wing mesoderm at various times after death should have occurred as dictated by the clock. The cells, once removed, died unless the protection was maintained *in vitro* for at least six days. At that point the clock was stopped entirely and the PNZ cells survived, even in the absence of the dorsal wing mesoderm. The pattern of cell death observed in the proximal portions of the wing is therefore a product of two factors: a cellularly autonomous clock that is triggered in a specific group of cells, and the position of those cells with respect to the remainder of the mesoderm of the limb.

Analysis of the cell death in the interdigital regions reveals another level of regulation. As shown in Figure 7–16, there is little cell death in these regions of the duck foot. This lack of death results in the webbing present between the second, third, and fourth digits. In the interval between the first and second digits where there is no web, there is necrosis. Saunders and Fallon constructed chimeras of duck-limb mesoderm with chick ectoderm and the reciprocal. These hybrid leg buds were then grafted onto the flank of host chick embryos, and the pattern of interdigital necrosis observed. In both cases the duck pattern was seen. Therefore, it is possible to conclude that duck ectoderm, like the dorsal mesoderm of the wing, can inhibit the necrosis normally seen in chick foot development. Chick ectoderm, however, does not induce necrosis in duck mesoderm. Thus, it is possible that the regulation of interdigital necrosis is similar to that in the PNZ; that is, it is an autonomous character of the interdigital mesoderm that may or may not be expressed, based on the environment in which the presumptive necrotic cells come to lie.

That the pattern of cell death in limb development is indeed genetically controlled was shown by Hinchliffe and Thorogood in an analysis of the *talpid* (*ta*) mutation in the chicken. The *ta* mutation is inherited as a simple autosomal recessive and in the homozygous condition causes osseous polydactyly and soft tissue syndactyly of both the wing and leg. The observed polydactyly results primarily from a fusion of the radius and the ulna, and of the tibia and the fibula. The resulting limbs, especially the leg, are broad and paddlelike and bear six or seven digits rather than the normal four (Fig. 7–16). The development of these limbs is characterized by an almost total lack of cell death. The absence of the anterior and posterior necrotic zones accounts for the breadth of the mutant appendage because the contouring, which normally narrows this portion of the limb, does not occur. The fusion of the forelimb bones results from the absence of the opaque patch. This failure in separation may also result in the broadening of the distal portions of the appendage and the seeming mirror-image symmetry seen in the distal limb (Fig. 7–16). There is also an absence of the normal interdigital necrotic zones and the subsequent failure in separation of the digits. Interestingly, this same pattern of syndactyly has been observed by Johnson in the *polysyndactylous* mutant of the mouse, but in the absence of osseous polydactyly. What remains to be determined is whether the *talpid* mutation affects the death clock in the various mesodermal regions, or if it is allowing the rescuing factor produced by dorsal mesoderm to reach the cells destined to die. However, whatever the proximate cause, it is clear that there is genetic regulation of the process of cellular necrosis, and that cell death is important in the production of the final morphology of the vertebrate limb. With respect to this final point, it is interesting to note, as do Hinchliffe and Thorogood, that the simple bifurcating pattern seen in the limbs of the *talpid* mutants is reminiscent of the pattern of elements found in the paddles of crossopterygian lungfish, such as *Eusthenopteron* and *Sauripterus*. Therefore, it is possible that the complex pattern of development observed in the limbs of higher tetrapod vertebrates may have evolved through a process of genetically regulated and patterned cellular necrosis.

There are also mutations that increase regions of necrosis. Thus, mutations in *Drosophila* such as *Bar* eyes or *vestigial* wings, which dramatically reduce the eye and wing, do so by an increase in the regions of cell death normally present in the development of these two structures. Additionally, the *wingless* and *rumpless* mutations in the chicken seem to act by a similar mechanism, allowing necrosis to excess. One can envision this type of alteration, in a less drastic form, as being important in the reduction or elimination of structures necessary during one stage of development but unnecessary at a later time. A simple example is the tail of the tadpole larvae of frogs. Again, as for the mutations that suppress cell death, there are two conceivable levels at which the process could be regulated. The intrinsic clock could be activated in more cells or extrinsic factors, such as the diffusing, protective substance found in dorsal wing mesoderm, may be eliminated.

Genes Involved in Late Events and Growth

It is clear that mutations in genes directly controlling developmental pathways, particularly those functioning early in development, may have cataclysmic effects. However, there are also late-acting genes that, while affecting the overall morphology of the organism, often have no obvious deleterious effects. These include genes that control the growth characteristics of the organism subsequent to the production of basic morphology and organogenesis. Such genes have been discovered as a result of mutations that alter hormone function to produce giantism or dwarfing. Alterations in form (e.g., relative size of appendages) can result from changes in patterns of growth introduced as pleiotropic effects of the basic hormonal change. Although this type of alteration has certainly led to evolutionary changes, for example, giantism in European cave bears during the Pleistocene and dwarfing in elephants discussed in Chapter 2, these kinds of late developmental changes have not resulted in basic reorganizations in morphology. After all, a pygmy elephant is still undeniably an elephant.

There is one size affecting mutation, however, that demonstrates the plasticity of the developmental process and thus deserves some special comment. This is the *giant* (*gt*) mutation in *Drosophila melanogaster*. This sex-linked recessive trait was originally discovered by Bridges and Gaberchevsky in 1928. The entirety of *Drosophila* development, from fertilization to adult, normally takes place in 10 days. Individuals bearing the *gt* mutation take from two to five days longer to complete the same process. The resulting *gt* individuals are normal in morphology, but are nearly twice the size of normal flies. The manner in which this change is produced is quite interesting. The development of *gt* flies is normal from the embryonic through the late-third larval instar. However, at the point in time when normal individuals pupate and begin metamorphosis, the *gt* individuals continue on as larvae. It is during the larval instars and early pupal stage that those groups of cells, the imaginal discs, which are destined to form the tissues of the adult fly, proliferate, and it is the number of these cells that controls the size of the fly. The *gt* larvae, during their period of extended larval development, apparently go through at least one extra cell division. This is implicated by the fact that those cells in the larvae that are polytene (e.g., the salivary glands) participate in at least one, and in some cases, two extra rounds of DNA synthesis. The *gt* larvae, after this 2–5-day extra period of growth, pupate at about twice normal size. Metamorphosis then takes place and a morphologically normal, albeit double-sized, adult ecloses after a slightly protracted pupal stage. Therefore, the animal is able to regulate its development such that no extra elements are formed, despite the apparent proliferative event prior to differentiation. We have previously encountered this type of developmental plasticity (discussed in Chapter 4) in the evolution of early ontogenic events in the tunicates and lower chordates. In these organisms, as an adaptation to the relative importance of the larval stage in different species, the numbers and the times of the cleavage divisions have changed. This has resulted in differing portions of the early embryo of different species contributing to specialized larval organs. This demonstrates the ability of two quite different developmental systems to take a major alteration into stride and produce a completely integrated organism.

Other terminally acting genes include those that control patterns and amounts of pigment production, and the effects of genetic variation in these genes are obvious in natural populations of most organisms.

Although these kinds of alterations are clearly important to selective processes, and thus evolution, they are probably not important to alterations in morphology *per se*.

Our basic tenet is that genes control ontogeny. As we have shown, this control is exerted at several levels. Maternal effect mutations can be utilized to demonstrate the genetic control of the organization of the egg in organisms, such as *Drosophila*, that have a mosaic mode of early development. Other mutations, such as *tailless* in the mouse and the *cardiac* and *eyeless* defects in the axolotl, indicate the existence of genetic controls over subsequent stages of development, just as specific genetic information is necessary for the proper function of the cascade of ontogenetic events resulting in differentiation of the basic germ layers of the embryo and the proper inductive events necessary to organogenesis. Finally, genetic changes can alter later events in development, including growth and pigmentation, and thereby modify the final form of the adult organism. The evolutionary process may select alterations in expression of the class of developmentally important genes discussed in this chapter to produce new developmental pathways. What must be stressed, however, is the difference between this class of gene and the nature of the mutation (alterations in expression) important in evolutionary change. The mutations discussed in this chapter have been, for the most part, drastic and deleterious in their effects. They are important in revealing the genetic elements underlying certain developmental processes. In all likelihood, it is mutations that are more subtle in their effects that contribute to morphological evolution. Mutations that alter timing or duration of events or strengths of interactions will all result in evolutionary modification of developmental pathways. An altogether different class of genes and mutational changes are also crucial to pattern formation and morphogenesis. This is discussed in the remaining chapters.

Eight

Homoeosis in Ontogeny and Phylogeny

See! See!
What Shall I See?
A Horse's Head,
Where His Tail Should Be!

Rube Goldberg

Homoeosis and Homoeotic Mutations

Gene action is intimately involved in ontogeny and this involvement is revealed by mutations that drastically disrupt the development of the organism. There is, however, another class of mutations that alter but do not disrupt the process of ontogeny. These are the homoeotic mutations. The importance and theoretical utility of this type of developmental change was first recognized by William Bateson in his book *Materials for the Study of Variation* published in 1894. His logic in coining and definition of the term homoeosis is still valid and points up the salient features of the concept.

> The case of the modification of the antenna of an insect into a foot, of the eye of a crustacean into an antenna, of a petal into a stamen, and the like, are examples of the same kind. It is desirable and indeed necessary that such variations, which consist in the assumption by one member of a meristic series, of the form or characters proper to other members of the series, should be recognized as constituting a distinct group of phenomena....I therefore propose...the term Homoeosis...for the essential phenomenon is not that there has merely been a change, but that something has been changed into the likeness of something else.

Bateson then goes on to catalog various examples of homoeotic alterations in organisms as diverse as mammals and annelid worms. For example, among the mammals some rare individual ground sloths have been described as having thoraciclike vertebrae in the lumbar region of the spine. However, it is more common to find homoeotic variants among the arthropods, animals that are comprised entirely by a series of metameric segments; and within the Arthropoda the insects have

235

yielded the most information about the nature of patterns and mechanisms of homoeosis.

Some of the oldest fossil insects are found in the strata belonging to the Carboniferous Period. These insects, like modern pterygotes, had four wings similar in morphology to those found on extant species. Unlike modern insects there was also a pair of flaps or paranotal lobes projecting from the dorsum of the first thoracic segment. These flaps have been cited as possible evidence for the origin of wings as lateral integumental outfoldings. These lobes could have served originally as organs to aid in gliding. The primitive condition, wings on the second and third thoracic segments and paranotal flaps, is not seen in modern insects, but can be produced by means of a homoeotic mutation in the cockroach *Blattella germanica*. M. H. Ross has described a heritable trait *Pro-wings* in this primitive insect that causes winglike flaps to develop on the dorsal prothorax. A similar homoeotic alteration has been described by I. Herskowitz in *Drosophila melanogaster*. Another sex-linked recessive mutation, *labiopedia*, has been described in the beetle *Tribolium confusum*. Animals homo—or hemizygous for this mutation have their labial palps transformed into thoracic legs. This transformation is evident in both the larval and adult stages and can be seen in Figure 8–1. The transformation is complete and includes the formation of musculature normally found in the legs. Apparently, however, these muscles are not innervated as the labial legs are not observed to move.

A more extreme set of transformations is observed in the lepidopteran silkworm *Bombyx mori*. There exists in this organism a complex of mutants referred to as the *E*-allelic series (*Extra legs*). The caterpillar of this moth has a very distinctive morphology. The full-grown fifth instar larva has a darkly pigmented head, bearing eyes, followed by three thoracic segments each bearing jointed legs. Dorsally, the second thoracic segment has a darkly pigmented eye spot. The abdomen has eight segments followed by a terminal caudal segment. The third through sixth and the caudal segments all bear fleshy walking appendages. The second and fifth abdominal segments are marked dorsally by a pigmented crescent and star, respectively. The morphology of the wild-type caterpillar is presented at the top of Figure 8–2. Dominant mutant variants of the *E* series, as shown in Figure 8–2, cause segmental alterations in this pattern. Thus, individuals heterozygous for the *Extra Crescents and legs* (E^{El}) mutation have a normal head and thorax but have walking appendages on the normally legless first and second abdominal segments. In addition, the first abdominal and third thoracic segments have a pigmented crescent on their dorsal sides. The mutation *Extra legs—New additional crescent* (E^N) causes the production of a crescent on the third abdominal segment, the deletion of the star from the fifth abdominal segment, and the production of stars on the sixth, seventh, and eighth abdominal segments. These alterations can be interpreted as a change of segmental identities (both dorsal and ventral) in the abdomen and thorax. This alteration is most dramatically seen in homozygotes for the above mentioned E^N allele. This genotype is lethal and results in death in the early larval instars. The morphology of these

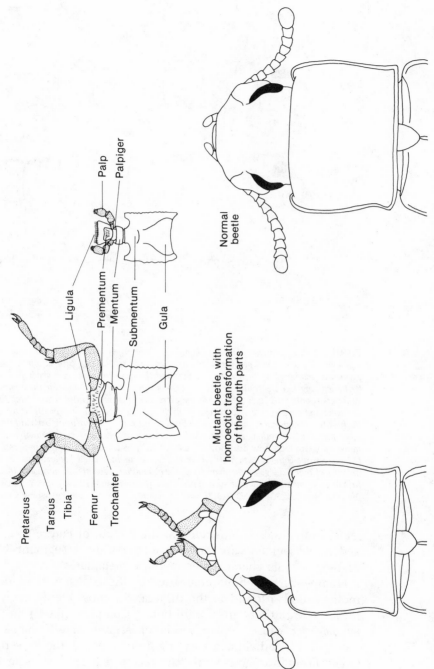

FIGURE 8–1. *Drawings of the heads of normal (right) and labiopedia mutant (left) individuals of Tribolium confusum from the dorsal aspect. Insets show the specific portions of the mouth parts that are transformed into leg in the mutant individuals. The distal portions of the normal labial elements (palp and palpiger) are transformed into a normal leg, including a trochanter proximally and tarsal claws distally. [Redrawn from Daly and Sokoloff, 1965.]*

Pretarsus
Tarsus
Tibia
Femur
Trochanter

Ligula
Prementum
Mentum
Submentum
Gula

Palp
Palpiger

Normal
beetle

Mutant beetle, with
homoeotic transformation
of the mouth parts

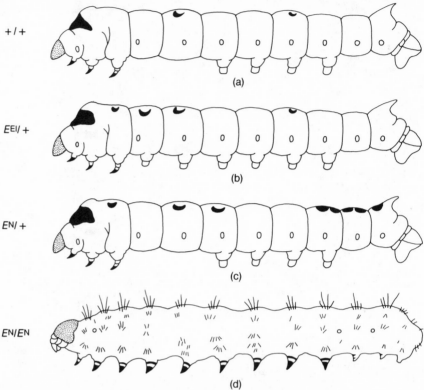

FIGURE 8-2. *Drawings of larvae of* Bombyx mori: (a) *Normal fifth instar—the first three body segments (thoracic) bear pigmented walking legs; the second and fifth abdominal segments have pigmented markings on their dorsal surface and the third through sixth abdominal segments have fleshy walking legs ventrally.* (b) *Heterozygous fifth instar individual for the dominant mutation* Extra crescents and legs (E^{El}). *Note the extra crescents on the third thoracic and first abdominal segments and the presence of walking legs on the first two abdominal segments. It would appear that the first two abdominal segments and the third thoracic segment are being partially transformed into more posterior abdominal segments.* (c) *Heterozygous fifth instar individual for the dominant mutation* Extra legs new additional crescent (E^N) *shows similar changes in dorsal pigmentation but no ventral transformation.* (d) *Late embryo of an individual homozygous for* E^N. *Note the presence of thoracic legs on the majority of the abdominal segments. In normal individuals only the first three segments would bear such appendages.* [Adapted and redrawn from Y. Tanaka, Genetics of the silkworm, *Adv. Genet.* **5**:239–317, 1953, by permission of Plenum Publishing Corporation.]

lethal individuals is illustrated at the bottom of Figure 8–2. All of the abdominal segments are transformed into thoracic segments bearing the segmented legs characteristic of these metameres.

Homoeotic mutations analogous to those observed in beetles and moths are also found in the diptera, the most highly evolved of the insects. Mutations in the Bithorax Complex (BX-C) in *Drosophila melanogaster* cause similar kinds of segmental alterations to those incurred by individuals carrying mutations of the *E* series. Adult *Drosophila* homozygous for the two recessive mutations, *bithorax* (*bx*) and *postbithorax* (*pbx*), demonstrate the remarkable transformation shown in Figure 8–3. Like all dipterans, *Drosophila* has a single pair of wings borne by the second thoracic segment of the adult. The second pair of wings found in other winged insects on the third thoracic segment has

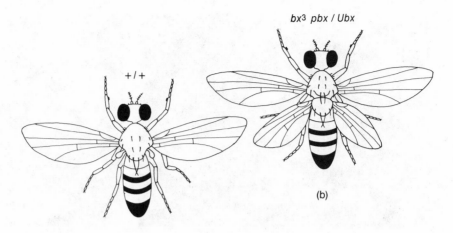

FIGURE 8–3. *Dorsal views of* (a) *normal and* (b) bx³ pbx/Ubx *mutant adult* Drosophila. *The homoeotic mutations result in the transformation of the normally small dorsal third thorax and haltere into the larger second thorax and wing.*

been reduced to a pair of small nubs called halteres or balancer organs. These halteres are transformed into wings in the *bx pbx* double mutants. Moreover, the adult third thorax, normally quite small in *Drosophila,* is transformed into the likeness of the large second thoracic segment. *Drosophila* also exhibit a series of mutations that transform the mouth parts into other structures. These genes are members of a second gene complex, the Antennapedia Complex (ANT-C). In the *proboscipedia* (*pb*) mutation in *Drosophila,* as for the labiopedia mutation of *Tribolium,* the labial palps are transformed into legs. However, some alleles of *proboscipedia* transform the palps to antennae (Figure 8–4). Another dominant mutation, *Antennapedia* (*Antp*), performs the transformation alluded to in Bateson's original definition of homoeosis. The antennae of the adult fly are transformed into perfectly formed legs that extend prominently from the head of the fly (Fig. 8–4). The task, of course, is to go beyond the mere cataloging of these defects and to attempt an explanation of how these seemingly magical alterations have transpired.

The most extensive genetic work on homoeotic lesions has been performed on *Drosophila melanogaster.* A large repertoire of mutations that alter or modify the segmental pattern exists in this organism, and several investigations on the developmental properties of these mutations have helped to define the genetic mechanisms of the specification of body pattern and segment identity as they are controlled by homoeotic loci. As a general class, homoeotic mutations result in the apparent expression of altered states of determination. This being the case, they have been regarded as mutations defective in the interpretation of the positional information supplied to the egg and early embryo by the maternal genome. Moreover, they have been envisioned as developmental (binary) switches that translate position in an either/or fashion. However, homoeotics comprise a rather heterogeneous group of mutations in terms of their genetic and developmental properties.

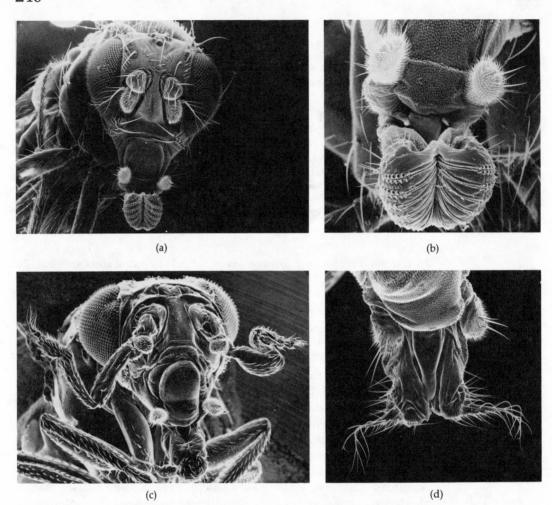

(a)   (b)

(c)   (d)

FIGURE 8–4. Scanning electron micrographs of the (a) head and (b) mouth parts of a normal D. melanogaster. An individual heterozygous for an Antennapedia mutation is shown in (c). The antennal segments are transformed into the majority of segments seen in a normal leg. The terminal portions of this homoeotically transformed element, however, still resemble the terminalia of a normal antenna. In (d) the mouth parts of an individual homozygous for a mutant allele of proboscipedia are shown. The normal pseudotracheal rows of the labium are replaced by structures that resemble the arista or terminal segment of the normal antenna. Other mutant alleles at this locus transform the labial lobes into legs. [Photographs courtesy of Dr. F. R. Turner.]

The homoeotic loci that have been identified in *Drosophila melanogaster* are distributed among the four chromosomes of this species, as diagrammed in Figure 8–5. They are found on all chromosomes and exist as both dominant and recessive alleles.

The wide range of properties displayed by these genetic lesions suggests that they may differ in the mechanism by which any particular transformation is effected. Put another way, some may reside in switch genes having a primary involvement in the determinative process, but others may not. This contention is pointed up by the fact that mutant alleles of several of these loci have pleiotropic affects, causing not only homoeosis, but other nonspecific phenotypic anomalies. A good exam-

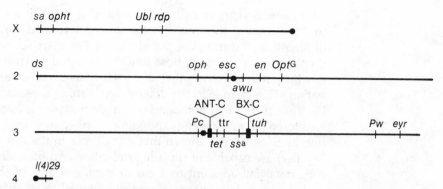

FIGURE 8–5. *A diagrammatic representation of the distribution of most of the known homoeotic loci on the four chromosomes of* D. melanogaster. *The solid circle represents the position of the centromere on the sex (X), second (2), third (3), and fourth (4) chromosomes. There are two clusters of this type of gene on the right arm of chromosome 3. These are the* Antennapedia *cluster (ANT-C) and the* bithorax *cluster (BX-C). The remaining abbreviations are as follows in their order on the map:* sa = sparse arista; opht = opthalmoptera; rdp = reduplicated; Ubl = Ultrabithorax-like; ds = dachsous; opl = opthalmopedia; esc = extra sex comb; awu = augenwulst; en = engrailed; Opt^G = Opthalmoptera of Goldschmidt; Pc = Polycomb; tet = tetraltera; ss^a = spineless aristapedia; tuh = tumorous head; Pw = Pointed wing; eyr = eyes-reduced; 1(4)29 = lethal-29 *on chromosome 4.* [Adapted from a compilation in Ouweneel, 1976.]

ple of this is the *spineless-aristapedia* (*ss^a*) mutation, which produces an antenna-to-leg transformation similar to that described for the dominant and nonallelic *Antp* mutation. In addition to the homoeotic transformation, the flies also have thin bristles, drooping wings, and fused tarsal segments. These latter effects would not be expected from a gene causing a switch between two alternate patterns of development. Therefore, we may conclude that *ss^a* causes its homoeosis by an indirect, perhaps metabolic, defect. The possible nature of this indirect effect is revealed by an analysis by A. Shearn and his colleagues of a group of disc-specific lethal mutations in *Drosophila*. These lesions have no direct effect on embryogenesis as it relates to the development of a normal larva. However, the imaginal discs that normally give rise to the adult integument during metamorphosis are not formed or do not proliferate. Temperature-sensitive alleles of several of these loci have been recovered. Transfer of mutant larvae to the nonpermissive temperature at various times during the proliferative stages of imaginal disc development does not result in the elimination of the discs, but in homoeotic transformations in the adults derived from these larvae. Because the discs have been shown to be already determined at the time of the temperature change, the mutation must be altering the determined state subsequent to its initial establishment. This same kind of change has been observed during the *in vitro* culture of imaginal discs by E. Hadorn and his students. Explanted discs cultured in the abdomens of adult females will continue to grow, and if cut into fragments, cell division will occur. When injected into a metamorphosing larval host, these cultured fragments will themselves metamorphose. A majority of the time, the cultured disc will develop into a structure pertinent to the disc originally cultured; that is, a wing disc will yield wing structures even after several years in culture. Indeed, it is just this sort of experiment that has shown that the disc tissue is determined early and achieves differentiation during the pupal stages. The discs so treated, however, do not always metamorphose into the expected structures but occasionally into derivatives of other discs. This has been termed transdetermination. This change in the determined state of these cells is not caused by, as in the case of homoeosis, a single mutational event but rather by an altered pattern of gene expression. Further analyses of this phenomenon by S. Strub has shown that not all cells in the disc (specifically the leg disc) are capable of transdetermination. Additionally, this group of cells must fall adjacent to a cut or damaged area of the disc and cell division must take place before the potential to transdetermine can be fulfilled. It is possible that the discless mutations of Shearn and the ss^a lesion cause their respective homoeoses by stimulating these events (i.e., localized damage or cell death followed by a compensatory cellular proliferation by adjacent cells) *in situ*. Therefore, the change in cell fate comes about by altering an already determined state, as does transdetermination, rather than affecting the initial determinative event.

Another indirect cause of homoeosis is demonstrated by mutations at the *Ultrabithoraxlike* (*Ubl*) locus mentioned in Chapter 7. Mortin and Lefevre have shown that this sex-linked dominant mutation mimics the autosomal *Ultrabithorax* (*Ubx*) mutation that is related to the *bx* and *pbx* mutations mentioned previously. *Ubx*, like *bx* and *pbx*, causes a transformation of the adult halteres, to wing, and indeed fails to complement both *bx* and *pbx* (Fig. 8–3). *Ubl* causes a similar transformation and interacts with *Ubx* to cause a more extreme transformation than either single mutation alone. Homozygous *Ubl/Ubl* individuals are lethal and other recessive lethal alleles of *Ubl* have been recovered that do not exhibit any homoeotic transformation or interaction with *Ubx*. Biochemical work on this locus by Greenleaf and his associates has shown that Ubl^+ encodes one of the subunits of RNA polymerase II, the polymerase necessary for the transcription of genes that produce messenger RNAs (mRNA). Because only a single allelic variant of the *Ubl* gene has homoeotic effects, although demonstrated null alleles do not, we may conclude that the unique homoeotic allele produces an altered enzyme, which in turn may affect the transcription of the *Ubx* gene. This indirect effect is supported by Mortin's recent demonstration that *Ubl* affects the functioning of several other nonhomoeotic loci throughout the genome—*Ubx* is only the most sensitive to the enzymic alteration.

The foregoing discussion points out that a thorough analysis of both the genetic and developmental properties of any locus must be obtained before any general conclusions as to the mechanism of homoeotic transformation, and therefore, the role of any particular gene in normal development can be found. This type of extensive analysis has been performed for two groups of loci and the conclusion reached is that both groups cause their effects through their involvement in the initial determinative events that control segmental identity. Before discussing the possible mechanisms by which transformations are caused by these loci, it is necessary to discuss the normal process of segmentation in *Drosophila*.

Normal Segmentation in *Drosophila*

Subsequent to fertilization and syngamy, the *Drosophila* embryo undergoes a series of syncytial cleavage divisions in which the nuclei divide without cellularization of the embryo. The first nine of these divisions take place throughout the cytoplasm of the egg, and after the ninth division the majority of the nuclei migrate to the cortical cytoplasm. Four more nuclear divisions follow, and after the 13th division is completed cellularization takes place, giving rise to a cellular blastoderm of about 6,000 cells. This stage can be seen in Figure 8–6, where the somatic blastoderm cells can be morphologically distinguished from the spherical pole cells (the presumptive germ line) at the posterior tip of the embryo. It is at this point in development that determination of cell fate takes place for larval as well as adult structures. Moreover, it is at this point that the zygotic genome first becomes detectably active in transcription. During gastrulation, a lateral anterior fold called the cephalic furrow forms. This furrow separates a majority of the head from the trunk. The pole cells are simultaneously carried from the posterior tip to the dorsal surface of the embryo by the elongation of the germ band. Three separate infoldings—the ventral furrow, posterior midgut, and anterior midgut invaginations—carry the mesoderm and anterior and posterior endoderm, respectively, to the inside of the embryo. Subsequently, at about eight hours of development, the surface of the germ band becomes invested with a series of lateral folds that physically divide the embryo into segments. These embryonic segments can be seen in Figure 8–6 and correspond to the metameric segments of the larva and adult. Thus, by 10 hours of development, the embryo has three germ layers and is divided into a head, thorax, and abdomen. The head is made up by the clypeolabral, procephalic, mandibular, maxillary, and labial segments. The latter three segments are collectively called the gnathocephalic segments. There are three thoracic segments: prothorax, mesothorax, and metathorax. The remaining nine segments are the eight abdominal and single terminal caudal segment.

Embryogenesis is completed by the elaboration of this basic metameric pattern by the formation of the cuticular covering and internal organs of the larva. The larva hatches at about 24 hours after fertilization. Each of the three thoracic and eight abdominal segments is marked at its anterior ventral margin by rows of denticles, the morphology of which allows the distinction between the larval abdominal and thoracic segments. The segments of the head have inverted so that the head has come to reside inside the larva where the gnathocephalic segments have given rise to the larval mouthparts. The imaginal anlagen or discs that were determined and segregated from the larval cells at the cellular blastoderm stage will grow during the three larval instars and then achieve their adult differentiated state during metamorphosis in the pupal state. The adult, like the embryo and larva, is metamerically organized. The various imaginal discs are likely derived from cells distributed among the various metameres of the embryo. We have presented this derivation in cartoon fashion in Figure 8–7.

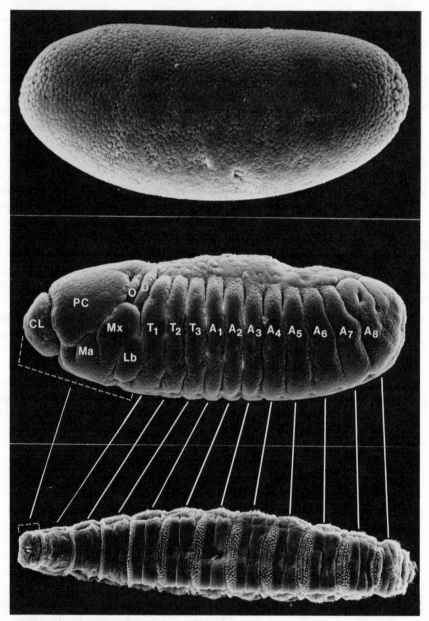

FIGURE 8–6. *Scanning electron micrographs of the cellular blastoderm (top), segmented germ band (middle), and first instar larval stages (bottom) of* D. melanogaster. *The lines show the segmental identities for the eight-hour embryo and the mature larva. The head of the embryo and larva are indicated by the dotted bracket. The identity of the segments is given by the following abbreviations: CL = clypolabrum; PC = procephalic; Ma = mandibular; Mx = maxillary; Lb = labial; O = optic; D = dorsal ridge; T_1, T_2, T_3 = first, second, and third thorax; A_1–A_8 = first to eighth abdominal.* [Photographs courtesy of Dr. F. R. Turner.]

The adult head consists of eyes, antennae, clypeolabrum, and maxillary and labial palps. The latter three adult mouthpart elements probably derive from the corresponding embryonic segments. In the higher diptera, such as *Drosophila*, there is no mandibular element in the adult. The eye and antenna in all likelihood originate in the procephalic

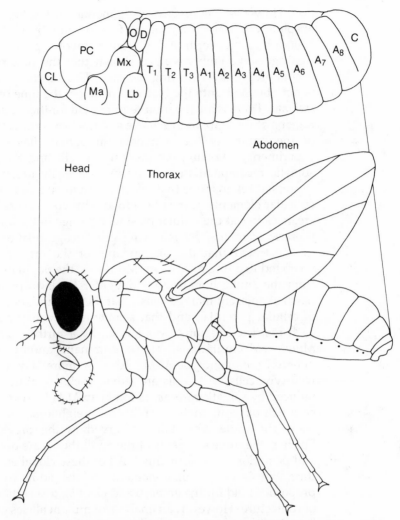

FIGURE 8–7. *Diagrammatic representation of the hypothesized segmental correlations between the eight-hour embryo and the adult of* D. melanogaster. *Abbreviations are as in Figure 8–6.*

lobe of the embryo. The entirety of the adult head derives from three discs, with the majority of the head capsule formed from the eye antennal disc which also gives rise to the maxillary palp and is therefore of mixed segmental origin. The other two discs derive from one segment each. Each of the three thoracic segments bears a pair of walking legs each derived from a single disc. Leg discs are present in each of the three embryonic thoracic segments. The dorsal aspect of the mesothorax makes up the majority of the adult thorax and is derived from the wing disc. Both the dorsal pro- and metathoraces are greatly reduced. The latter bears the halteres referred to earlier. Finally, there are the eight abdominal segments and the terminal genital structures having the appropriate derivations. Whether or not the genitalia are derived from the caudal segment is not known.

All of the segment boundaries appear simultaneously during gastrulation. However, the process of segmentation has been shown not to be absolutely mosaic in character. Experiments by G. Schubiger and W. Wood, and W. Herth and K. Sander involving the ligation of various cleavage-stage embryos have shown that the patterning of segments occurs progressively during cleavage. The full complement of segments is not obtained until the blastoderm stage, the time of determination of cell fate. There is an apparent requirement for the interaction of different regions of the embryo to achieve a normal pattern. A further elucidation of the nature of the segmentation pattern has been provided by experiments involving the marking of cells and their descendants by somatic recombination induced by x-rays. By utilizing autonomously expressed cell markers that affect the pigment and morphology of adult cuticular elements, Garcia-Bellido and his co-workers have been able to create clones at the cellular blastoderm stage in the anlagen of the adult thoracic elements. By following cell lineage relationships, they have demonstrated that the descendants of the original marked cell are restricted to a single segment. Moreover, the population of cells derived from the single homozygous daughter tends to remain together as a patch of adjacent marked cells. That these boundaries reflect restrictions of cellular potential, and that segments are actually divisible into two such restricted regions, was shown by a further experiment utilizing *Minute* (M) mutations. These dominant mutations cause cells of heterozygous genotype M/M^+ to grow more slowly than do M^+/M^+ wild-type cells: M/M cells are lethal. If clones of the M^+/M^+ cells are induced by somatic recombination in an M/M^+ background, the normal cells will outgrow their heterozygous neighbors. Large M^+/M^+ clones show the same restrictions observed in the previous experiments. Further, the three segments of the adult thorax are divided into anterior and posterior "compartments." All of these clonal experiments involve the analysis of cuticular elements of the adult fly. That the same properties hold for the embryo and larva is, to a certain extent, a matter of conjecture. However, an analysis of mutant alleles of the *engrailed* (en) locus by Kornberg is consistent with this hypothesis. Leaky mutations of the *en* locus transform the elements of the posterior compartments of the adult thoracic segments into anterior compartment structures. This is most strikingly seen in the wing blade of the adult where the mutant wing is comprised of two anterior halves in mirror image symmetry. More stringent alleles cause lethality in the late embryo. These dead individuals can be seen to have similar mirror-image duplications of all of the thoracic and abdominal segments. Therefore, it would appear that compartments are formed in the embryo and this segmental segregation of cellular potential is at work in both the larva and adult.

The Relation of Homoeosis to Normal Segmentation

There are two groups of homoeotic loci that are apparently involved directly in the determination of segmental identity. Interestingly, these two sets of genes exist as two clusters of tightly linked loci in the right arm of the third chromosome. Members of the more distal group have already been presented. The associated loci together are called the Bithorax Complex, and there are at least 10 complementing loci comprising this group. The other more proximal group has at least six members and is called the Antennapedia Complex. The physical relationship of these two complexes to each other and the other homoeotic loci is seen in Figure 8–5. Figure 8–8 shows the makeup of each group and its position in the polytene chromosomes. The members of the BX-C are responsible for specifying segmental identity

FIGURE 8–8. *Drawing of the regions of the polytene chromosomes of* D. melanogaster *in which the ANT-C and BX-C are located. The loci constituting the two complexes are given above the chromosome segments.* pb = proboscipedia; zen = zerknüllt; Dfd = Deformed; ftz = fushi tarazu; Scr = Sex combs reduced; Antp = Antennapedia; abx = anterio-bithorax; bx = bithorax; Cbx = Contrabithorax; Ubx = Ultrabithorax; bxd = bithoraxoid; pbx = postbithorax; Hab = Hyperabdominal; Uab = Ultraabdominal; Mcp = Miscadestral pigmentation; tuh = tumorous head; iab = infraabdominal.

Antennapedia complex

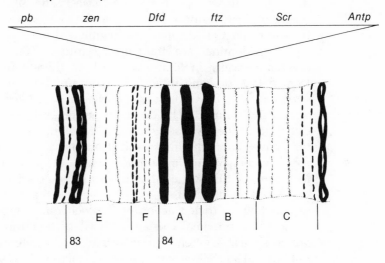

Bithorax complex

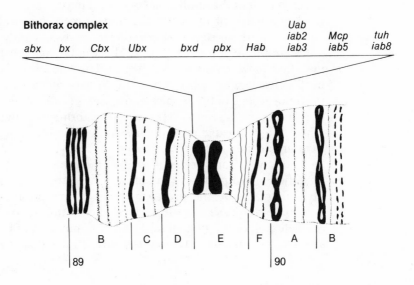

in segments posterior to the mesothoracic segment in both the larva and the adult. Thus, as we have already seen, in the absence of the *bx* and *pbx* loci, the adult metathorax is transformed into mesothorax. The *bithoraxoid* (*bxd*) mutation causes a like transformation of the first abdominal segment into a mesothoracic segment. In this latter case the transformation can also be seen in the larva. These three recessive mutations complement each other (i.e., *bx/pbx* or *bx/bxd*, for instance, are wild-type in phenotype) but all three fail to complement the dominant *Ultrabithorax* (*Ubx*) mutations. E. B. Lewis has concluded that the *Ubx* lesions represent mutations in a portion of the BX-C responsible for the regulation of the *bx, pbx,* and *bxd* loci. Individuals homozygous for a *Ubx* mutation are lethal in the late embryo and show transformations of the third thoracic and first abdominal segments into a second thorax, a result consistent with Lewis' interpretation of the function of *Ubx*. As more of the distal sites in the BX-C are deleted (Fig. 8–8b), more posterior segments are transformed into thoracic segments. Thus, deletion of those loci labeled *iab-2, iab-3,* and *iab-8* in Figure 8–8(b), results in the transformation of abdominal segments into mesothorax. If the entire BX-C is deleted, the segments from the metathorax to the eighth abdominal are transformed into mesothorax, as presented diagrammatically in Figure 8–9. All of these transformations result from a lack of gene function. There are, however, other dominant variants in the BX-C that result in the abnormal activation of functions within the complex. The dominant mutation *Contrabithorax* (*Cbx*) produces adult flies whose wings are transformed into halteres. Genetic tests have shown that this results from the derepression of the bx^+ and pbx^+ functions in the mesothorax where they are normally not active. Mutations of this sort exist at at least three points in the BX-C and are presumed to affect *cis*-regulatory regions that control the normally segmentally restricted action of the loci comprising the complex.

The more proximal cluster (ANT-C) of homoeotic loci (Fig. 8–8a) causes similar segmental transformations in anterior portions of the embryo and adult. The *proboscipedia* (*pb*) locus is necessary for the normal development of the adult labial segment. In the absence of pb^+ function, the labial palps of the adult are transformed into prothoracic legs. A deletion of the *Sex combs reduced* (*Scr*) locus results in the transformation of the prothorax into mesothorax. This can be seen most readily in adult male flies in which the sex comb, a structure normally found on only the prothoracic leg, is absent. Moreover, the overall pattern of bristles of the prothoracic legs of both males and females is indicative of a transformation of first leg to second. The same transformation can be seen in larvae where the distinctive morphology of the prothoracic segment is altered to resemble the mesothoracic segment. Finally, certain alleles of the *Antp* locus allow a transformation of the mesothorax and portions of the metathorax into prothorax in both the larva and adult. This can be seen in adult males by the presence of sex combs on all six legs. A total deletion of the ANT-C results in lethality in the late embryo just before the hatching of the first larval instar. Examination of these individuals, as well as earlier stages before lethality, shows that segments from the

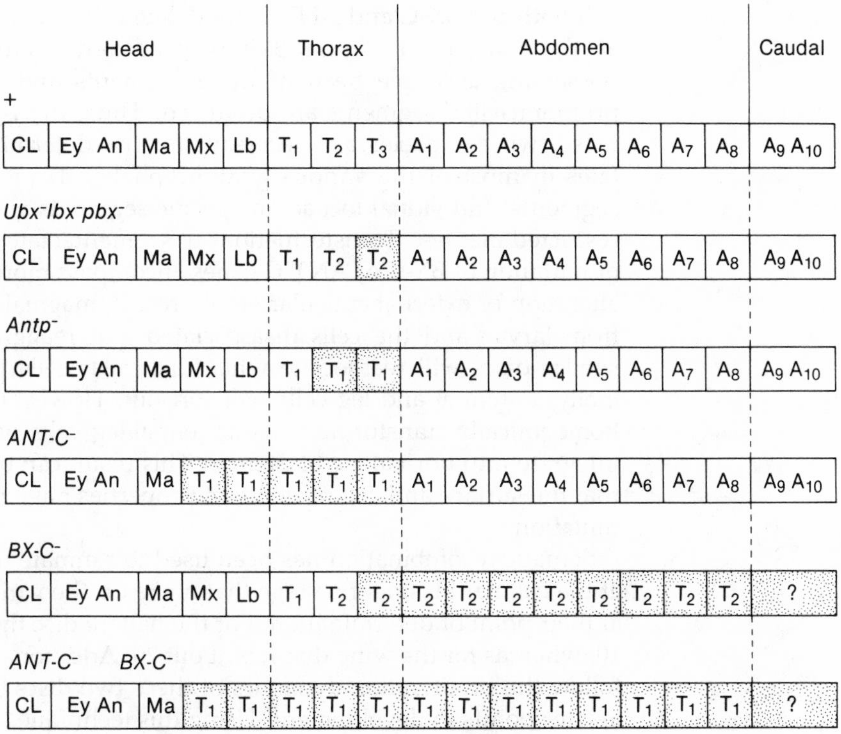

	Head					Thorax			Abdomen								Caudal

+

CL	Ey An	Ma	Mx	Lb	T₁	T₂	T₃	A₁	A₂	A₃	A₄	A₅	A₆	A₇	A₈	A₉ A₁₀

Ubx⁻lbx⁻pbx⁻

CL	Ey An	Ma	Mx	Lb	T_1	T_2	T_2	A_1	A_2	A_3	A_4	A_5	A_6	A_7	A_8	$A_9 A_{10}$

Antp⁻

CL	Ey An	Ma	Mx	Lb	T_1	T_1	T_1	A_1	A_2	A_3	A_4	A_5	A_6	A_7	A_8	$A_9 A_{10}$

ANT-C⁻

CL	Ey An	Ma	T_1	T_1	T_1	T_1	T_1	A_1	A_2	A_3	A_4	A_5	A_6	A_7	A_8	$A_9 A_{10}$

BX-C⁻

CL	Ey An	Ma	Mx	Lb	T_1	T_2	T_2	T_2	T_2	T_2	T_2	T_2	T_2	T_2	T_2	?

ANT-C⁻   BX-C⁻

CL	Ey An	Ma	T_1	T_1	T_1	T_1	T_1	T_1	T_1	T_1	T_1	T_1	T_1	T_1	T_1	?

FIGURE 8–9. Diagram of the segmental transformations observed after the deletion of portions or the entirety of the ANT-C and BX-C. The transformed segments are indicated by stippling and the new identity is given within the stippled area. The various genotypes are given at the left, above each box diagram. The indicated transformations are a simplification and compilation of effects on adults, larvae, and embryos. In most cases, however, the transformations presented are those hypothesized to be present in the embryo at the initial point when segmental identity is attained. Ey = eye; An = antenna. The remaining abbreviations are as in Figure 8–6.

maxillary through the third thoracic segment all resemble thoracic segments. The embryonic pattern of these last two ANT-C lesions is presented in Figure 8–9.

Like the BX-C, dominant gain of function mutations also occur in the ANT-C. The most obvious of these are the *Antp* lesions causing a transformation of the antenna to leg. The function of *Antp*⁺ based on the phenotype of deletions of this locus is to elicit the proper development of the second and third thoracic segments. In dominant *Antp* lesions this function is active in the more anterior antennal segment and antennal anlagen. That *Antp*⁺ has no function in normal antennal development is demonstrated by the normal morphology of the antenna, even when homozygous clones of *Antp*⁻ cells are created in this organ by somatic cross-over events. The same, however, is not true of these clones in the second and third legs, where *Antp*⁻ clones can produce either a prothoracic or antennal pattern of development. Thus, it would appear that *Antp*⁺ function is necessary for normal ventral meso- and metathoracic development, and the transformation of antenna to leg results from the abnormal activation of this function in the antennal anlagen.

If both the BX-C and ANT-C are deleted, the segmental pattern seen at the bottom of Figure 8–9 is produced. Lethal embryos each possessing a simple head of three segments and a series of similar prothoraciclike segments are produced. Thus, the genes that comprise these two complexes are necessary for the determination of separate fates in most of the various gnathocephalic, thoracic, and abdominal segments. Individual loci act in specific segments and in a segmentally restricted manner. Transformations of segmental fate caused by deletion or mutation of BX-C or ANT-C genes encompass more than a superficial alteration of external cuticular structures. If imaginal discs are removed from larvae and the cells disassociated and reaggregated, cells from similar discs will reassociate while dissimilar cells will sort out. Normally, antennal and leg cells will sort out. However, if the antenna is homoeotically transformed, the antennal/leg cells will reassociate and integrate into normal leg blastemas. This result can be taken to indicate that the surface and cell recognition properties have been altered by the mutation.

Somatic recombination has been used to estimate the number of cells at the cellular blastoderm stage that make up the various imaginal discs at their point of determination. For the haltere disc this number is about 10, whereas for the wing disc it is about 20. Additionally, it is possible to follow the proliferation dynamics of these two discs by inducing clones at later points in development. Using this technique, Garcia-Bellido and his associates have been able to show that *bx pbx* double mutants alter both the growth dynamics and the number of cells recruited into the haltere disc at the cellular blastoderm stage. Indeed, both of these properties, like the morphology of the disc, are altered to resemble a wing, and therefore, a mesothoracic derivative.

The result that the program of development of both the larva and the imaginal discs is altered at a very early point (apparently at cellular blastoderm for the anlagen) can be taken to indicate that the homoeotic genes are important in the actual determination of cell fate. This contention is supported by the fact that bithorax transformations can be phenocopied by heat treatment of embryos at cellular blastoderm. Two-hour-old embryos, shocked by high temperature, produce nonheritable developmental alterations that mimic *bx* mutants. Garcia-Bellido has shown that this same effect can be produced by treatment of this same embryonic stage with ether. The efficacy of the treatment is significantly increased by altering the dose of BX-C genes with heterozygous deletions of *bx* and *pbx*, increasing the frequency of phenocopies. These observations strongly suggest that the early determinative events are the target of the homoeotic loci. However, it is also the case that these loci are necessary for the maintenance of that state. Temperature-sensitive mutations at the *bx* and *pb* loci have temperature-sensitive periods in the third larval instar, four to five days after the cells of the discs have been determined. Therefore, the state of determination of these cells can be altered after its initial setting but before differentiation takes place during metamorphosis.

The results of genetic and developmental studies on the Antennapedia and bithorax homoeotic gene complexes have demonstrated that these loci do act as switches that control segmental fate. This control is effected at the time of determination and apparently results from the positive action of the loci in the complexes on batteries of other genes. It should be noted that the alternate pattern of development that occurs in the absence of these genes is not chaotic but is normal to some other portion of the animal and these genes can indeed be viewed as switches that select among a series of alternate states. The change in response caused by the homoeotic mutation does not result, as Goldschmidt may have wanted, in the production of hopeful monsters that bear the potential for macroevolutionary events. The changes are, in actuality, atavistic and may provide evidence of the history of the gentic controls acquired in evolution. This is perhaps most clearly demonstrated through a consideration of the evolutionary history of the insects.

Arthropod Phylogeny, or Homoeology Recapitulates Phylogeny

The phylogenetic relationships of the major arthropod groups are given in Figure 8–10. This phylogeny represents an interpretation derived from paleontology, embryology, comparative anatomy, and a variety of other investigations. The origin of arthropods from an annelidlike ancestor most probably occurred late in the Precambrian, and the first great radiation of arthropods began in the Cambrian. Representatives of all of the arthropod groups except insects are present in the early Paleozoic fossil record. The insects appear in the later Paleozoic, or about 350×10^6 years ago. While there is some unanimity of opinion about the origin of the arthropods from annelids, or an annelidlike group, there is less of a consensus as to whether the major arthropod groups arose in a mono- or polyphyletic set of events. We do not propose to enter into this argument but to concentrate on the portion of the phylogeny represented by the uniramous arthropods, the line of ascent that ultimately gave rise to the insects. The changes in segmental pattern in this group of arthropods can be analyzed in the light of the homoeotic mutations of *Drosophila*.

The features that suggest that annelids are ancestral to the arthropods are their basic metameric organization, ventral nervous system, and dorsal heart, which are characteristics shared by the arthropods. Despite these common adult characteristics, early development of the two groups is quite different. The annelids, like mollusks, have spiral cleavage, whereas arthropods in general have the intralecithal type of cleavage we have previously described for *Drosophila*. Therefore, one of the major events in the origin of the arthropods was a dramatic alteration in early development while maintaining a basic adult body plan. Figure 8–11 shows the later developmental stages of the marine polycheate worm *Polydora ciliata*. Subsequent to cleavage and gastrulation, a free-swimming trochophore larva is formed. This larval form is

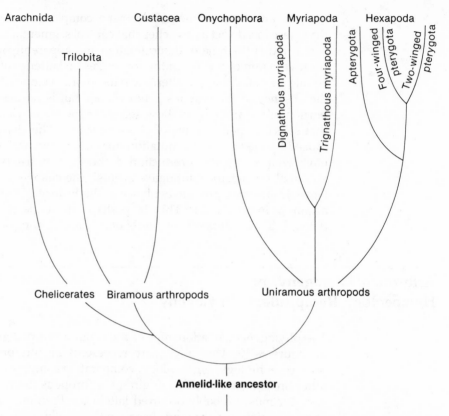

FIGURE 8–10. *The phylogenetic relationships of the major arthropod groups, with an emphasis on the line that gave rise to the modern insects (Hexapoda).*

divided by ciliary bands into three discrete regions. The most anterior region is the prostomial area, separated from a medial trunk area by the prototroch. The posterior pygidium is, in turn, separated from the trunk by the telotroch. Larval development proceeds by the delineation of the trunk area into three discrete trunk segments. At this stage it is possible to discern a total of six segments: the prostomium, the mouth segment, three trunk segments, and the caudal pygidium. The larva continues to grow by the subterminal addition of more segments in a growth zone that resides between the last trunk segment and the pygidium. This posterior growth is accompanied by a continued elaboration of structures particular to each species on the more posterior segments. After

FIGURE 8–11. *(Opposite) Drawings of the segmental pattern of a marine polychaete of the genus* Polydora *(a–c). (a) Early larval stage; (b) later larval stage after addition of several segments at the posterior growth zone; (c) early adult. Below are drawings of the embryo (d and e) and adult (f) of representatives of the onychophora. (d) Ventral view of the early germ-band stage of* Peripatoides novaezealandiae. *Like the annelids, segments are added in a posterior growth zone. (e) Ventral view of the embryonic head of* P. capensis. *The head is composed of three segments: the antennal, jaw and mouth, and slime palp-bearing elements. (f) A generalized adult onchyphoran. Each of the similar trunk segments bears a single, jointed walking leg with a claw distally.* [Parts a–c are redrawn from D. P. Wilson, 1928. Part d redrawn from Sheldon, 1889. Part e drawn after A. Sedgwick, 1888. Part f drawn from preserved specimens.]

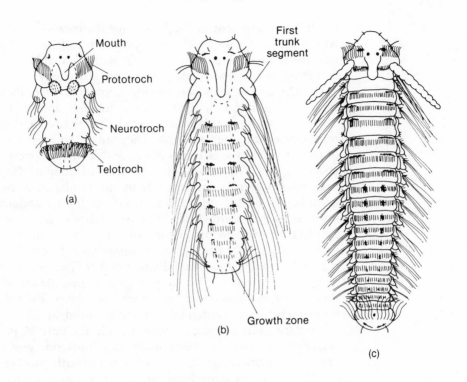

Mouth
Prototroch
Neurotroch
Telotroch
(a)
First
trunk
segment
Growth zone
(b)
(c)
Antennal somite
Ventral
extraembryonic
ectoderm
Growth
zone
Dorsal
extraembryonic
ectoderm
(d)
Antenna
Labrum
Mouth
Jaw
Slime papilla
Embryonic
ectoderm
Ventral
extraembryonic
ectoderm
(e)
(f)

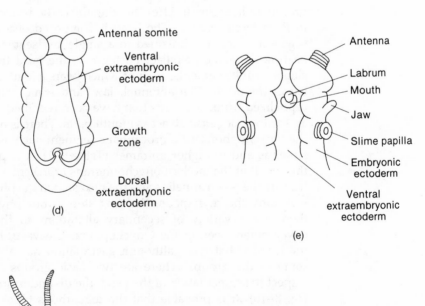

the full development of the larva, metamorphosis produces the adult worm. This latter process need not concern us here, as it is sufficient to note the segmental nature of the animal and how this segmental pattern is altered in the "next" stage in insect phylogeny.

The Onychophora, when initially discovered, were thought to represent the "missing link" between the annelids and higher arthropods. On first appearance this would seem to be the case, and they indeed do represent an intermediate form. They are not, however, on a direct line of ascent. Cleavage is intralecithal, and in those species having large yolky eggs the blastoderm forms by cellularization on the surface of the yolk mass. Gastrulation proceeds by the formation of two ectodermal germ bands on the midventral surface of the blastoderm and by the invasion of mesodermal cells into the interior of the blastoderm. Major gastrulation movements occur only for the anterior midgut. These movements appear first at the presumptive anterior of the animal and proceed caudally. After the initiation of these movements the process of segmentation begins. This early stage of segmentation, as it appears in *Peripatoides novaezealandiae*, is shown in Figure 8–11(d). Anteriorly, there is a prominent antennal segment followed, at this stage, by a series of similar somite-bearing segments. As in the polycheate larval forms, segments are added subterminally in a posterior growth zone. The entirety of segmentation is completed before birth (most onychophorans are viviparous or ovoviviparous). A later stage of embryogenesis is presented in Figure 8–11(e) showing the early development of the head in *Peripatopsis capensis*. The head is composed of an anterior antennal segment, a mouth contained in a single jaw segment, and a segment bearing the slime palpillae. Posterior to the head there are a series of similar trunk segments, each of which bears a jointed, clawed, walking leg in the adult. The antennal, jaw, and slime palpilla segments are cephalized trunk segments that have been recruited, in an evolutionary and ontogenic sense, to act as mouth parts. There is no good evidence in the Onychophora for a procephalic segment such as that found in the annelids and in higher unirames (Table 8–1). It is partially because of this fact that the modern onychophorans represent a relict rather than a true link between annelids and the next step in the phylogeny of insects.

Among the myriapods, the next step in our phylogenetic journey, there are a variety of secondary alterations to the general plan of development seen in the Onychophora. Cleavage, however, is still of the intralecithal type, although gastrulation has altered somewhat in some of the groups. There are two basic groups of myriapods with respect to segmentation of the head: the di- and trignathous (Table 8–1, Fig. 8–10). It is probable that the dignathous types are actually more recently evolved, despite their less complex head structure. Subsequent to gastrulation the germ band is invested by a series of furrows, creating a group of segments that in the trignathous types includes the cephalic, antennal, premandibular, mandibular, maxillary, labial, and three to six trunk segments. This stage is shown in Figure 8–12(e) and (f) for *Hanseniella*. As in the Onychophora, the full complement of adult segments is added during embryogenesis in a growth zone between the

TABLE 8–1. Segmental Identities in Annelids and Uniramous Arthropods

| Segment No. | Annelids | Onychophora | Myriapods | | Hexapods | |
			Dignathous	Trignathous	Apterygotes	Pterygotes
1	Prostomium	Antennal	Procephalic	Procephalic	Procephalic	Procephalic
2	Mouth	Mouth	Antennal	Antennal	Antennal	Antennal
3	Metatrochal	Jaw	Premandibular	Premandibular	Premandibular	—
4	Trunk	Slime palp	Mandibular	Mandibular	Mandibular	Mandibular
5	"	Trunk	Maxillary	Maxillary	Maxillary	Maxillary
6	"	"	Collum	Labial	Labial	Labial
7	"	"	Trunk	Trunk	Thoracic$_1$	Thoracic$_1$
8	"	"	"	"	Thoracic$_2$	Thoracic$_2$
9	"	"	"	"	Thoracic$_3$	Thoracic$_3$
10	"	"	"	"	Abdominal$_1$	Abdominal$_1$
11	"	"	"	"	"	"
12	"	"	"	"	"	"
13	"	"	"	"	"	"
14	"	"	"	"	"	"
15	"	"	"	"	"	"
.						
.						
n	"	"	"	"	"	"

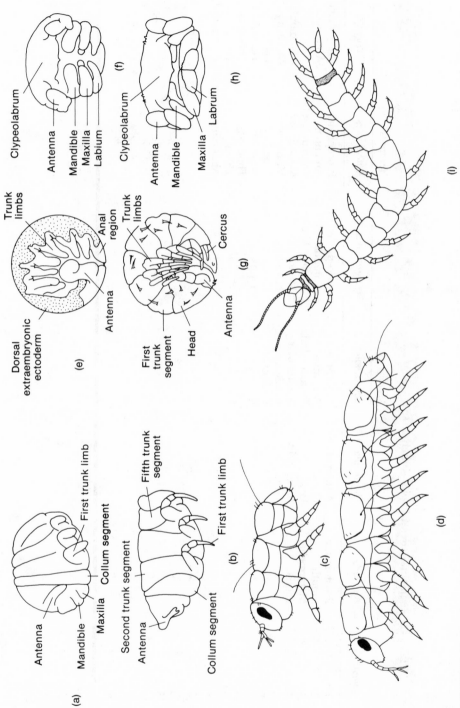

FIGURE 8-12. *Drawings of the embryonic, larval, and adult stages (a–d) of a dignathous myriapod Pauropus, and the embryonic stages (e–h) of Hanseniella and the adult (i) of Scutigerella, both trignathous myriapods. (a) 10-day-old embryo showing early basic body segmentation. (b) Early pupoid stage. (c) First instar, shortly after hatching. (d) Adult after addition of trunk segments at posterior growth zone. (e) Entire germ-band lateral view of 8-day-old embryo. (f) Ventral view of head and mouth part segments of embryo, as in (e). (g) Embryo just prior to hatching, lateral view. (h) Ventral view of head and gnathal elements in embryo at same age as (g). (i) Adult dorsal view. [Parts a, b, e–h redrawn from D. T. Anderson, 1973; Parts c, d, and i adapted from R. E. Snodgrass: A Textbook of Arthropod Anatomy. Copyright © 1952 by Cornell University. Used by permission of the publisher, Cornell University Press.]*

penultimate posterior segment and the anal region. Further development results in the elaboration of the walking legs on each of the trunk segments and the formation of the mouth parts from the mandibular, maxillary, and labial segments (Fig. 8–12g and h). The premandibular segment bears only a transitory appendage and does not contribute markedly to the adult head. The more anterior antennal and cephalic segments bear the antennae and clypeolabrum, respectively. Thus, relative to the Onychophora, there are six rather than three segments composing the head. But like the Onychophora, these specialized segments are derived by recruitment from post-oral trunk regions of the early embryo. The dignathous myriapods show a singular difference in the pattern of segmentation. The segment that corresponds to the labial in the trignathous myriapods does not contribute to the adult mouth parts. Instead, it forms the specialized collum segment that forms a kind of neck between the head and the first trunk segment (Figure 8–12a and b). Other than this, the pattern of segmentation is similar. In the particular example presented in Figure 8–12 (a-d), *Pauropus silvaticus,* there is another striking and significant difference. This class of myriapod (the Pauropoda) develops anamorphically rather than epimorphically, like the chilopod example presented previously; that is, the full complement of segments is not present at hatching. The newly hatched individual has only three trunk segments, as shown in Figure 8–12(c). The remaining segments found in the adult (Fig. 8–12d) are added by subterminal addition in a growth zone at the posterior end of the animal. This is not a characteristic of dignathy *per se* because the Symphyla, a trignathous group, also exhibits anagenesis. The existence of a larval form possessing only three trunk segments has been taken as partial evidence that the insects arose from a myriapodlike ancestor by a process of paedogenesis. That this is possible can be seen in an examination of the apterygotes, the primitive wingless insects.

Among the apterygotes there are groups that show either myriapodlike or more pterygotelike patterns of early development. Anderson has interpreted these variations as functional adaptations within each of the groups and not as significant differences belying a link between myriapods and insects. The most striking similarity among myriapods, apterygotes, and pterygotes and among the apterygotes themselves is in the manner by which segmentation of the germ band takes place. Like the trignathous myriapods there are (early in the segmentation process) six head segments followed by three trunk segments. During subsequent development, more segments are added posteriorly in a subterminal growth zone (Figure 8–13d and e). This addition occurs anterior to the terminal caudal segment. The gnathocephalic segments are recruited from embryonic trunk elements and their appendages modified to mouth parts. The major departure in the apterygota from the myriapoda is that the trunk is divided into a three-segment thorax and an eight-segment abdomen. Only the thorax carries the walking legs, whereas the abdomen has greatly reduced appendages (Figure 8–13f). A comparison of Figures 8–12c and d and 8–13f will demonstrate the striking similarity of body patterns seen in the anamorphic myria-

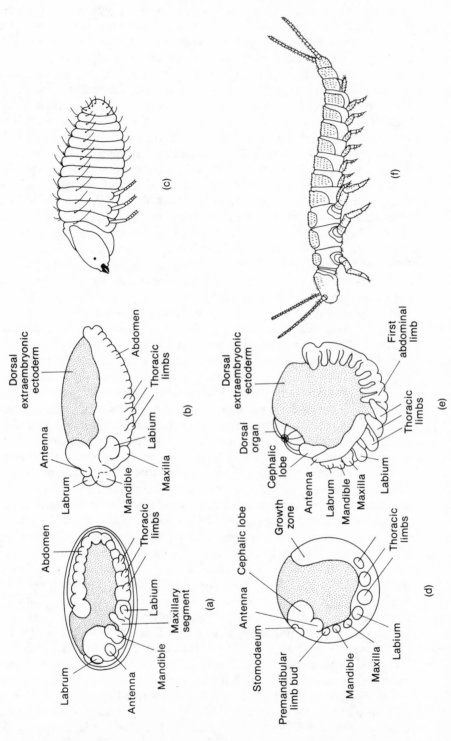

FIGURE 8–13. *Drawings of the lateral aspects of the segmented embryos and hatched forms of Campodea (d–f), an apterygote insect, and Bruchidius (a–c), a pterygote form. (a) Fully segmented germ band; (b) shortening of the germ band; (c) hatching stage; (d) segmenting germ band—segmentation occurs progressively at the posterior of the band; (e) fully formed and segmented germ band; (f) adult. Compare this body form to that of the myriapod in Figure 8–12c and d. It is the striking similarity between these forms that has resulted in part of the hypothesis that the primitive insects arose from a myriapodlike ancestor through paedogenesis. [Parts a–c redrawn from D. T. Anderson, 1973. Part f adapted from R. E. Snodgrass: A Textbook of Arthropod Anatomy. Copyright © 1952 by Cornell University. Used by permission of the publisher, Cornell University Press.]*

258

pods and the primitive apterygote insects, the essential difference residing in the suppression of limbs on the segments posterior to the third trunk segment in the apterygotes.

The final stage in the phylogeny is the pterygote or winged insects. We have already seen the aspects of early development of this group, as exemplified by *Drosophila melanogaster* (Fig. 8–6). The early embryonic events described for the particular case are found throughout the pterygotes and need not be repeated. As an example of a more primitive form, however, we have included in Figure 8–13 the pattern of segmentation of the germ band of *Bruchidius*, a beetle. Like *Drosophila*, there is no apparent posterior growth zone and segmentation occurs as a seemingly single event. There are three gnathocephalic and three thoracic segments forming the mouth parts and thorax, respectively. Unlike the apterygotes, the three thoracic segments are differentiated such that the second and third segments bear the wings. This latter morphology has been further altered in the Diptera, where the metathoracic wings are reduced to halteres and the dorsal prothorax is greatly diminished in size, forming in *Drosophila* only a narrow band of cuticle in the adult (Fig. 8–7). In all pterygotes, the reduced abdominal limbs of the apterygotes are completely absent in the adult. Moreover, the premandibular head segment is difficult to discern or totally absent. This reduction of head segments is carried one step further in *Drosophila*, where the mandibular segment contributes to the formation of the larval mouth parts but not to the adult. A summary of the segmental changes seen in the unirames is given in Figure 8–14. As we follow the phylogeny from annelid to pterygote we can observe an elaboration of segmentally arrayed structures and subsequent loss of some, but not all, of these segmental elements. A comparison of Figures 8–14 and 8–9 presents a rather interesting parallel. A deletion of the bx^+ and pbx^+ functions transforms a dipteran *Drosophila* into a more primitive four-winged condition. The $Antp^+$ function is necessary to the development of the meso- and metathoracic segments. In the absence of this gene, the embryonic thorax develops as three similar prothoracic or nonwing-bearing segments, thus partially mimicking the apterygote condition. The posterior end of the embryo develops as a series of similar thoracic or trunk segments if the entire BX-C is missing. This pattern of segmental identities is reminiscent of the trignathous myriapods. Finally, if both the ANT-C and BX-C are removed, an embryo with three head segments and a series of similar trunk segments is produced. This pattern resembles the onychophoran condition. Thus, by sequential deletion of a relatively small amount of genetic material we have managed to traverse a rather large amount of evolutionary ground.

We do not intend that these analogies be taken too literally and it must be stressed that the atavisms presented are not actual. What has changed by deleting these genes is the pattern of segment identity. The segments produced are still undeniably drosophilid in character. This indicates that what these loci represent are genes involved in the specification of pattern— the control of ontogeny *per se*. They act as switches controlling the fate of cells by regulating the genes expressed in

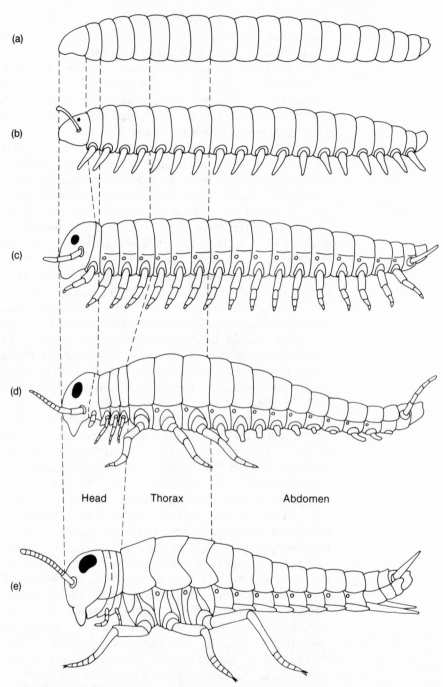

(a)

(b)

(c)

(d)

Head   Thorax   Abdomen

(e)

FIGURE 8–14. *Diagrammatic representation of the hypothesized evolution of segmental pattern in the phylogeny of the Insecta. (a) annelid; (b) onychophoran; (c) myriapod; (d) apterygote insect; (e) pterygote insect. A comparison of this figure with Figure 8–9 demonstrates that the progressive deletion of the homeotic loci results in a progressive simplification of the segmental pattern of* Drosophila, *which to a certain extent mimics the phylogeny of the insects.* [Redrawn from Snodgrass, 1935.]

any particular segment. The structural genes under their control must have evolved concomitantly with, but at the same time in some sense, separately from the controlling homoeotic loci.

Based on the phylogeny of the insects presented and the fine structures of the two clusters of loci, we view the evolution of the homoeotics as a stepwise duplication and subsequent divergence of function. The easiest place in the phylogeny where one may envision this process is in the myriapod-to-apterygote-to-pterygote transitions. The *iab* genes serve to differentiate the thorax and abdomen. In the absence of these genes, abdominal segments become thoracic in character. Therefore, an *iab* function arose that could suppress the formation of limbs on the posterior segments. E. B. Lewis has shown that there are several of these loci in *Drosophila* perhaps one for each of the eight abdominal segments. However, in the apterygotes these segments are added sequentially during embryogenesis, possibly requiring only a single gene that is sequentially activated as new abdominal segments are added. With the advent of the pterygote mode of simultaneous formation of all segments a serial duplication of this *iab* gene may have occurred as a necessary element to this alteration in development.

The members of the ANT-C can be envisioned to have arisen at an earlier point in insect evolution. The transition from annelidlike to onychophoran to myriapod involved the recruitment and elaboration of post-oral trunk segments into the gnathocephalic segments that give rise to the mouth parts. This process is at least partially accomplished by the *proboscipedia* locus. Moreover, the Scr^+ and $Antp^+$ functions (Fig. 8–8a) are necessary switches for determining the identity of the first and second thoracic segments. It is conceivable that these functions arose, duplicated, and diverged in much the same manner as the BX-C loci. Indeed, it may be the case that the two complexes are themselves related by duplication, elements of the ANT-C giving rise by duplication to a precursor of the BX-C.

We have to this point presented the kinds of transformations produced by the homoeotic genes and when and where they function, but not how they perform as switches. This last question actually contains two elements. How are the homoeotic genes themselves regulated, and how do they in turn regulate the pattern of gene expression within each segment? Before attempting an answer to these two questions, first it is necessary to discuss the manner in which pattern formation is thought to occur.

Nine

Pattern Formation

From the moment of my birth to the instant of my death
There are patterns I must follow just as I must breathe each breath

Paul Simon, "Patterns"

Pattern and the Origin of Form

The basic problem of the developmental biologist is to account, in a mechanistic sense, for the manner in which the single-celled zygote is elaborated into the morphologically more complex multicellular adult. From the geneticist's point of view, this entails an explanation of how the two-dimensional information encoded in DNA is translated into the three-dimensional organism. The developmental program actually consists of two classes of related phenomena, cytodifferentiation and pattern formation. That these two aspects of ontogeny are separable is demonstrated by a rather simple example. If a biochemical analysis were performed on the hand and arm of a human and the component parts cataloged, the muscles, tendons, bones, and so on of the right hand would be identical to the left. Yet a simple inspection of these two organs reveals that they are not identical. It is not possible to substitute one for the other. This same comparison is, of course, even more striking if made between the hand and the foot. It is the manner in which these differences arise, the development of pattern and form, that concern us in this chapter. It is ultimately within the framework of pattern formation that the evolution of both morphology and cytodifferentiation must be understood.

One way of visualizing the ontogenetic process was presented by C.H. Waddington in the 1940s. A cell proceeds through development by traversing what he referred to as the epigenetic landscape. This construct is presented in Figure 9–1 as a plane invested with a series of valleys. These valleys begin at the high end of the plane and proceed downhill, diverging to produce a group of unique endpoints at the low end of the plane. The cell moves from the high to the low end of the plane through the valley system. Each branch point represents a developmental decision made by the cell and a restriction in the potential of the cell. At the top of the landscape, the cell can theoretically reach any of the unique endpoints. However, once an initial decision is

262

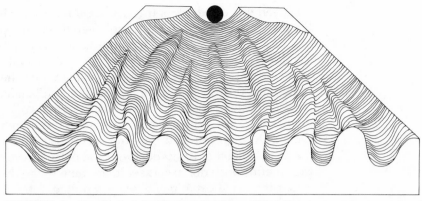

FIGURE 9–1. The epigenetic landscape of Waddington. The ball at the top represents a cell and the valleys below the various developmental paths the cell may take. [Redrawn from C. H. Waddington, *Principles of Development and Differentiation,* Macmillan Publishing Co., Inc., New York, 1966. Copyright © 1966 by C. H. Waddington.]

made at, for example, the first branch point, only a subset of the valley endings is possible. The cell begins in a totipotent state and becomes more and more restricted by determinative events. This process was referred to by Waddington as canalization. The general properties of this process can be applied to both cytodifferentiation and pattern formation. The decision points, the valley forks, can be influenced by extrinsic forces, such as hormonal stimulae or induction, and the decision will be dependent on the genetic response of the cell to that stimulus. As we shall see, the decisions one actually observes cells making in ontogeny have an either/or character and can be viewed as a series of right, left, or binary decisions. The downhill grade of the landscape represents time of development. As such, the model can be applied to either mosaic or regulative development by simply moving the valley branch points toward either the high or low end of the hill for each respective type of development. Although the model as presented does give us a way of viewing development, it does not explain the process. Moreover, the model is static in that it does not formally incorporate the process of cell division. Nevertheless, the idea of a series of canalizing events is an important one and is reflected in the development of the vertebrate limb as it relates to the determination of the axes that differentiate right from left.

The Determination of Limb Axes

Given that the left and right arms are different, how do they get that way? A description of this process was developed from the embryological studies of R. G. Harrison on the limbs of the axolotl. The limbs of *Ambystoma* first appear as lateral thickenings of the mesoderm at the sites on the flank where the legs will eventually develop. The development of the leg is dependent on this mesoderm because transplanting it to other sites on the flank will cause legs to develop in odd locations.

The overlying ectoderm of the limb regions does not have this property. If a presumptive limb region is explanted *in toto* and placed on the flank of a host in the same polarity on the same side as it was removed from the donor, the explant will produce a complete limb of polarity similar to the host. If, however, the explant is placed on the contralateral side of the host relative to the donor, the donor appendage has the reverse anterior-posterior polarity of the host's normal appendage. The dorsal-ventral polarity is the same. If the same donor limb region is instead rotated through 180° and then transplanted to an isolateral position in the host, again the correct dorso-ventral polarity results. Therefore, only the anterior-posterior axis is set (determined); the dorsal-ventral is not. When the same experiments were repeated with slightly older donor limb regions, Harrison found that the dorsal ventral axis had also become determined. Similar experiments involving transplants in which the proximal-distal sides of the leg mesoderm were inverted showed that this axis is the last to be determined. Therefore, the determination of the axes of the limb are set in a manner quite like the canalization events described in the epigenetic landscape. There are a series of decisions, anterior versus posterior, dorsal versus ventral, and proximal versus distal, that occur as a temporal series of restrictions of developmental potential. What remains to be seen, however, is how these restrictions take place. What is the nature of the instructive event at the decision point and what is the nature of the cellular response? Although it is not possible to answer these questions precisely, it is possible to gain some insight by a consideration of two related concepts: embryonic fields and positional information.

Pattern Formation and the Specification of Position

The leg region that was transplanted by Harrison can be regarded as an embryonic field. More generally, a field is a region of an embryo or a group of cells within which regulation can occur. In the early limb field of the amphibian, this is demonstrated by the fact that a portion of the presumptive limb can be removed and the remaining cells will still generate a leg with all of the normal parts; that is, the cells that remain, if the surgery is performed early enough, can respond to the deletion and replace the elements that have been removed. T. H. Morgan, before his founding of the American school of genetics, realized that this process could occur in two fundamental ways: the first, in which the missing structures are replaced subsequent to cell divisions that replace the extirpated material; the second, in which regeneration of the field occurs in the absence of cell division. The first he called epimorphosis, which is exemplified by limb regeneration in the amphibia. The latter he termed morphollaxis, the best example of which is the entire early embryo. What both types of regulation have in common is that the cells within the field have the ability to read and register their positions within the field.

To account for the regulative behavior of cells within fields, Lewis Wolpert and his associates have formulated the idea of positional information. This information is supplied to the cell as a cue to its location in relation to other cells and it is this information that regulates the pattern within which the embryo or field develops. Most commonly, the positional information is viewed as a gradient of some substance (sometimes termed a morphogen) which is distributed in the field from a high to a low concentration.

Many of the concepts of positional information have come from developmental analyses performed first by Saunders and his associates and subsequently by Wolpert and his colleagues on the chick wing field. The wing bud of the chick develops initially in a similar fashion to that of the limb bud of amphibians. As the bud grows out from the main axis of the body to form a paddlelike appendage, the skeleton of the limb is formed as cartilage that is derived from the mesenchyme in the interior of the limb. The first elements to appear are the proximal ones, and subsequently, the distal digits. The final form of the elements is given in Figure 9–2. A fate map of these elements was developed by marking experiments using carbon particles. These showed that, as the limb bud grew in the proximal-distal direction, more and more distal structures were added. Early in bud growth the humerus was present, later on the radius and ulna, then the carpals, and finally the digits. Experiments performed by Wolpert and his co-workers demonstrated that as this proximal-distal growth proceeded there was little regulation. If they removed the overlying ectoderm (which does not induce limb morphogenesis) plus some of the underlying mesoderm at the distal tip of the bud, no further limb growth occurred. If the tissue was removed early, only the humerus was formed. Slightly later humerus, radius, and ulna were formed, but no carpals or digits. This temporal sequence coincided nicely with the fate map derived from the marking experiment. The temporal map of the chick wing is given below the normal wing at the top of Figure 9–2. It should be noted that the removal of tissue from the developing bud occurred long before any discernible morphogenesis or gross cytodifferentiation had taken place. Therefore, the pattern of the limb was being affected before any overt signs of differentiation had taken place. To account for this result, Wolpert proposed that the cells in the mesoderm underlying the terminal apical ectoderm of the limb bud form what he termed a progress zone. The mesenchymal cells in this region are undetermined, but through division and distal growth, leave behind on their proximal boundary cells that are determined to form specific wing structures; however, which structures are formed depends on the point in time the daughter cells leave the progress zone. The later a cell's departure from the zone, the more distal the structure formed. Cells therefore have their fates determined as a function of a time coordinate rather than strictly as a function of position.

This type of regulation is in the class of epimorphosis because it requires cell division. In order to test this hypothesis, reciprocal grafts of the putative progress zone were made between "young" and "old" limb

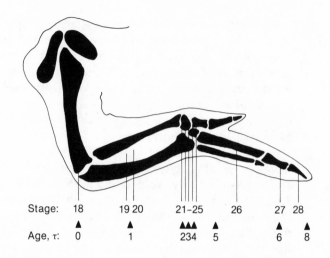

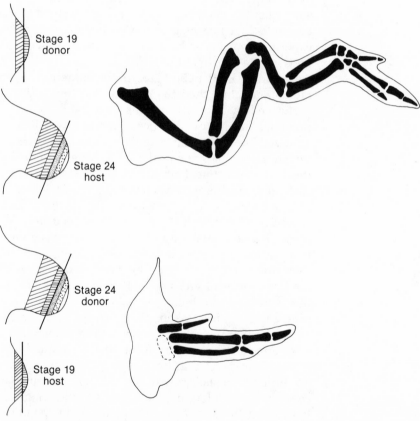

FIGURE 9–2. *Drawing of the structure of the bony elements of the chick wing and the results of transplantation experiments between wing buds of different ages. The embryonic stages given below the limb delineate the time at which the varius limb elements are specified during development. 18 = humerus; 19 and 20 = radius and ulna; 21–25 = the wrist region; and 26–28 = the phalanges. The four marked regions of the limb buds shown below indicate the fate-mapped positions of the more proximal elements of the limb. The limbs that develop from the indicated host are shown at the right. The older host (stage 24) capped with young bud (stage 19) produces a tandemly duplicated wing, whereas the reciprocal graft yields a deficient limb.* [Adapted from Wolpert, Lewis, and Summerbel, 1975.]

buds. Representative results from those experiments are diagrammed in Figure 9–2. If at the time when only the humerus and proximal portions of the radius and ulna have been determined, a wing bud is used as a donor and an older bud, which is determined to the level of carpals, is used as recipient, a limb with tandemly duplicated humerus, radius, and ulna is formed. The reciprocal graft results in a deletion of the previously duplicated elements. The resulting limb has carpals projecting directly from the lateral flank. This result is entirely consistent with the progress zone hypothesis and presents the possibility that position may be specified in time as well as space. A further point about the specification process is that positional information can cue more than just cellular identity in terms of pure morphology. This conclusion is reached through a consideration of the wrist region of the wing. Inspection of Figure 9–2 will show that the rather small wrist appears to take a rather long period of time for specification relative to the larger bones in the upper arm. Studies designed to find the relative sizes of the primordia for each segment of the arm demonstrated that initially they are similar in size; that is, the migration of the progress zone is uniform through time. What is different is the proliferation dynamics of the various regions after specification. The wrist grows rather less than do the other regions of the wing. Therefore, one of the apparent aspects of specification is a determination of relative growth as well as different morphology.

Specification of Anterior-Posterior Polarity

The specification of anterior-posterior polarity of the developing wing can be envisioned to be set by positional information in a manner similar to the proximal-distal axis. However, although there is an interaction with the progress zone in this specification, the elaboration of position is more in the form of a gradient from a high point in the posterior of the bud to a low point at the anterior. As such, this axis is thought to be determined by a system similar to that present in morphallactic regulation. The origin of the positional information specifying the anterior-posterior axis is thought to reside in a group of mesodermal cells at the posterior edge of the limb bud adjacent to the posterior necrotic zone mentioned in Chapter 7. This group of cells was discovered by Saunders and his collaborators and called the zone of polarizing activity, or ZPA. If the ZPA is transplanted to the anterior edge of a developing limb bud such that there are two ZPAs, the resulting wing will possess a mirror-image duplication of distal structures that will be specified by the progress zone at points after the ZPA transplant. Elements already specified at the time of transplant are not duplicated. This result is presented diagrammatically in Figure 9–3. Also presented in Figure 9–3 is the result of a transplant of the ZPA to the middle of the limb bud. This also produces a duplication, but not a mirror image. These results have been interpreted as demonstrating the

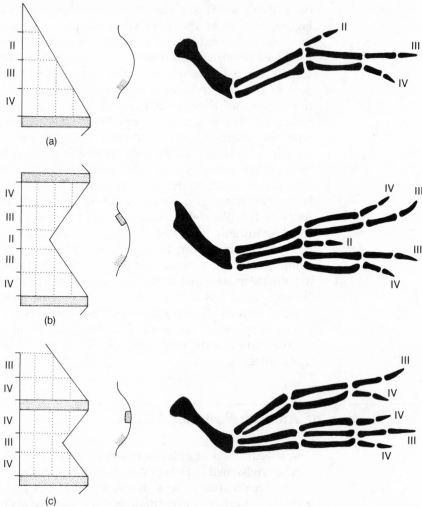

FIGURE 9-3. *The determination of the anterior-posterior axis of the chick limb by the zone of polarizing activity (ZPA). The normal limb is shown in (a) along with a hypothetical gradient of positional information that originates in the ZPA (stippled area) at the posterior edge of the limb bud; (b) and (c) show the results of transplantation of the ZPA to the anterior and medial portions, respectively, of an otherwise normal wing bud. The resulting limbs are shown at the right and the hypothesized alterations in the gradient are shown at the left.* [Redrawn from Ede, 1978.]

ZPA as a source of positional information. The anterior/posterior array of, for example, the digits in the wing is specified by a cell's reading of the concentration of that information. High concentration specifies digit 4; low, digit 2. As diagrammed in Figure 9–3 the transplant to the anterior edge of the limb bud produces a gradient with two high points and a valley in the middle; thus, the mirror-image duplication. The transplant to the center of the bud produces a double gradient in tandem, thus the resulting roughly tandem duplication of digits. The small valley between the ZPAs produces a mirror duplication of digit 4 and no digit 2 because no value low enough in the gradient exists to specify that structure.

The number of structures duplicated is dependent on the age of the limb bud at the time of transplant, with only those elements that have not emerged from the progress zone being duplicated. Therefore, the anterior/posterior axis is actually specified by two interacting parameters of positional information, distance from the ZPA and time in the progress zone. This dual nature of positional cues is not unique to this particular system but has been utilized in an attempt to explain the patterns of epimorphic regulation seen in the regeneration of cockroach and newt limbs and the imaginal discs of *Drosophila*. A theory put forward by French, Bryant, and Bryant has the interesting property in that it can account in large part for the phenomena associated with all three systems.

Pattern Formation and Polar Coordinates

Hadorn and his students have discovered that if an imaginal disc is removed from a third instar larva of *Drosophila* and transplanted by injection to a new larval host, it will differentiate with the host. Moreover, the disc can be cut into pieces and each fragment will differentiate autonomously into a portion of those structures that would normally be produced by that disc, each specific fragment producing a specific set of structures. By this technique it is possible to produce a fate map of each disc and project the determined states of the various cells onto the surface of the undifferentiated disc. A portion of this type of map is presented for the wing disc in Figure 9–4. These experiments also demonstrate that the imaginal discs do not show morphallactic regulation. They do, however, show epimorphosis. If the fragments are cultured in the abdomens of adult female *Drosophila* instead of larval hosts, they will grow by cell division but will not differentiate. Disc fragments cultured like this for seven days and then transplanted to a metamorphosing larval host do show regeneration in very consistent patterns. When the disc is cut into two fragments, one smaller than the other, the larger fragment will regenerate the structures that would normally be formed by the cells in the smaller piece. The smaller fragment, on the other hand, forms a mirror-image duplicate of itself. The results of two such experiments are presented diagrammatically in Figure 9–4.

Cells can be envisioned as being spatially determined on a set of radial and circular coordinates, each cell being specified by a unique pair of signals. If a cut is made, the cut surface juxtaposes normally nonadjacent positional values. Given the proper conditions (i.e., culture in the abdomen of an adult female), growth and cell division are induced by this juxtaposition. This growth will result if the positional confrontation occurs for either the circular or radial values. The cells that are produced by this process will recreate missing positional information in much the same manner as the progress zone model for the chick limb bud. To account for the result observed with the reciprocal fragments, French,

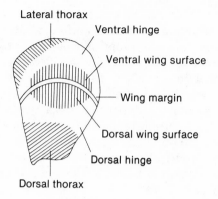

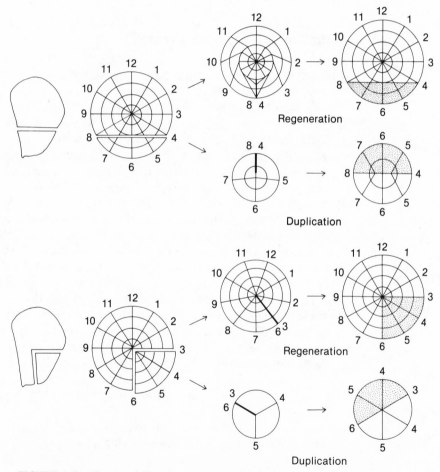

FIGURE 9–4. *The use of the polar coordinate model to account for the results of wing disc regeneration and duplication in D. melanogaster. At the top of the figure a rough fate map of the wing disc is presented. Below, at the left, two specific cuts of this disc are shown. In both cases the smaller fragment will duplicate, whereas the larger will regenerate the missing elements. This result is presented at the right of each cut disc in terms of the polar coordinate model. The rules of regeneration in this model can be found in the text.* [Adapted and redrawn from V. French, P. Bryant, and S. Bryant, Pattern Regulation in Epimorphic Fields, *Science* **193**:969–981, 1976. Copyright 1976 by the American Association for the Advancement of Science.]

Bryant, and Bryant have proposed the shortest intercalation rule. This is most easily seen in the example on the bottom of Figure 9–4. The small fragment intercalates the positional values 4 and 5 because that is the shortest route relative to the remaining circular values on the fragment. It is conceptually similar to a linear gradient model in which a fragment of gradient can regenerate only lower positional values. The large fragment will intercalate the same missing information, position 4 and 5 on the circle, but by doing so, will complete the circle and produce all the necessary positional information. The small fragment will duplicate and the large will regenerate; thus, the result. A second rule is also necessary. This is the complete circle rule, which states that in order to produce distal structures (the areas specified in the center of the circle) it is necessary to regenerate all of the circular positional values. It is to this last rule that some exceptions have been found. However, these do not take away from the basic utility of the model as a way of thinking about the manner in which the ontogenic process elicits pattern from apparent lack of form. Indeed, we must remember that these models represent abstractions of very complicated cellular and genetic events and are at best approximations of the reality of the process. It would be wise to keep in mind Waddington's advice that "...an embryological field is therefore essentially a concept appropriate to the realm of discourse which deals in multidimensional spaces. Any attempt to reduce it to three or even four dimensions plus one field variable must be recognized as a drastic abstract simplification, which may be justified for certain particular purposes but must always be regarded with great caution."

Bryant and his collaborators have found that the positional information resident in discs is apparently the same in all discs. The normally duplicating fragments of wing discs can be induced to undergo regeneration if they are cocultured with reciprocal fragments that have been killed by irradiation. The killed fragments do not contribute living cells to the resultant regenerate but apparently can contribute positional information. This information apparently can come from any disc, not only the same disc. More specifically, a proximal duplicating fragment of wing disc like that shown at the top of Figure 9–4 can be induced to regenerate distal wing structures by growing it in the presence of either killed distal wing or haltere disc fragments. This result is also consistent with the observation that the homoeotic mutations cause transformations that are serially homologous, that is, distal antenna transforms to distal leg. This indicates that positional information is the same in all imaginal anlagen; what differs is how it is interpreted. This apparent universality of positional specification has been shown to cross even species boundaries. Tickle and collaborators have shown that the ZPA taken from the posterior of the mouse limb bud is capable of specifying anterior/posterior axial information in the chick. Mouse ZPA can induce extra chick digits if grafted to the anterior border of the chick limb bud. A similar result was obtained by Fallon and Crosby using ZPAs taken from turtle embryos. Therefore, it would appear that the specification of position in the developing limb field of vertebrates has been conserved and evolutionary change has occurred in the cellular response to that information. The nature of that response is almost certainly a result of differential patterns of gene expression or, put more directly, in the throwing of genetic switches.

Binary Switches in the Interpretation of Position

The possible nature of the genetic response to positional information has been presented in a formal sense by S. Kauffman. The genes that react to position are viewed as a set of on/off switches that act in cascades and networks. This model is presented diagrammatically in Figure 9–5. The first gene in the pathway or cascade is derepressed by the input of positional information. This activation can be envisioned most easily as a simple response to concentration. If the relative concentration of morphogen is high enough, the switch is thrown. The product of this first gene in the network then causes subsequent binary switch events to occur in other genes, and the products of these genes in turn can activate other loci to form the cascade. It is not necessary for the initial input to the system to be a single entity, and indeed, multiple inputs can be incorporated into the model to produce redundancy, thereby affording homeostatic properties to the system. To put this in terms of Waddington's epigenetic landscape, we can view the progress of a cell through a series of valley forks, where at each branching a switch is thrown and a decision is made. Thus, at the output stage each cell has imprinted on it a hierarchical set of switch decisions that are a unique reflection of its ontogenetic history. That this is a reasonably accurate

FIGURE 9–5. *A diagrammatic representation of a hypothetical set of gene switches incorporated into a forcing structure. An external inducer or inducers (morphogenes) activates gene A, which in turn activates B, B in turn C, and C back to A, forming the forced loop. Once accomplished, the external inducer can be removed and the system remains "on." In this particular construct genes B and C also act externally to the loop to activate other loci.* [Adapted from S. Kauffman, 1972.]

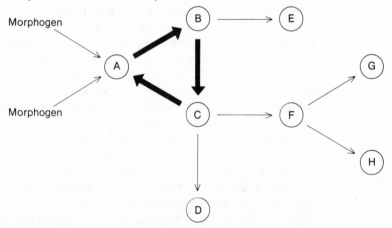

reflection of fact can be seen in the development of the imaginal wing disc of *Drosophila*.

Garcia-Bellido and his collaborators have shown that disc development proceeds as a series of compartmentalization events; that is, in a manner similar to the amphibian limb field, the axes of the imaginal disc are determined as a series of sequential events. Like the limb field, first the anterior-posterior axis is set, followed by the dorsal-ventral and a set of three sequential proximal-distal restrictions. Each of these compartments is separate from the others and cells in one compartment are not capable, in normal development, of transcending the compartmental boundary. Each cell in the adult wing and thorax can be seen to have a unique address that is related to the series of canalizing events through which it has passed during development. Kauffman views these events as reflective of a series of binary decisions. Thus, a cell at the distal front tip of the wing would be labeled as anterior/not posterior, dorsal/not ventral, wing/not thorax, distal wing/not proximal wing. Using a chemical wave model originally postulated by Turing, it is possible mathematically to show that an original wave form can be subdivided sequentially by a series of nodal lines, the position of which will be dependent on the shape of the original container of the wave and any growth dynamic that would alter the shape or size of the container. The shape of the wing disc is roughly an ellipse. If the growth properties of this disc are considered, a prediction of the position of the nodal lines can be made. As can be seen in Figure 9–6, the similarity between the

FIGURE 9–6. *Diagram of the actual and theoretical compartmental boundary lines of the imaginal wing disc of* D. melanogaster. *At the left the lines are superimposed on the fate map of the disc. The numbers associated with each line indicate the order in which the compartment boundaries are observed to restrict the potentials of cells in the disc (dotted line 4 has not been observed). On the right the lines are formed on an elipse using Kauffman's theoretical application of the Turing model. The similarity between the two is striking.* [Adapted and redrawn from S. Kauffman, R. Shymko, and K. Trabert, Control of sequential compartment formation in *Drosophila*, *Science* **199**:259–270, 1978. Copyright 1978 by the American Association for the Advancement of Science.]

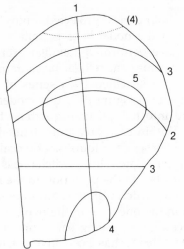 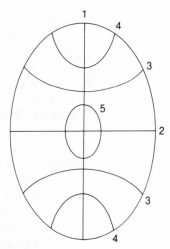

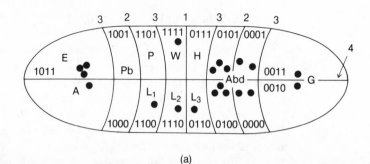

(a)

(b)

FIGURE 9–7. The series of compartmental boundary lines superimposed on the cellular blastoderm stage of the Drosophila *embryo. Within each defined compartment are the binary code assignments proposed by Kauffman that specify the identity of each segment. E = eye; A = antenna; Pb = proboscis; P = prothorax; L_1, L_2, L_3, = leg 1, 2, and 3; W = wing; H = haltere; Abd = abdomen; and G = genitalia. In (b), a scanning electron micrograph of a segmented embryo is shown for comparison to the model embryo in (a). The labeling on the micrograph indicates the proposed segmental origins of the adult structures indicated by the abbreviations in (a).* [Part a redrawn from S. Kauffman, R. Shymko, and K. Trabert, Control of sequential compartment formation in *Drosophila, Science* **199**:259–270, 1978. Copyright 1978 by the American Association for the Advancement of Science. Part b photograph courtesy of Dr. F. R. Turner.]

actual compartmental boundaries projected onto the disc and the position and sequence of the nodal lines is quite striking. Kauffman, Shymko, and Trabert have performed a similar exercise by projecting nodal lines on the early embryo at cellular blastoderm. In this case the regions of the embryo are specified by a series of sequential binary decisions. As diagrammed in Figure 9–7, this model predicts that the embryo is first divided into halves, then quarters, then eighths, and finally dorsoventrally. Within each area thus defined, the cells can be given a binary address by a series of four hypothetical switches that can exist in either a one or a zero state. Thus, switch one is in the one state in the anterior and the zero state in the posterior. In the next set of nodally defined zones, the second switch is in the zero state in the anterior and posterior quarters and in the one state in the two central quarters. This combinatorial coding continues, giving each zone a unique label.

An identical labeling scheme has been invoked for the manner in which the identity of each segmentally derived imaginal structure is specified. Thus, the various imaginal discs can be labeled by using a five-switch set of codes. It is, of course, inherent to this model that a hierarchy exists and that certain segments and discs should be more closely related in that hierarchy than others. This aspect of the model agrees very nicely with Hadorn's observations on transdeterminative events among the imaginal discs; that is, when disc fragments in culture change from one determined state to another, they do not do so randomly. Like the combinatorial code, they demonstrate a hierarchy. Moreover, there are some transdeterminative events which have never been observed, for example, wing disc changing to proboscis. Kauffman would account for this by saying there are two switches that differentiate the two structures and both must be thrown to cause the desired event, which is highly unlikely. In addition to accounting for the transdetermination relationships, the combinatorial code is also consistent with the kinds of homoeotic transformations observed in *Drosophila*. By utilizing the code words derived for the zones of the blastoderm in Figure 9–7 and superimposing the fate map positions of imaginal discs on this map, it is possible to assign a combinatorial code word to adult structures. Most observed homoeotic transformations among these structures can be accounted for by alterations in single switches. Thus, to change antenna (1010) to second leg (1110) requires a single $0 \to 1$ switch. This is in effect what the *Antennapedia* mutation does, and Kauffman proposes that some homoeotic loci may represent switch genes important in the interpretation of positional information.

There is one further property of these switches that is also put forward by Kauffman. Once thrown or activated, the gene must remain in that state. The gene must have memory. This memory is hypothesized to exist as a series of "forcing circuits." Once activated by some extrinsic signal (e.g., positional cues), the first gene in the circuit activates a second locus which in turn can activate a third. Both the second and third loci may function themselves as activators that derepress other loci. As diagrammed in Figure 9–5, one of the loci derepressed by the third gene in this simple scenario would be the first gene. Gene 3 would then serve to establish a feedback loop that would reinforce the original activation of gene 1. Moreover, genes 2 and 3 can themselves be envisioned to be maintained in an active state by other loci, thereby reinforcing the action of the loop even further. A simple concrete example of this type of system is offered by the lactose operon of *E. coli*. The enzyme β-galactosidase is inducible in *E. coli* by providing the cells with lactose. In order to be induced, however, some substrate must get into the cell. This occurs initially by the molecule simply leaking into the bacterium and inducing the synthesis of the catabolic enzyme and a permease. The permease subsequent to synthesis is incorporated into the plasma membrane of the cell where it facilitates the transport of lactose into the organism. The initial activation of the system is then dependent on the entry of the lactose molecule, which in turn is dependent on concentration. At suboptimal concentrations, the enzymes will not be induced and the cells will not utilize lactose as a carbon source. If a population of *E. coli* is grown in the presence of a concentration of lactose just above the inducing level, the cells in the population will gradually gain the ability to utilize the sugar. This occurs by some cells being induced while others are not. If, at a time when roughly half of the cells in the population are induced, the concentration of lactose is diluted to suboptimal levels, the previously induced cells remain in that state while the noninduced cells stay in that state. The cells in the population maintain this situation even through division; that is, the daughters "remember". The reason for this resides in the permease incorporated into the membranes of the induced cells. The permease facilitates transport of the lactose even though it is in low concentration. This in turn induces the formation of more permease, which reinforces the induced state. Therefore, in this population there are cells in two alternate states—one activated, one not. However, in order to make this situation similar to the one hypothesized by Kauffman, there should be an internal inducer of the lactose operon that would maintain activity even in the absence of the external inducer.

As with the theory of positional information, these ideas on gene switches and combinatorial coding offer a theoretical framework in which to contemplate the genetic regulation of development. Indeed, results obtained recently on the genetic control of the basic pattern of segmentation in the *Drosophila* embryo have shown that the specifics of the combinatorial code are not correct; however, the idea of a sequential partitioning is to the point.

The Origin of Segments

The basic segmental pattern of the *Drosophila* embryo appears as a series of nearly simultaneous lateral invaginations of the germ band during gastrulation. Despite the seeming mosaic nature of this event it can be shown that the patterning of segments occurs progressively during cleavage before the actual blocking out of segments. Ligation of cleavage-stage embryos performed in several laboratories by Vogel, Herth, and Sander, and by Schubiger and Wood has demonstrated that this patterning can be interrupted with varying degrees of severity, depending on the age of the cleavage-stage embryo. The earlier the ligation is performed, the more severe the defect. The full complement of segments is not determined until the cellular blastoderm stage is reached. During cleavage there is an apparent requirement for the interaction of different regions of the embryo to achieve a normal pattern. The possible nature of this interaction and the genetic regulation of the process of segmentation has been revealed in a series of elegant developmental and genetic analyses carried out by Nüsslein-Volhard and Weischaus.

As was pointed out in Chapter 7, Nüsslein-Volhard has characterized two maternal effect genes that severely disrupt the basic polarity of the entire embryo. Mutations at the two loci, *bicaudal* (*bic*) and *dorsal* (*dl*), cause homozygous mutant females to produce, respectively, embryos with two posterior ends or two dorsal sides. It is possible that these two loci represent genes whose products supply the initial positional information specifying the basic anterior-posterior and dorsal-ventral axes of the embryo. Whether these loci encode the information itself or are necessary to the proper elaboration of that information is not known. However, because these genes show a maternal pattern of inheritance, it is suggested that this information is built into the egg during oogenesis. Based on the results of the ligation experiments it would appear that the full pattern of segments results from a further refinement of these basic axial cues. However, it is not simply a matter of mechanical partitioning of the maternally supplied information. Segmental number, position, polarity, and identity must all be specified. One aspect of this refinement involves the homoeotic loci discussed in Chapter 8. These loci are involved in the specification of segmental identity and apparently mainly act subsequent to fertilization and not maternally. Other genes establish segmental number and polarity. The recent study by Nüsslein-Volhard and Weischaus has revealed genes that function in the progressive refinement of the basic axes of the egg into a defined number of segments. Like the homoeotic specification of segment identity, these genes are expressed during embryonic development. By genetic and developmental tests a group of 20 such segment pattern loci have been identified. Mutations in these loci result in zygotic but not maternal effect lethality, and produce phenotypes in which the formation of segments is disrupted. A genetic and developmental characterization has not been completed on these mutations, so it is not clear if all 20 genes are specifically involved in segmentation or if some affect the process in a pleiotropic fashion. It appears, however, that the 20 loci identified are close to the total number of such genes in the *Drosophila* genome.

Despite the preliminary nature of the analysis of these loci, the terminal phenotypes of the mutants do offer some possible clues to the nature of the genetic involvement of segmental patterning in the embryo. Moreover, these mutants offer an intriguing insight into the possible integration of segment patterning and the specification of segmental identity by the homoeotic genes. The segment pattern genes fall into three broad categories, based on their terminal phenotypes. These are the "gap" loci, in which groups of contiguous segments are missing; the "pair-rule" loci, in which a pair-wise association of normally adjacent segments resulting in half the normal number of segments is observed; and the segment polarity loci, in which embryos with the normal number of segments but containing mirror-image duplications of anterior portions of each segment are produced. All such segments consist of two anterior halves.

Mutants of two of the gap loci produce phenotypes pertinent to the present discussion. These are *knirps* and the *Regulator of postbithorax* [*Rg(pbx)*]. Mutations at *knirps* produce embryos having a normal anterior and posterior end but only one large abdominal segment where the first through seventh abdominal segments should be. This phenotype is presented diagrammatically in Figure 9–8. Deletions of the *Rg(pbx)*

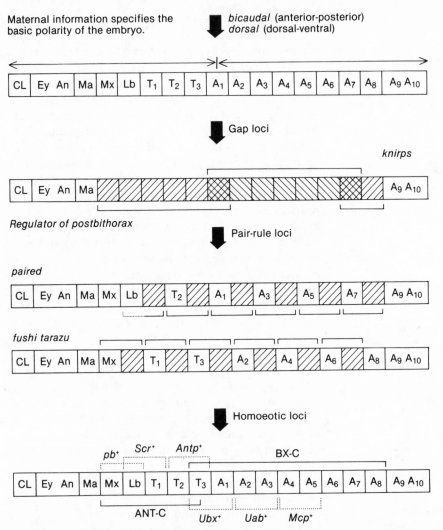

FIGURE 9–8. *Hypothetical flow of segmentation decisions made by the* Drosophila *embryo in early development. The abbreviations indicating segment identity are as in Figure 8–9. The brackets above and below the box figures indicate the domains of action of the indicated genes. The cross-hatched regions shown for the gap and pair-rule genes indicates the regions of the embryo that fail to segment properly. The bottom figure shows the proposed domains of action of the homeotic loci. The genes indicated are presented as examples and do not represent an exhaustive listing of the genes known to affect the pattern of segmentation. As such, this model represents a simplification of the genetic regulation of a complex process that is currently under intensive investigation.*

locus produce a kind of reciprocal "deletion" of segments relative to *knirps*. All three thoracic segments plus a portion of the gnathocephalic segments fail to form, and one large abdominal-like segment replaces them. In addition, the seventh and eighth abdominal segments fail to separate. In both cases large areas of the cellular blastoderm and germ band that would normally form several segments produce only a single large entity. A lesion of this kind can be broadly interpreted as resulting from a defect in the genetic program of the cells in specific regions of the embryo to react to the maternally supplied information in the egg. This contention is supported by the phenotype of embryos deleted for both the $Rg(pbx)$ locus and the bithorax gene complex (BX-C). A deletion of the BX-C results in a transformation of the abdominal segments into thorax. The double deletion has the same number of segments as a $Rg(pbx)^-$ embryo, but they are all thoracic in character. Therefore, the gap observed in the $Rg(pbx)^-$ embryos results in a regionalized deficiency in the early embryo and does not depend on the segments having a thoracic identity. Put another way, $Rg(pbx)^+$ is necessary for the segment-forming capacity in the anterior portion of the embryo, and the gene acts prior to or independently of the attaining of segmental identity. With this last point in mind, it is interesting to note that the deleted anterior segments in the $Rg(pbx)^-$ embryo are those in which the *Antennapedia* complex (ANT-C) is active in determining segmental identity. Therefore, the proper functioning of this particular gap locus is a prerequisite to the expression of the ANT-C.

There are two types of mutants found among the pair-rule genes. These are exemplified by the *paired* (*prd*) and *fushi tarazu* (*ftz*) loci. The *prd* mutation results in the "fusion" of the anterior of one segment with the posterior portion of the segment normally following it. The germ band of the embryo forms only one-half the normal number of segments, each of which are double the width of a normal segment. These double segments are shown diagrammatically in Figure 9–8 and are comprised in part of mesothorax and metathorax, first abdomen and second, third and fourth abdomen, and so on. It is also possible that the labial and first thoracic are combined as well. The other locus, *ftz*, is a member of the ANT-C (Fig. 8–8) mapping between the homoeotic *proboscipedia* (*pb*) and *Sex combs reduced* (*Scr*) loci. Like *prd* there is a "fusion" of segments; however, the frame of the fusion is offset relative to *prd*. This pattern is presented in Figure 9–8, and results in the association of the maxillary and labial segments, the first and second thorax, the third thorax and first abdominal segments, second and third abdominal segments, and so on. Observation of early embryos has revealed that this altered pattern is evident, even at the initial stage of segmentation. The germ band is divided into one-half the normal number of segments, all of which are roughly double the size of a normal segment. Temperature-sensitive alleles of both the *prd* and *ftz* loci have been analyzed, and their temperature-sensitive periods (TSP) occur in very early development. For *ftz* the TSP begins at about two hours of development, a time corresponding to the cellular blastoderm stage. The TSP ends at about four hours, before the germ band is physically segmented. Thus a genetically regulated event takes place in the early embryo that is necessary for the proper completion of segmentation.

The absence of this genetic activity also demonstrates that at some point in the determination of segmental pattern the germ band is divided into a series of units the width of two segments. Like the $Rg(pbx)$ locus the function of *ftz* is necessary to the proper expression of the homoeotic loci. In the pairwise "fusion" of segments observed in *ftz*⁻ embryos, only the segmental character of the anterior segment is evident in the double-wide segment. This is most easily seen in the fused third thorax and first abdominal segments. In the mutants only thoracic structures are found in this large segment; no abdominal character remains. Because the *bithoraxoid* (*bxd*) locus functions to specify the identity of the first abdominal segment, one may conclude that the absence of ftz^+ function precludes the expression of the *bxd* locus. However, the activity of *ftz* does not rely on segmental identity. Embryos in which both the *ftz* locus and the BX-C are deleted have one-half the normal number of segments, but they are all thoracic in character. This phenotype would be expected if ftz^+, like $Rg(pbx^+)$, acted prior to the homoeotic loci which in turn function within the segmental pattern specified by the gap and pair-rule loci.

With this latter point in mind, it should also be noted that the paired segments define associated functions within the ANT-C and BX-C. As diagrammed in Figure 9–8, the third thoracic-first abdominal region defines the range of action of the bx^+, pbx^+, and bxd^+ loci, all of which fail to complement the dominant *Ubx* mutations. Similar realms can be seen for other BX-C sites in the remaining posterior paired segments. With respect to the ANT-C, mutations at the *pb* locus affect the maxillary and labial segments. The *Scr* mutations produce transformations of the labial and prothoracic segments, and null alleles of the *Antp* gene indicate that $Antp^+$ is necessary for normal development of the second and third thoracic segments. Therefore, the effects of the genes of the BX-C and ANT-C indicate that these loci act in regions the width of two metameres. Some of these regions are defined by the *ftz* phenotype; others are revealed by the alternate pattern of segmental associations seen in the *prd* mutant. In a manner similar to the gap loci, the pair-rule genes also define domains of action of the homoeotic loci and it is this definition that associates the two kinds of segment-regulating genes. Mutants at the segment polarity loci have the normal number and identity of segments but produce pattern duplications. These are likely to be most closely related in function to the homoeotics, in the sense that they can be envisioned as interpreters of positional information which in turn specify identity rather than number of metameric elements. Based on the above observations, it is possible to construct a coherent and not entirely fanciful picture of the manner in which an insect embryo is patterned into segments.

A Model of Gene Action in the Specification of Segmentation

A diagrammatic representation of our view of the process of segmentation is given in Figure 9–8. The basic axes of the embryo are determined by maternally supplied positional information that specifies anterior versus posterior and dorsal versus ventral. These basic coordinates are then elaborated on by the various segmentation-controlling loci. The first event involves the partitioning of the presumptive germ band into a group of large domains, some of which are defined by the gap loci and encompass several segmental anlagen. Subsequent to the action of the gap loci, the pair-rule genes further subdivide these larger regions into domains covering the width of two segmental anlagen. These dual-segment domains occur in two separate but overlapping frames. It should be noted that the same overlap of domains is observed between the two gap loci pictured in Figure 9–8. The overlap could serve several purposes. The first would be to reinforce the boundary between the two adjacent domains, because the cells in this region are unique in having both genes active. A second function could be in the actual demarcation of segments, each segment being defined by the overlapping paired frames seen in the *prd* and *ftz* mutants. Finally, these segmental decisions are reinforced by the activity of the segment polarity loci that maintain the proper anterior/posterior relationship within each segment as it was originally specified by the maternally supplied information for the entire embryo. These temporally restricting embryonic fields in turn represent the domains of activation for the homoeotic loci of the ANT-C and BX-C.

At present, the most plausible role in ontogeny for the BX-C and ANT-C genes is as interpreters of positional information. The mechanism by which position may be interpreted has been most explicitly presented by Garcia-Bellido. In his model the homoeotic loci, specifically in the BX-C, act as selector genes or switch genes. The positional information specified during oogenesis and refined during embryogenesis activates or inactivates these loci in a sequentially patterned fashion. Once active, the products of the homoeotic loci select or repress batteries of other genes. The pattern of activity of both the homoeotic loci and the genes they in turn regulate is unique to each particular segment. E. B. Lewis has presented a model in which the activation of the BX-C is polarized in a proximal-to-distal direction along the chromosome. That is, there is a sequential activation of the BX-C genes that corresponds to the anterior-to-posterior order of segments in which the loci are active. Lewis further envisions the proximal-to-distal orientation of loci in the complex as corresponding to a gradient of repressor binding along the chromosome. The more anterior the segment, the greater the concentration of repressor, and consequently, fewer genes are active. Recent analyses of the *Polycomb* (*Pc*) locus by Lewis and the *extra sex combs* (*esc*) locus by Struhl have lent some validity to this model of BX-C activation. Both of these mutations result in a transformation of the second and third thoracic legs into prothoracic legs. This is readily seen in males where sex combs are found on all six legs, a phenotype similar to that produced by some lesions at the *Antp* locus. However, embryos bearing lethal alleles of either of these two loci have a very different phenotype relative to the *Antp* mutations. All of the thoracic and abdominal segments 1–7 are transformed into eighth abdominal segments. This result, coupled with the fact that both *Pc* and *esc* mutations are sensitive to the number of genomic copies of BX-C, (that is, the more BX-C genes present the more extreme the transformation to eighth abdomen of the entire animal), indicates that the *Pc*[+] and *esc*[+] products may act as the hypothetical repressor of BX-C.

The demonstration of loci such as *Pc* and *esc,* as well as the gap and pair-rule mutations, shows that the attaining of proper segmental identity is regulated on at least two separate levels. There is a direct regulation of homoeotic genes at the level of transcription, but this requires the attainment of proper segmental patterning events before the derepression of the homoeotic loci can have an impact on segmental identity.

The model, as presented in Figure 9–8, bears a striking resemblance to the theoretical considerations of Kauffman. He envisioned the segmentation process as a series of binary decisions that sequentially subdivided the egg, and as a consequence, assigned specific addresses to different portions of the embryo (Fig. 9–7). Kauffman's formal model saw the embryo as divided first in half, then quarters, then eighths, and so on. Based on the mutant phenotypes observed thus far, that view is too simplistic. However, this should not be taken as a rejection of the basic form of the model because there does appear to be a sequential partitioning of the early embryo, and this partitioning is integrated with the loci that function to regulate segmental identity.

In a formal application of the model, one might assume that the binary set of switches, like the ball descending on Waddington's epigenetic landscape, would be linearly ordered and dependent; that is, the throwing of one switch depends on the prior event in the pathway. In some cases this is indeed true. However, this is not necessary. There are instances in which genes that regulate basic morphogenetic events are dissociated in their function. This fact was illustrated in Chapter 5 by the uncoupling of morphogenesis from cytodifferentiation in the developing pancreas. It can also be seen in the example of segmentation of the *Drosophila* embryo. Despite the fact that segment number- and identity-controlling loci are spatially integrated, the two functions can be uncoupled in mutant individuals; that is, the gap loci do not depend on the action of the homoeotic loci nor do the homoeotic loci require the activity of the gap loci. This dissociation is what allows the ontogenic process to be "tinkered" with in evolution. This fact is quite apparent in the evolution of insects, as presented in Chapter 8. The changes observed have been involved in segmental identity and have been independent of segment number.

Phenotypic Choice and the Morphogenetic Potential of the Genome

There is another important aspect of genomic control of development. Phenotypes are not rigidly determined by a single genetic program. Morphologies also reflect nongenetic environmental influences, although, as extensively discussed by Waddington, the possible range of phenotypic responses is determined by the nature of the genetic program. And indeed, choice of alternate phenotypes is an almost universal part of ontogeny. Phenotypic plasticity is a very large and complex topic; we thus confine ourselves to alternate phenotypic choices that are part of the programmatic developmental pattern of a species, and that provide the basis for evolutionary changes as typified by some of the classic examples of heterochrony. Programmed phenotypic choices include sexual dimorphism, different larval and adult morphologies, and alternate adult morphologies, as exemplified by environmentally determined neoteny in amphibians or by hormonally elicited castes in social insects. In all of these cases, switches determine which of the possible alternate programs for morphological development will be selected from the repertoire of a single genome.

That the genome has a greater potential than the production of a single invariant morphology is nicely and dramatically illustrated by a consideration of the castes produced by the black-mound termite *Amitermes atlanticus*. Like most social insects, a colony of these termites is founded by a queen after her nuptial flight. The subsequent brood of siblings produced by this queen form the population of termites in the colony. In his book *Dwellers in Darkness*, Skaife pointed out that five castes can be found in a mature colony of these organisms: workers; soldiers; and primary, secondary, and tertiary reproductives. All of these are illustrated at the bottom of Figure 9–9. As can be seen, there is a wide range of morphology among the castes. Only the reproductives have wings and eyes, whereas the workers and soldiers are blind and wingless. The soldiers differ from workers and reproductives in that they have enlarged heads and powerful mandibles. In a nest producing several of these kinds of adults, no differences can be discerned among the different types at hatching. Moreover, it should be remembered that all members of the colony are siblings. When the termites are about half grown, differences begin to appear in their morphologies, depending on the caste to which they are destined to belong. It is unlikely that this is caused by genetic differences because in young colonies only workers are found, and as the colony matures first soldiers and then different classes of reproductives are produced. In addition, there are seasonal fluctuations in the number and type of reproductives produced by a mature colony. Thus, it would appear that a group of sibling organisms with similar genotypes are capable of producing a rather varied series of phenotypes, all of which are functional integrated organisms.

More is understood of the mechanisms of caste differentiation in the more-primitive subterranean termites, largely through the work of Luscher. Caste differentiation in these species is under the control of a pheromone that inhibits production of secondary reproductives, and of the principal insect developmental hormones. Apparently, ecdysone causes undifferentiated nymphs to transform into reproductives, whereas high levels of juvenile hormone cause nymphs to transform into soldiers. Thus, the endocrine control of caste differentiation may involve the competing actions of ecdysone and juvenile hormone.

The role of hormones in triggering developmental switches in insects is further illustrated by the analysis by Wheeler and Nijhout of the production of soldiers by the ant *Pheidole bicarinata*. Like the termites, there are workers and soldiers as well as reproductives in colonies of

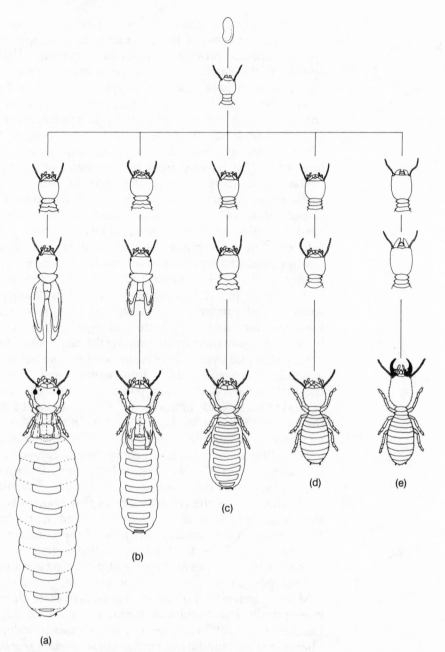

FIGURE 9–9. *Developmental differentiation producing caste distinctions in the black-mound termite* Amitermes atlanticus. *All castes begin life with a similar morphology, but complete development with distinct morphologies.* (a) *primary queen;* (b) *secondary queen;* (c) *tertiary queen;* (d) *worker;* (e) *soldier.* [Adapted and redrawn from Skaife, 1955.]

these ants. It was found that soldiers could be produced from undifferentiated female larvae if an insect juvenile hormone analog was topically applied during a specific period in the last larval instar. This application results in a longer period of growth and a change in the size of larvae at the time of metamorphosis. The larger adults produced by the longer growth period are soldiers. As shown by J. S. Huxley and by

E. O. Wilson, the workers and soldiers of ants typically fall on the same allometric curve (see Fig. 2–10). The extreme size of the jaws and heads of soldiers are a consequence of their greater size. In some ant species the situation is more complex because allometric relationships change slope with increasing size. Di- and triphasic allometric relationships are known. Adjustment of hormone levels allows thresholds to be set whereby the ant colony can "choose" the adult morphology a larva will assume.

These examples are only two of a great number in which phenotypic choices occur in the course of an ontogeny directed by a single genome. The potential for evolutionary changes in morphology is obvious, and examples of evolution involving the flexibility offered by developmental switches leading to different functional morphologies have been presented in Chapter 6.

One final example deserves mention here to show that programmatic changes in ontogeny can also be observed in fossil organisms. Figure 9–10 illustrates two heteromorph ammonites that once graced the Late Cretaceous seas of western North America. These shells record the growth history of the animals that secreted them, and each shell

FIGURE 9–10. *Two heteromorph ammonites of the Cretaceous sea of the American western interior. Both* Didymoceras nebrascense *(left) and* D. stephensoni *(right) initially grew straight shells. This mode of growth was followed by a period of helical shell growth, and finally, at maturity, development of a final coil in a different plane.* [Redrawn from J. R. Gill and W. A. Cobban, Stratigraphy and geologic history of the Montana group and equivalent rocks, Montana, Wyoming, and North and South Dakota. *U.S. Geol. Surv.* prof. paper no. 776 (1973); and G. R. Scott and W. A. Cobban, Geologic and biostratigraphic map of the Pierre Shale between Jarre Creek and Loveland, Colo. *U.S. Geol. Surv.* (1965), Map I-439.]

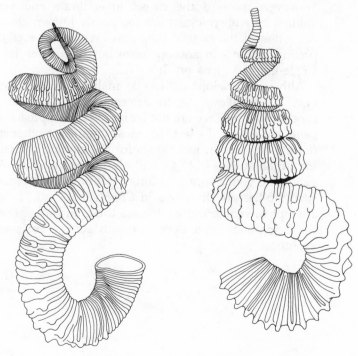

indicates three sequential modes of growth. The initial shell is straight until a discrete point in development at which the growth program switched to generate a torticonic shell. As maturity approached, growth of the shell changed direction to produce a terminal, U-shaped living chamber. Aside from the aesthetic appeal of these ammonite shells, they indicate that a single species had a considerable degree of morphogenetic flexibility for evolutionary change in shape. The rapid evolution of heteromorphs, as exemplified in Figure 2–4, suggests that the flexibility provided by switches in growth mode was exploited in ammonite evolution.

Chapters 7 through 9 have been intended to demonstrate that genes indeed control ontogeny in very specific ways; that is, that there is such a thing as a genetically determined developmental program. Probing of the genetic program by analysis of mutations that affect development has revealed that whereas mutants of many genes disrupt development, there is a smaller set of genes in which mutations produce entirely novel effects. These genes, typified by the homoeotic clusters of *Drosophila*, act as switches specifying alternate patterns of morphogenesis. We have explored the characteristics of the homoeotic genes in detail because they represent the most thoroughly studied developmental switch genes. Other, less well-understood switch genes exist in other organisms, where they presumably play crucial roles in making developmental decisions.

An exciting feature of the genes in *Drosophila* that make the switching decisions involved in determination of segment number, polarity, and identity is that these genes act as interpreters of positional information. Positional information, as discussed in Chapter 4, is an important feature of development in a wide phylogenetic spectrum of organisms. However, most of the classic invertebrate and vertebrate systems studied by embryologists are too poorly known genetically to allow a dissection of the genetic programs involved in establishing and interpreting patterns in any organism but *Drosophila*; thus, our seemingly single-minded focus on flies.

Although development can be analyzed by classic genetic methodology, which allows us to determine which genes have regulatory functions *in vivo*, we are not limited to this approach in the analysis of gene expression. In fact, for most organisms other methods must be used. Advances in techniques for the cloning of genes and for high-resolution studies of DNA and RNA have made it possible to examine genes and gene expression directly during development. The results of such studies are discussed in Chapters 10 and 11. In Chapter 12 we attempt to synthetically integrate these and other chapters to provide the beginnings of a developmental genetic basis for morphological evolution.

Ten

Adaptations for Gene Expression in Development

Life is a force which has made innumerable experiments in organizing itself;...the mammoth and the man, the mouse and the megatherium, the flies and the Fathers of the Church, are all more or less successful attempts to build up that raw force into higher and higher individuals...

George Bernard Shaw, Man and Superman

Embryonic Adaptations

The evolutionary history of a species determines what structures and processes are available on which selection may act. History and adaptation are intertwined. The result has been a marvelous diversity of larval morphologies. In some groups similar patterns of development and larval stages are retained despite very dissimilar adult morphologies. Thus, barnacles and other Crustacea retain the presumably primitive nauplius larva, and mammalian embryos possess structures resembling the gill arches of fish. These are the cases that led to von Baer's law and later to Haeckel's recapitulation doctrine. But the opposite also occurs. Certain insects in which the adults are very similar produce quite dissimilar larvae, owing to divergent larval specializations. These cases and the existence of specialized structures associated with development, such as the extraembryonic membranes and placenta of mammals, represent purely larval adaptations that may be totally divergent from ancestral or even related forms.

Embryonic or larval stages, while having to meet the inescapable requirements of being integrated and viable organisms, are nonetheless stages in a dynamic process involving differentiation and growth within the embryo. If we agree that changes in these processes provide the mechanistic basis for achieving morphological evolution, then the nature of the adaptations that underlie the expression of genes controlling embryonic stages of development are of central interest.

That genes indeed control morphological development was documented in Chapters 7 through 9. In this chapter we consider two major aspects of this expression: the genetic cost of development in terms of

287

the proportion of the genome devoted to development, and specializations in genomic organization to support development.

How Many Genes Are Necessary for Development?

Fortunately, there are methods of estimating the amount of genetic information available to higher organisms. One of the finest of these tools is classic Mendelian genetics. The major difficulty with this method is that surprisingly few organisms have been subjected to sufficiently detailed genetic analysis to yield the desired data. In fact, only one organism, the common fruit fly *Drosophila melanogaster*, presently fulfills the criterion of sufficient genetic familiarity. Since *Drosophila* was first taken into the laboratory by T. H. Morgan and his students in 1910, a monumental amount of information has been gathered on this single modest insect. Thousands of mutations scattered throughout the genome of *Drosophila* have been induced by use of chemical mutagens and x-rays. The current catalog of mutations in *D. melanogaster*, a testament to the assiduousness of *Drosophila* geneticists, is now over 500 pages in length.

This vast collection of mutations, as well as the ability to recover more, makes possible a dissection of the relationship between genes and the morphological, metabolic, developmental, or behavioral characteristics of the organism. Mutations can be used to gain two basic kinds of information about development. Analysis of the ways in which mutations interfere with individual ontogenetic processes can provide major insights into the control and function of such processes. The second kind of information, with which we are concerned here, is the numbers of genes controlling the developmental processes, which in turn allows us to estimate the fraction of the entire genome devoted to ontogeny *per se*.

Because any single gene can yield a number of different mutant alleles exhibiting different phenotypes, the number of genes involved in any process cannot be estimated by the appealingly simple procedure of counting up mutations. Consider two mutations that both affect a particular phenotypic trait. The decision as to whether the two mutations in question are in one gene or two is made by means of a standard complementation test.

The number of complementation groups regulating development of *Drosophila* can be estimated by taking a sample of a particular class of mutation and by extrapolating the total number of genes in that class in the entire genome. The extrapolation is based on two assumptions. The first is that genes of similar function are randomly distributed either in a particular chromosome or in the entire genome; that is, there is no clustering of genes of related function in development. With certain important exceptions discussed in Chapter 8, this assumption appears reasonable. The second assumption is that genes of like function are equally mutable, and that these mutations are equally recoverable. This assumption is violated by some genes that turn out to be more mutable than most, but the assumption appears valid if applied to a large sample of genes. Mutations can be recovered that block development in a variety of ways and can be grouped into general categories. As many mutations as possible are gathered for each category, but an exhaustive sample in which the genome is saturated for the desired class of mutants is, of course, impossible. Thus a nonsaturating sample is taken and must be extended statistically. This requires that the genes actually sampled by mutation represent a random sample of the class of interest.

What one chooses to consider as developmentally significant classes of genes depends on the particulars of development in the organism being studied. It is important to remember that *Drosophila* is a holometabolous insect and that the larva represents a morphological, physiological, and behavioral stage completely distinct from the adult. Larval structures do not transform directly into equivalent adult structures. The fly is something of an insect phoenix. Most larval tissues are broken down and absorbed by the adult tissues that develop from the imaginal discs present within the larva. There are then actually two developmental systems in this organism in which gene function crucial to adult development can occur. The first is during embryogenesis when the component parts and morphology of the larva are produced. The second is the establishment, maintenance, and proliferation of the imaginal discs within the larva, and their differentiation to produce the adult.

The first of these systems, larval development, actually represents two temporally distinct systems of gene action. Information present within the egg that is necessary to early development represents the products of gene action during oogenesis. Subsequent to its period of dependence on the gene products of oogenesis, the embryo begins active transcription to provide the information required for the remainder of development. Thus in *Drosophila*, a tally of developmentally important genes will have to include genes that exhibit a maternal pattern of inheritance (i.e., function during oogenesis), genes whose function is vital during embryonic or larval development, and genes that specifically affect the development of imaginal discs. Estimates of all three classes have been made for *Drosophila*.

Both Gans and co-workers and Mohler have isolated a large number of maternal effect mutations in genes located on the X chromosome (see Chapter 3 for a discussion of *Drosophila* chromosomes), and similarly, Rice and Garen have isolated maternal effect mutations located on the third chromosome. There are actually two classes of such mutations. In the first, females produce morphologically abnormal eggs (e.g., with defects in the egg shell), whereas females of the second class produce ostensibly normal eggs that fail to complete development. It is only this latter class that is of interest to us. The tally presented in Table 10–1 indicates that Gans recovered 42 mutations of the interesting class; Mohler, 146; and Rice and Garen, 6. Subsequent genetic complementation tests showed that all of these maternal effect mutations fell into 30, 60, and 5 separate complementation groups, respectively. That is, some of the mutations represented repeat hits within the same gene.

Before embarking on the analysis that follows, it is worth recalling

TABLE 10–1. Estimation of Total Maternal Effect Genes in the *Drosophila* Genome

Investigator	Chromosome	Total No. of Mutations	No. of Genes Mutated	n_o	Estimated Total No. of Genes
Gans et al.	X	42	30	53	83
Mohler	X	146	60	38	98
Rice and Garen	3	6	5	8	13
Our extrapolation	All	—	—	—	117

Thomas Huxley's admonition that "Mathematics may be compared to a mill of exquisite workmanship, which grinds you stuff of any degree of fineness, but, nevertheless, what you get out depends upon what you put in..." Making use of the assumptions of a random arrangement of genes and their equal mutability, the number of unmutated genes of any class can be estimated by use of the Poisson distribution. This procedure assumes that the probability is high that most of the genes of a class have not yielded any mutations at all, that a smaller number have yielded a single mutation, and that two mutations per gene is even less probable. The quantitative relationship between the no-hit, one-hit, and two-hit groups is given by the expression

$$n_o = \frac{(n_1)^2}{2n_2}$$

where the terms are defined as follows: n_1 is the number of genes that have been mutated once; n_2 is the number that have been mutated twice, and n_o the number of genes for which no mutation has been detected. The complementation data for any class yield numerical values for n_1 and n_2.

The data of Gans and coworkers yield a value of n_o equal to 53 and the data of Mohler yield 38 for total estimates of 83 and 98 maternal effect genes on the X chromosome. Considering the nature of the estimates, the two numbers are in remarkably close agreement and suggest an average of 90 genes on the X chromosome vital to embryogenesis whose products are supplied as a result of gene action during oogenesis. The same exercise can be carried out using the data of Rice and Garen, yielding an estimate of $n_o = 8$ and $n_{total} = 13$ for this class of genes located on the third chromosome. The disparity in the totals between the X and the third chromosome are interesting but not readily explicable because the third chromosome contains roughly twice the DNA of the X chromosome and the discrepancy cannot be accounted for by differences in sample sizes examined. It thus appears that the maternal effect genes are predominantly located on the X chromosome.

If we assume that the second chromosome, an autosome similar in DNA content to the third chromosome, is similar to the third chromosome in organization it too would have 13 maternal effect genes located on it. The last autosome, the very small fourth chromosome, can be estimated to have one maternal effect gene, based on its relative size. These estimates taken together suggest that there are a total of about 117 maternal effect genes in the *Drosophila* genome with 90 located on the X chromosome, and based on results obtained by examination of the third chromosome, roughly 27 more on the autosomes.

In order to estimate the number of genes necessary for embryonic development but not in the maternal effect class, we will have to use a slightly different approach. It might be possible to collect a random sample of lethal mutations and determine the number of these that are lethal during embryonic development. An extrapolation of the total number of embryonic lethals in the entire genome could then be made by applying the fraction of embryonic among all lethal mutations in the random sample to a determination of the total number of genes in the genome capable of mutating to lethality. Our method will essentially follow this outline except that our random sample of lethal mutations will come from two small but representative regions of the genome that have been saturated with lethal mutations.

The first requirement is, of course, that we have an estimate of the total number of genes in the *Drosophila* genome. Judd and his co-workers found in an intensive genetic analysis of a small segment of the X chromosome that there is a nearly 1:1 correspondence between the number of bands in the polytene chromosomes of *Drosophila* and genes as defined by genetic complementation. The total number of bands visible in the polytene chromosomes of *D. melanogaster* shown in Figure 10–1 is 5,000, and a reasonable estimate of the total number of genes is also 5,000. Needless to say, this number has been hotly debated by geneticists since it was originally proposed in 1972, but subsequent studies have continued to substantiate the approximate correspondence between bands and genes.

The proportion of these 5,000 genes capable of mutating to lethality comes from two studies: one by Shannon and her co-workers, and the other by Hochman. Shannon et al. analyzed the mutations in a small region of the X chromosome between the genes *zeste* and *white*, a region that, as illustrated in Figure 10–2, contains 13 bands and 13 identifiable genes. Hochman's study entailed a similar analysis of the small fourth chromosome, which contains only 50 bands altogether, on which he was able to define 43 genes. Of the 13 defined genes on the region of the X chromosome studied by Shannon and co-workers, only two failed to mutate to lethality, whereas six of the 43 genes identified by Hochman on the fourth chromosome yielded no lethal mutations and are thus presumably not vital to the fly. Extrapolating from this limited sample, approximately 85% of the 5,000 genes of *Drosophila* appear to be capable of producing lethal mutations.

In developmental analyses of the lethals thus recovered and mapped, Shannon et al. and Hochman found that 1/10 and 5/37, respectively, were embryonic lethals. If the frequencies of embryonic lethals to all lethals are an accurate reflection of the total in the genome then, as summarized in Table 10–2, there are 425–550, or approximately 500 genes capable of yielding mutations that are not only lethal but are lethal

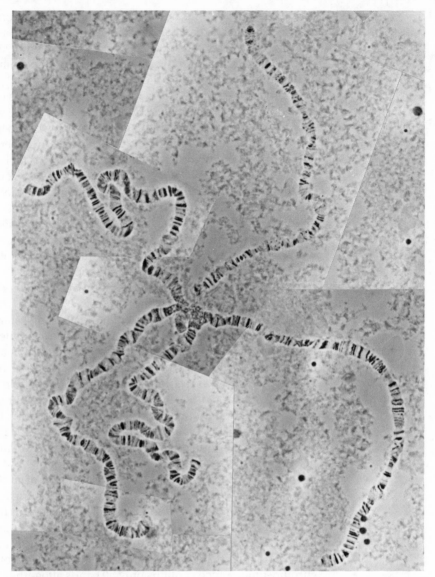

FIGURE 10–1. *Photomicrograph of the salivary gland polytene chromosomes of* D. melanogaster. [Courtesy of R. Lewis.]

specifically during embryonic development. Combining this estimate with that for maternal effect lethals, there are 617 genes necessary for early embryogenesis and for the formation of a functioning larva.

An estimate of the number of genes necessary for the development and maintenance of the imaginal discs in the absence of any larval defects can be calculated from the results of a study by Shearn and Garen. These workers selected for lethal mutations that permitted ostensibly normal embryogenesis and larval development, but in which death occurred during metamorphosis. These mutations were found to cause imaginal discs either to be defective or lacking altogether. A total

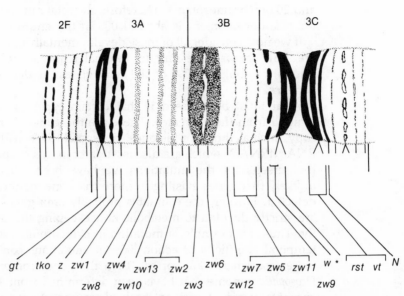

FIGURE 10–2. *Identification of genes in a small region, between the genes* zeste *and* white, *or the x chromosome of* D. melanogaster. *Thirteen genes, corresponding to the thirteen identifiable bands, have been identified.* [Redrawn and adapted from T. C. Kaufman et al., 1975.]

sample of 57 mutations of this type was isolated on the third chromosome. Complementation tests of the 57 mutations revealed 52 complementation groups or genes. By the same procedure used in the calculation of maternal effect genes, $n_o = 384$ for this class of gene on the third chromosome. The total theoretical number of imaginal disc function genes on chromosome 3 is therefore 436. Mutations of this class have also been discovered on the X chromosome by Kiss and co-workers. Because these mutations are sex-linked lethals with which it was impossible to do a standard complementation test, Kiss and co-workers were forced to do a more complicated analysis that need not concern us here. They arrived at an estimate of a total of 118 genes vital for imaginal disc development and function on the X chromosome. If we assume that the second and fourth chromosomes have a number, relative to DNA content, of these genes similar to that of the third chromosome there should be 436 genes of this class on chromosome 2

TABLE 10–2. Estimation of Total Embryonic Lethals in the *Drosophila* Genome

Investigator	Region	No. of Bands	No. of Genes	No. of Lethals	No. of Embryonic Lethals	No. of Embryonic Lethals in Genome
Shannon et. al	zeste white	13	13	10	1	425
Hochman	Fourth chromosome	50	43	37	5	550

and 20 on chromosome 4. Therefore, the total number of genes needed for disc development is about 1,000 in the entire genome.

If we now consider all of these developmentally important genes there are 1,617 genes necessary for normal embryogenesis and metamorphosis, or roughly 30% of the total genes, as defined by Mendelian genetics.

What must be stressed is that certain types of genes are entirely omitted from the estimates. These are the genes that exist in multiple copy, such as those for histones, ribosomal RNAs (rRNAs) and transfer RNAs (tRNAs), which are discussed later in this chapter. The nature of the screens for the mutations we have been considering makes it unlikely that genes existing in more than one identical copy would be detected. Further, the estimates probably omit genes whose functions are strictly devoted to metabolic housekeeping through the life of the fly. Thus, for example, a mutation in cytochrome *c* would not show a temporal specificity of action like the mutations considered before. If these assessments are correct, then the genes comprising 30% of the *Drosophila* genome that have been determined from screens for developmentally important genes probably represent a reasonable estimate of the genetic cost of development from egg to fly.

Genetic Cost of Development at the Molecular Level

Two problems arise in applying estimates of the number of genes required for the development of *Drosophila* to other organisms. *Drosophila* appears, at least in some major respects, to have a genome organized rather differently from that of most other animals. It is entirely possible that genetic inferences drawn from *Drosophila* may be misleading when applied elsewhere. The second difficulty lies in the definition of a gene. The classic geneticist's definition is largely operational, with a gene being defined by its effect on viability or visible phenotype. Unfortunately, genetic analysis has been made in only a most rudimentary form for most organisms studied by embryologists. For these animals an alternative molecular approach to estimates of numbers of active genes has been used. To molecular biologists an operational definition of a gene has generally meant a sequence of DNA that encodes the amino acid sequence of a protein. Although techniques exist for separating and estimating the number of relatively common protein species present in a cell, it is nearly impossible experimentally to estimate the number of species of rare proteins. Fortunately, it is feasible to determine the number of different mRNAs present and thus yield an estimate of the number of functioning genes. The molecular biologist's definition of a gene will be ultimately congruent with the geneticist's definition in many instances. However, it is not at all clear that all genes necessarily produce mRNAs, nor does a mRNA contain all of the information present in the gene from which it is transcribed.

Embryos have two sources of mRNAs available: those stored in the egg during oogenesis, and those synthesized by embryonic cells during the course of development. The major functional role for the large and diverse store of mRNAs of the egg is to allow the newly fertilized egg to begin a massive program of protein synthesis to provide the proteins needed for assembly of nuclei, membranes, and other subcellular structures needed during the period of rapid cleavage. Once cleavage has produced sufficient nuclei, the embryo itself can sustain a sufficient rate of mRNA synthesis to provide for the protein synthetic requirements of further development.

Each individual species of mRNA present in the embryo represents the coding sequence portion of a particular structural gene. Reference to Table 3–1 leaves little doubt that such messenger sequences provide a reflection of only a portion of the genome. It is also evident that many of the regulatory genes important in morphogenesis may not produce mRNAs at all. Thus a crucial class of genes that manifests itself to the Mendelian geneticist because mutations in it result in altered phenotypes or in developmental lethals of the sort tallied in the previous section of this chapter will simply be missed if we use mRNA diversity as an index of the number of genes required for development. It is not possible to gauge accurately the quantitative effect of this discrepancy, but fortunately there is not a total incongruity between regulatory genes and genes represented as mRNA copies. Genes that function as regulators of morphogenesis are, at least in some cases, transcribed. The RNAs detected by Kalthoff in eggs of the insect *Smittia*, discussed in Chapter 4, seem to represent transcripts of this class of gene because these RNAs provide the determinant for differentiation of the anterior end of the embryo. Further, many structural genes do have important roles in development and morphogenesis. The chorion proteins of the insect egg shell, which will be discussed in some detail in the next section of this chapter, serve as useful examples of this category of genes. Individual chorion genes are switched on and off in a concerted pattern as morphogenesis of the egg shell proceeds. The functions of chorion proteins are of course as structural components, and not as regulatory molecules, yet failure of these genes to switch or presence of a defective gene can result in assembly of a modified or defective final structure.

There is another source of ambiguity in using mRNA diversity to reveal gene numbers. Purely genetic methods allow the isolation of mutations that affect specific developmental processes or periods. These mutations reveal genes specific to ontogeny rather than genes basic to the metabolism or structural maintenance of cells at all stages of development. Messenger RNAs of embryos contain both stage-specific and housekeeping sequences, and both groups of sequences contribute to the overall diversity of the mRNA population. A distinction between groups of sequences can only be made by experiments in which mRNAs from embryonic stages are compared to those from adult tissues to reveal the proportion of sequences in the embryo common to all stages. These can be assigned to the housekeeping category.

Measurements of mRNA diversity are made by use of nucleic acid hybridization techniques similar to those introduced in Chapter 3. However, instead of annealing complementary strands of DNA to each other, RNA is annealed to the coding strand of genomic DNA. Therefore, the proportion of the DNA that hybridizes to mRNA from any stage can be used to calculate the number of genes represented.

This kind of estimate has been made with varying degrees of completeness and success for several organisms. Values of mRNA diversity, and thus numbers of genes active in production of proteins in eggs and embryos of several protostomes and deuterostomes, are presented in Table 10–3. Estimates of mRNA diversity in this table are derived from either total cytoplasmic RNA or, preferably, from RNA associated with polysomes and thus presumably actually engaged in directing protein synthesis. Polysomal RNA is more likely to include only *bona fide* mRNA than is total cytoplasmic RNA, but a note of caution is necessary in considering even studies made with polysomal RNA. Most of the RNA sequences revealed by hybridization studies are too rare to be identified by their presumptive protein products: They are mRNAs by inference only.

The numbers of active genes estimated for *Drosophila* eggs and larvae based on nucleic acid hybridization data have generally been found to be high with respect to genetic estimates for genes specifically needed for larval development. This is not surprising, because housekeeping as well as development-specific genes are expected to be expressed. On the average, published mRNA diversity values have been similar to genetic estimates of the total number of genes in *Drosophila*. But, one recent study by Zimmerman et al., using methods designed to detect all classes of mRNA, revealed that there are approximately 14,500 mRNA sequences present in larvae. This is over twice the number predicted by the genetic analysis of Judd and his co-workers. Resolution of the discrepancy is difficult without more precise data on the specific identities of the presumptive genes counted by these two very different approaches.

The other organisms listed in Table 10–3 generally exhibit diversities of genes expressed as mRNAs in development higher than those of *Drosophila*. It is not clear what this really means, but differences in diversity may well be an aspect of the C-value paradox (the discrepancy between morphological complexities and amounts of DNA in the genomes of organisms).

The most detailed studies of mRNA diversities in development have been performed by E. H. Davidson and his collaborators with the sea urchin *Strongylocentrotus purpuratus*. This group, in a paper by Galau et al., compared the diversity of genes expressed in the mRNAs of developmental stages and adult tissues. They also addressed a perhaps more significant question than mere gene numbers; that is, how many genes are stage-specific and how many are expressed in several stages and in adult cells? Their results, summarized in Table 10–3, reveal a curious fact. The number of genes being expressed as proteins actually declines during sea urchin development, despite a pronounced increase in morphological complexity and differentiation between early cleavage and the larval pluteus stage. Thus, egg mRNAs share the sequences

TABLE 10–3. Diversity of Cytoplasmic RNAs Present in Eggs and Embryos[a]			
Organism	Stage[b]	Estimated No. of Different Genes Represented	Source of Complexity Values
Protostomes			
Drosophila melanogaster	Egg	8,000	Hough-Evans et al., 1980
(fruit fly)	Larvae[c]	3,100	Bishop et al., 1975
	Larvae[c]	5,400	Levy and McCarthy, 1975
	Larvae[d]	14,500	Zimmerman et al., 1980
Musca domestica	Egg	16,000	Hough-Evans et al., 1980
(house fly)			
Urechis caupo	Egg	21,000–31,000	Davidson, 1976
(echiurid worm)			
Deuterostomes			
Xenopus laevis	Egg	18,000–27,000	Davidson, 1976
(frog)	Tadpole[d]	20,000	Perlman et al., 1977
Arbacia punctulata	Egg	20,000	Davidson, 1976
(sea urchin)			
Strongylocentrotus	Egg	24,000	Davidson, 1976
purpuratus	16-cell[d]	18,000	Hough-Evans et al., 1977
(sea urchin)	Blastula[d]	15,000	Hough-Evans et al., 1977
	Gastrula[d]	11,000	Galau et al., 1976
	Pluteus[d] (larval stage)	10,000	Galau et al., 1976

[a]Estimated gene number is the number of RNA species present assuming an average RNA length of 1,500 nucleotides. The expression used for such an estimate (assuming transcription of only one strand of DNA) is size of single copy part of genome in nucleotide pairs × fraction represented in RNA × 2 = RNA complexity in nucleotides, and RNA complexity in nucleotides/average number of nucleotides per mRNA = number of mRNA species.

[b]Diversity in eggs is for total RNA, presumably mRNA, because too little protein synthesis occurs in eggs to allow fractionation of functioning mRNA from other cytoplasmic RNAs.

[c]Total cytoplasmic RNA.

[d]mRNA isolated from polysomes engaged in protein synthesis.

present in the gastrula and possess an equal number of sequences not present in gastrula at all. Blastula and pluteus stages share most of the gastrula sequences, but have other sequences as well. Three adult tissues were also examined, which is no mean feat because an adult sea urchin is essentially a limestone box filled with gonads and little else. Adult tissues exhibit a lower diversity of mRNAs, with values ranging from about 2,000 to 4,000 genes expressed as mRNAs in any tissue. Even though these tissues have a much lower mRNA diversity than gastrula, they nevertheless share about 1,500–2,000 sequences with gastrula. Interestingly, these appear to be the same subset of gastrula sequences in all three tissues.

Do these represent housekeeping genes whereas the rest of the diversity is required for embryonic development or maintenance of adult cell types? That this is not an improbable conclusion is suggested by measurements of mRNA diversity in organisms of extremely simple morphology, such as bacteria or fungi. In these forms it is reasonable to expect that very little gene activity is devoted to morphological organization, but rather to synthesis of proteins involved in metabolism, maintenance of cell structure, and cell replication. The bacterium *E. coli* has been found by Hahn and his colleagues to contain a mRNA diversity of 2,300 sequences, essentially the full capacity of the genome, assuming that only one strand of DNA is transcribed. Hereford and Rosbash similarly have found that yeast, one of the most simple of eucaryotes, expresses 3,000–4,000 mRNA sequences. Another simple eucaryote, the fungus *Achlya ambisexualis*, studied by Timberlake and his collaborators, has a similar mRNA diversity of about 2,000–3,000 sequences.

There are other eucaryotes of equally simple morphology that exhibit higher mRNA diversities and differences in gene expression in the small number of different cell types they possess. The fungus *Neurospora crassa*, studied by Dutta and Chaudhuri, expresses about 10,000 genes in mycelial growth, but only about 5,000 in conidial cells. Similarly, Firtel observed that of the approximately 16,000 genes expressed in the life cycle of the slime mold *Dictyostelium discoideum* about 11,000 were specific to stage of differentiation, and about 6,000 were expressed at all stages. These results contrast with the finding of Zantinge and coworkers that of the roughly 10,000 genes expressed in the fungus *Schizophyllum commune* over 90% are shared between morphologically different mycelial types. Taken together these observations of diversities of genes expressed in simple organisms are difficult to interpret. Part of the high diversities observed probably results from the fact that these studies examined total cell RNA. That much of this RNA diversity is never translated into proteins is suggested by recent observations by Firtel and his collaborators on *Dictyostelium*. Only half of the RNA sequences measured by Firtel appear to serve as mRNA. If this finding also applies to the fungi *Neurospora* and *Schizophyllum*, it will serve to bring estimates of numbers of genes expressed closer to the mRNA diversities of yeast and *Achlya*. The most meaningful approach may be to regard the lowest gene numbers as representing the minimum numbers of genes required for housekeeping functions. In some forms, large changes in numbers of genes expressed accompany a rather simple morphological differentiation, but in others not greatly different in their morphogenetic abilities very small gene numbers seem to be sufficient. Again, in the absence of any solid data on the functions of the genes expressed, we suggest that the cases in which smaller numbers of genes are expressed may have more value in revealing the minimum number of genes actually involved in morphological differentiation.

The comparatively low mRNA diversities of adult sea urchin tissues suggest that a rather small number of genes represented by proteins are required for maintenance of the differentiated state in animals. This conclusion also follows from studies made with differentiated tissues of higher organisms, such as those on the mouse by Hastie and Bishop and on the chicken by Axel and co-workers. Mouse kidney, liver, and brain, like chicken oviduct and liver, were found to contain about 12,000 different mRNAs. In both organisms only about 10–15% of the mRNA species was unique to any tissue: The rest were shared. That only

1,000–2,000 genes expressed as proteins are sufficient to define individual tissues agrees with the result for sea urchins, but the question of why there should be such a large number of sequences, about 10,000 common to several very disparate tissues, remains open. The situation may in fact be even more complicated than is suggested by these studies. Hahn and his collaborators have examined the diversity of mRNAs in mouse brain by a hybridization technique capable of detecting rarer mRNA sequences than those observed by Hastie and Bishop, and Van Ness et al. detected the presence of as many as 170,000 different mRNA species. It should be kept in mind that brain is an incredibly complex tissue composed of perhaps hundreds of cell types and subtypes. Thus, high mRNA diversity may reflect a high degree of cellular diversity in this tissue.

Large numbers of genes have also been found by Kamalay and Goldberg to be expressed as mRNAs in the tissues of a higher plant, tobacco. Roughly 25,000 sequences are present in leaf, stem, root, petal, anther, and ovary. About two-thirds of the sequences are shared between tissues, leaving 6,000–10,000 sequences tissue-specific.

But our concern here is with the genetic cost of ontogeny *per se*. However, it may be risky to generalize from the only two organisms for which estimates have so far been made. In both the fly *Drosophila* and the sea urchin *Strongylocentrotus,* a relatively large proportion of the genes expressed at some time during the life cycle are expressed in a specific manner during ontogeny. The crucial question of how many of these genes control morphogenesis is simply unanswerable at present. The overall proportion of genes concerned with morphogenesis may be great, but paradoxically the number of genes that actually regulate morphogenesis may not be. Many structural genes required for morphological ontogeny provide essential products without which particular morphological entities could not be assembled. Yet these genes provide little in the way of regulatory information: They are instead regulated in their action. Genes of this type should not be thought trivial, however, because the products of some of them, as for example, tubulins, actins, or cell surface proteins, provide the actual machinery for cell shape-change and cell movements directly underlying morphogenesis. Much of the control exerted by regulatory genes, those genetic gray eminences, must be devoted to orchestrating the expression of ontogeny-specific structural genes. If regulatory genes were very large in number, interactions between them would be so complex as to render viable evolutionary changes nearly impossible. That large numbers are not involved, at least in several instances for which quantitative estimates of regulatory gene numbers are available, comes from evidence such as that discussed in Chapter 3, in which it was shown that 10 or fewer genes determine head shape in two species of Hawaiian *Drosophila*. The genes discussed in Chapter 8 that effect basic segmental commitment in *Drosophila* are also relatively few in number (with 15 involved in determination of most of the head segments and defining the fates of body segments), as are the genes discussed in Chapter 5 that regulate the number of toes in mammals or the number of ambulacra in echinoderms.

Gene Switching and Multigene Families

Simply tallying the number of genes whose activity is required for normal development provides an estimate of the complexity of ontogeny. However, such an estimate will be a misleading one unless it is understood that structural genes specific to development are not expressed everywhere in the embryo at all times. Both spatial and temporal controls regulate gene expression. The patterns of localization discussed in Chapter 4 are crucial to the establishment of the initial distribution of groups of determined cells within the embryo. Although such localization phenomena are a *sine qua non* for ontogeny, the process of development is fundamentally one of a series of cascades of events of ever-growing complexity. The initial localization patterns established during cleavage only serve to rough out a simple early embryonic morphology that generally changes dramatically with the morphogenetic events initiated by gastrulation and subsequent organogenesis. The cells whose fates are determined during cleavage, or later, are distinguished by their distinct locations within the embryo, and by their unique patterns of gene expression. The diversity of mRNAs extracted from a whole gastrula or larva represents the sum of mRNA diversities in several cell types.

Increasing complexity in ontogeny requires that the processes producing differentiation and morphogenesis have two characteristics: Events must occur in correct temporal relationship to one another, and differentiating regions of embryos must interact with one another.

Many structural genes expressed only at specific stages in development are subject not only to temporal control, but also have the interesting property of belonging to multigene families. Naturally, not all genes switched on or off during development are members of such families; many must be genes represented only once in the genome. However, a surprisingly large number of multigene families exists, and their expression, in almost all cases, involves developmentally regulated gene switching. Multigene families provide the embryo with a means of meeting its changing needs for proteins with similar but not identical functions as development proceeds and metabolism as well as cellular and embryonic architecture change. Because many multigene families include genes whose products are quantitatively important and easily isolated for study, these families have already contributed a great deal to our understanding of gene switches.

Our definition of a multigene family is slightly modified from that of L. E. Hood and his coauthors. A multigene family is a group of genes that exhibits close sequence homology, and has related or overlapping phenotypic functions. The degree of multiplicity can vary from a few copies, as in the case of globins, to several hundred copies, as in the case of histones and structural RNAs. The multiplicity and other characteristics of several well-documented multigene families are presented in Table 10–4.

A large number of multigene families are known. Some of these, the

TABLE 10–4. Some Multigene Families[a]

Family[b]	Gene Product	No. of Genes per Family[c]	Information Content/Family	Gene Organization (see Fig. 10–3)
18 and 28S	Ribosomal RNAs	100–600	Repeat of two genes	Tandem repeat
5S	5S RNA	2,000–24,000	Repeat of a single gene	Tandem repeat
tRNA	tRNA	6–400	Repeat of a single gene (many families)	Tandem repeat
Histone	Histones	10–1,200	Repeat of five genes (several subfamilies)	Cluster of five different genes with clusters repeated n times in tandem
Antibody v and c regions	Antibodies	Hundreds	Very large	Tandem repeats subject to rearrangement
Hemoglobin	α-Globins	1–3	1–3 related genes	Group of related genes in tandem
	β-Globins	2–7	2–7 related genes	Group of related genes in tandem
Tubulin	α-Tubulins	3–5 (possibly 10 or more in some organisms)	3–5 related genes	Groups of related genes or scattered
	β-Tubulins	3–5 (possibly 10 or more in some organisms)	3–5 related genes	Groups of related genes or scattered
Actin	Actins	5–20	5–20 related genes	Group of genes in tandem or scattered
Chorion	Insect egg chorion proteins	20–200	Groups of related genes in related subfamilies	At least three neighboring clusters of genes
Ovalbumin	Bird egg albumin and two related gene products	3	Three related genes	Groups of related genes in tandem
Keratin	Keratins	≥ 6	Several related genes	?
Cytochrome c	Cytochrome c	2	2 related genes	?
Vitellogenin	Yolk proteins	≥ 4	4 groups of related genes	?
Chymotrypsin	Chymotrypsin A and B	2	2 related genes	?
Haptoglobin	Haptoglobin α	2	2 related genes	?

(continued)

TABLE 10–4. (continued)

Family[b]	Gene Product	No. of Genes per Family[c]	Information Content/Family	Gene Organization (see Fig. 10–3)
Crystallins	α, β, γ crystallins of eye lens	> 10	Groups of related genes in subfamilies	?
Fibrinogen	Fibrinogens	3	3 distantly related genes	?
Preproinsulin	Insulin precursors	2	2 related genes	?
Amylase	Amylases	5	5 related genes	?

[a]Data presented in this table are drawn from the following: Bloemandal (1977), Brown et al., 1977; Brown and Dawid, 1968; Chambon et al., 1979; Childs et al., 1979; Cleveland et al., 1980; Cohen et al., 1976; Efstratiadis et al., 1979; Fuchs and Green, 1978; Goldsmith and Basehoar, 1978; Goldsmith and Clermont-Rattner, 1979; Hardison et al., 1979; Hennig, 1975; Hood et al., 1975; Jones et al., 1979; Kemphues at al., 1979, 1980; Kindle and Firtel, 1978; Kitchen, 1974; Kitchen and Brett, 1974; Lacy et al., 1979; MacDonald et al., 1980; Newrock et al., 1977; Raff et al., 1982; Schaffner et al., 1978; Scheller et al., 1981; Schuler and Keller, 1981; Sim et al., 1979; Wahli et al., 1979; Wood et al., 1977.

[b]Range in numbers of genes results from phylogenetic differences and do not represent uncertainty. For example, mammals have about 10 copies of the major histone gene cluster, whereas sea urchins have up to 1,200; thus, the tabulated figure of 10–1,200.

[c]Refers to individual family. There are, for example, many families of tRNA genes representing the 62 codons for the 20 amino acids found in proteins. Multiplicity of related families varies. There are many families of tRNAs, histones have a few subfamilies and most of the genes listed fall into one or a few subfamilies.

highly repeated satellite sequences, are not transcribed at all. Others, including many moderately repeated families, are transcribed to yield RNA sequences that apparently are not translated to produce proteins. The function of these RNA sequences is still to be determined. However, many multigene families consist of *bona fide* structural genes of well-defined function, which include the genes for the structural RNAs, ribosomal, 5S, and tRNAs, and the genes for a wide spectrum of proteins. Proteins encoded by multigene families include species important in cell motility and shape, such as actins and tubulins; structural proteins important to morphogenesis, such as collagens, keratins, and chorion proteins; some serum proteins; the oxygen carrier proteins, hemoglobins; some membrane proteins; the histones important in chromosome structure; the storage proteins of yolk; and antibodies. The arrangements of several of this multitude of multigene families on chromosomes are known, and there are several quite distinct multigene arrangements. These are diagrammed in Figure 10–3. The genes for several structural RNAs are arranged in the tandem repeat pattern, in which a series of identical genes for a particular product are linked together as gene—spacer—gene—spacer. The regions between the structural genes are sometimes transcribed, and sometimes not: They serve no known codogenic function, and are thus reasonably thought of as spacers between tandem structural genes. Genes in a multigene family are not necessarily identical. For instance, the family of genes coding for the related but not identical β-globins expressed in an orderly sequence during the development of some mammals are linked in the second pattern shown in Figure 10–3. A cluster of related genes can

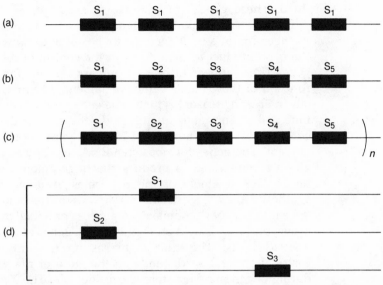

FIGURE 10–3. *Patterns of organization of multigene families. (a) Identical genes linked in tandem, as seen in ribosomal RNA genes; (b) related, but nonidentical genes linked in tandem, as seen in globin genes; (c) a gene cluster of nonidentical genes, as seen in sea urchin embryonic histone genes in which the clusters are tandemly linked; (d) related genes dispersed over several chromosomes, as seen for actins or tubulins.*

itself be the basic unit of a set of tandem repeats. The genes for the histones of sea urchins and other higher organisms form a cluster of genes organized as shown in the third diagram in Figure 10–3. The best-known cluster is that which accounts for the bulk of histone synthesis in the sea urchin embryo. This cluster has the structural genes for the five individual histone species arranged in the order

—(spacer—H2A—spacer—H3—spacer—H2B—spacer—H4—spacer—H1—spacer)—

The major histone types are very distantly related to one another, and each is in fact composed of several subtypes, making five histone multigene families. Thus, there are at least four or five distinct genes each for histones belonging to the H1, H2B, H3, and H4 families, and seven or more members of the H2A family. Some members of the major histone families are organized in clusters as above, but it is clear that not all histone genes are organized in this way.

Multigene families have their evolutionary origins in duplications or higher-order replications of single-copy genes. The initial duplication or multiplicative replication yields tandem genes. In some cases a large number of identical tandem genes is retained. The correction mechanism that keeps identical such genes as the 18 and 28S ribosomal RNA genes probably depends on the maintenance of a tandem arrangement. In some of the smaller tandem families duplication has been followed by gene divergence to produce related but nonidentical genes. In the case of the β-globin genes these have been maintained over considerable evolutionary time. In other cases, related genes, for example, those for tubulins or actins, are organized as shown in the last diagram of Figure

10–3. These genes are scattered over one or several chromosomes.

The diversity of multigene families thus far recognized suggests that membership in a multigene family has little to do with the function of the gene product *per se*, only with the control of the expression of the gene. There appear to be two primary reasons for the existence and function of these families in development. The first, and less intrinsically interesting reason, is that some gene products are needed in a short time and in enormous quantities. In these cases the multigene family consists of a larger number of identical gene copies, usually linked in tandem. The genes for ribosomal RNAs are organized in this manner, and are transcribed to produce the huge amount of ribosomal RNA needed for assembly of the ribosomes used by the cell in protein synthesis. A good, if extreme, example of the demand for the function of large numbers of ribosomal RNA genes is provided by the work of D. D. Brown and I. B. Dawid on the ribosomal genes of the oocyte of the frog *Xenopus laevis*. The embryo makes no new ribosomes until after gastrulation, and so depends on the store of ribosomes accumulated during oogenesis. This store is considerable: The *Xenopus* egg contains about 10^{12} ribosomes. Somatic cells of *Xenopus* contain 450 copies of the ribosomal RNA genes per haploid DNA complement. This is sufficient to provide for the needs of the relatively small somatic cells, but even this multiplicity of genes is inadequate to produce the ribosomal RNA needed by the egg. Brown and Dawid found that these genes were amplified a further 4,000 times in oocytes. A somewhat different strategy is used in oogenesis to provide the 5S RNA also required for ribosome assembly. There are 24,000 copies of the major oocyte-specific 5S RNA gene per haploid DNA complement in *Xenopus laevis*. A different, and smaller, 5S family is transcribed in somatic cells.

The second function of multigene families is to provide gene switching in development. Essentially, multigene families containing related but not identical genes produce similar products specifically required in distinct cell types or at different times in development. The best-understood example is provided by the small multigene families containing the genes for the globins. The evolutionary relationships of human globins, based on protein sequences, are shown in Figure 10–4. The ancestral hemoglobin diverged from myoglobin near the time of the origin of chordates late in the Precambrian. The ancestral β-globin in turn diverged from the ancestral α-globin about 500×10^6 years ago, early in the history of vertebrates in the early Paleozoic, and the fetal γ-chain diverged from the β-chain at most 200×10^6 years ago, early in the history of mammals. Finally, the δ-chain, a variant of β-globin found as a minor component of normal adult hemoglobins, diverged from the β-chain about 40×10^6 years ago. The genes for α-chains comprise a small multigene family of three members, and the genes for β-chains a family of seven members.

The functional hemoglobin molecule is a tetramer composed of two subunits of the α-type and two of the β-type, thus $\alpha_2\beta_2$. In humans there are seven clustered genes of the β-globin type, arranged as shown in Figure 10–4. Two of these code for the $^A\gamma$ and $^G\gamma$ chains found in the fetal hemoglobin $\alpha_2\gamma_2$: Two genes, δ and β, code for chains expressed

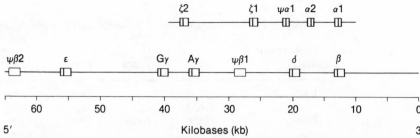

FIGURE 10–4. *Arrangement of human α-like and β-like globin genes. The scale indicates lengths of chromosomal DNA in kilobase units. Globin genes are shown as boxes. Black segments indicate coding sequences, whereas white segments indicate introns. Pseudogenes (ψ) are indicated as entirely open boxes. The 5' to 3' direction of transcription is from left to right.* [Redrawn from A. Efstratiadis et al., The structure and evolution of the human β-globin gene family, *Cell* **21**:653–668, 1980. Copyright by the Massachusetts Institute of Technology.]

after birth in major ($\alpha_2\beta_2$) and minor ($\alpha_2\delta_2$) hemoglobin species. Humans also possess another β-type globin, ϵ, and an α-type globin, ζ, which are expressed only in the embryonic hemoglobin $\zeta_2\epsilon_2$. Two other members of the β family, $\psi\beta_1$ and $\psi\beta_2$, are unexpressed pseudogenes. The time course of hemoglobin switching in human development reveals changes in which hemoglobin genes are expressed, and in site of expression. In the early embryonic portion of development, ζ- and ϵ-chains are synthesized by nucleated megablast cells produced in the yolk sac. This synthesis falls rapidly, and by the sixth week of development is replaced by a pattern of synthesis in which nonnucleated red cells derived from stem cells in liver and spleen produce the α- and γ-globin chains characteristic of the fetus. During late fetal development bone marrow becomes the preponderant site of globin synthesis. Shortly following birth there is a second switch in globin synthesis, and the adult pattern is assumed. The switch exhibits a very significant characteristic. The transition is one involving gene regulation within individual stem cells rather than a replacement of γ-producing stem cells by β-producing stem cells, because during the switch single red cells produce both γ- and β-globin chains.

While hemoglobin gene switching is the rule in vertebrates, there is a surprising diversity of switch patterns, even within the mammals. Humans express distinct embryonic, fetal, and adult globins. Fetal hemoglobins have been shown to have a higher oxygen affinity than the adult hemoglobins of the mother, thus facilitating oxygen transfer across the placenta to the fetus. However, as shown in Table 10–5, not all mammals possess distinct fetal hemoglobins. Rodents, carnivores, and horses, for example, pass through a direct transition from embryonic to adult globins in the fetus. In these cases the affinity for oxygen of the adult hemoglobin in the fetal red cells appears to be modulated by small molecules in the cytoplasm so that the fetal blood has a higher oxygen affinity than the maternal blood.

Organization of the β-globin genes may be very similar, even in organisms exhibiting significant differences in switching patterns. Table 10–5 shows that rabbits produce the embryonic β-like globins called

TABLE 10–5. Hemoglobin Gene Switches in Some Mammals[a]					
Developmental Stage	Rabbit	Sheep	Horse	Stump-Tailed Macaque	Human
Embryonic	$\chi_2\epsilon(y)_2$ $\chi_2\epsilon(z)_2$	$\alpha_2\epsilon_2$	$^S\alpha_2\epsilon_2$ $^F\alpha_2\epsilon_2$	None	$\zeta_2\epsilon_2$
Fetal	None	$\alpha_2\gamma_2$	None	$\alpha_2{}^1\gamma$ $\alpha_2{}^2\gamma$	$\alpha_2{}^A\gamma_2$ $\alpha_2{}^G\gamma_2$
Adult	$\alpha_2\beta_2$	$\alpha_2\beta_2$	$^S\alpha_2\beta_2$ $^F\alpha_2\beta_2$	$^1\alpha_2\beta_2$ $^2\alpha_2\beta_2$	$\alpha_2\beta_2$ $\alpha_2\delta_2$

[a]Based on Hardison et al., 1979; Kitchen, 1974; and Kitchen and Brett, 1974.

$\epsilon(Y)$ and $\epsilon(Z)$, but have no fetal β-globin equivalent to human γ-chains. Instead, the fetus synthesizes the adult β-chain. Lacy and her collaborators and Hardison et al. have found that the β-globin genes of the rabbit are organized in a cluster very similar to that of humans, and that the individual genes in the cluster undergo developmental switches. Two of the genes corresponding in position to the γ-genes in the human β-gene cluster shown in Figure 10–4 are expressed in the embryo, presumably to produce the ϵ-chains. The gene corresponding in position to the human δ-gene appears not to be expressed at any stage, and the gene corresponding to the human adult β gene serves the same function in the rabbit.

While the globins illustrate the role of switching in the expression of a succession of genes directly involved in metabolic functions, there are more complex multigene systems whose products are directly involved in gene expression or morphogenesis. Some of these families include a large number of members under developmental regulation.

The histone genes, best studied in sea urchins, nicely illustrate the adaptation of using multigene families to satisfy both the necessity of producing a large amount of protein in early development and the need for switching to express a sequence of related proteins in development. During cleavage the embryo doubles its nuclei and the chromosomes they contain as rapidly as every 10–20 minutes. Sufficient histones to support the assembly of chromosomes during rapid cleavage can come either from stores of proteins within the egg, as Woodland and Adamson have shown for the frog *Xenopus,* or from massive synthesis of histones in the cleavage-stage embryo, as has been found by Kedes and his collaborators to be the case in sea urchins. These two strategies make different quantitative demands on histone genes. The histones of the frog egg are accumulated slowly during weeks or months of oogenesis, whereas those of the sea urchin embryo are synthesized over a period of a few hours. The sea urchin egg contains only about 25% of the histone mRNA it will need to produce histones during cleavage; the remainder is transcribed during cleavage. As a consequence, the main histone gene family of sea urchins is much more highly repeated than that of frogs (or for that matter humans). Whereas sea urchins possess 300–1,200 copies of these genes, Birnstiel et al. and M. C. Wilson and co-workers have found *Xenopus* and human histone genes to be repeated only 10–20 times.

The histones of sea urchins have been found to be the subjects of a very elaborate set of switches, including both temporal and tissue-specific changes. Newrock and his collaborators have documented in intricate detail a sequence of changes, first observed by Ruderman and Gross, in which of the five major histones, three (H1, H2A, and H2B) are represented early in cleavage by a short-lived synthesis of cleavage-stage-specific subtypes. This synthesis is followed by synthesis of a sequence of other histone subtypes in a stage-specific manner, as diagrammed in Figure 10–5. The cleavage-stage subtypes, cs, of each histone type are succeeded by α-subtypes during cleavage, then by β-, γ-, and other subtypes in the blastula. Experiments by Newrock and co-workers, Kunkel and Weinberg, and Childs et al. all show that the histone protein switches result from a gradual succession of changes in histone mRNA synthesis. Each subtype is the product of a distinct gene related to other subtype genes in a family. The switching of histone subtypes results in a change in chromosome protein composition as development proceeds. Such changes may result in a "remodeling" of chromatin, potentially of importance to differentiation of cells within the embryo.

Switches are not limited to proteins functioning, like globins or histones, primarily in the internal economy of cells. Some are intimately involved in morphogenesis. For example, the microtubules so central to cell movement and cell shape are composed of α- and β-tubulins, which are the products of small multigene families. The tubulins have been found by E. C. Raff and her co-workers to be regulated by switches during the development of *Drosophila*. Some tubulin species are synthesized throughout development, but at least one β-tubulin is switched on and then off during a restricted period of embryogenesis, and Kemphues et al. have demonstrated that there is a tissue-specific β-tubulin expressed only in the testis. The testis-specific β-tubulin is required for the assembly of a very specialized microtubule structure, the axoneme of the sperm tail.

The study of the role of switches of multigenes in morphogenesis has been developed by F. C. Kafatos and his collaborators in their studies of the chorion proteins that make up the shell of the silkmoth egg. Seemingly a humble object, the egg shell can be seen with use of scanning and transmission electron microscopy to have an elegant structure and surface geometry, which would certainly have pleased D'Arcy Thompson. The surface of the eggshell of the silkmoth *Antheraea polyphemus* is shown in Figure 10–6. The predominant surface features are the hexagonal pavement, which marks the former sites of follicle cells, and the tall, chimneylike respiratory structures evocatively called aeropyles. The pit in the foreground is the micropyle, through which the sperm enters at fertilization. In cross-section the shell can be seen to be mechanically strong but light, with internal vaulting. The layers of the shell are secreted by a sheath of follicle cells that synthesize the chorion

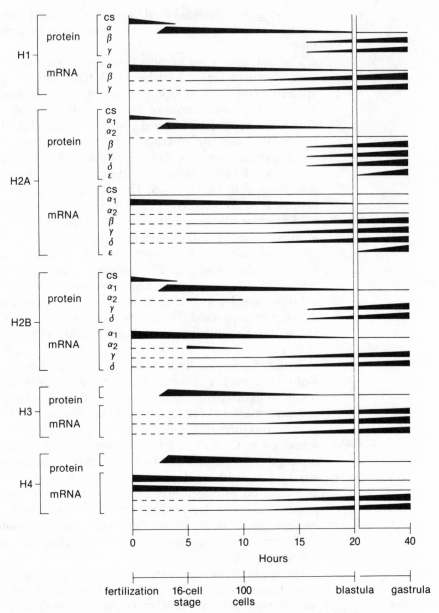

FIGURE 10–5. *Expression of members of histone families during the development of sea urchin embryos. Thickness of bars is a schematic indication of level of synthesis of protein or amount of mRNA present. Note that translational regulation as well as mRNA synthesis control histone protein synthesis because mRNAs stored in the egg are present in early development but not immediately translated.* [Modified from K. M. Newrock et al., Histone changes during chromatin remodeling in embryogenesis, *Cold Spring Harbor Symp. Quant. Biol.* **42**:421–431, 1977; and G. Childs, R. Maxson, and L. H. Kedes, Histone gene expression during sea urchin embryongenesis: Isolation and characterization of early and late messenger RNAs of *Strongylocentrotus purpuratus* by gene-specific hybridization and template activity, *Dev. Biol.* **73**:153–173, 1979. Additional data from D. E. Wells et al., 1981. Unpublished data of D. E. Wells et al. and of L. H. Cohen were used to define the times of onset of translation of histones from stored mRNAs.]

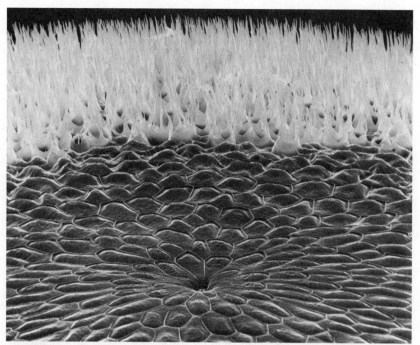

FIGURE 10–6. *Surface structure of the eggshell of the egg of the silkmoth* Antheraea polyphemus. *This view shows the micropyle (site of sperm entry) surrounded by a pavement of concentric cell imprints in the foreground and the aeropyles (respiratory structures) standing in the background.* [A complete description of the chorion has been presented by J. C. Regier, G. D. Mazur, and F. C. Kafatos, The silkmoth chorion: Morphological and biochemical characterization of four surface regions, *Dev. Biol.* **76**:286–304, 1980. Photograph courtesy of G. D. Mazur and F. C. Kafatos.]

proteins, and there are close to 200 different chorion protein species synthesized by the silkmoth. These fall into five broad classes defined by molecular weight and amino acid sequence relationships. Each class comprises a family of related genes, but as C. W. Jones et al. have shown there is also a considerable degree of relatedness between restricted regions, or domains, of proteins belonging to different families within the chorion superfamily. The relatedness of the genes for these proteins is further indicated by the finding of Marian Goldsmith and her collaborators that in the silkmoth *Bombyx mori* a large number of the genes coding for members of these families are organized into three clusters of genes on a single chromosome.

Expression of chorion genes is controlled by a series of switches. Sim et al. have found that synthesis of members of the three higher-molecular-weight families predominates early in choriogenesis, with synthesis of members of the two lower-molecular-weight families predominating throughout middle and late choriogenesis. By use of appropriate probes consisting of recombinant DNA clones of specific chorion genes, Sim et al. were able to detect the mRNA species corresponding to the cloned genes. Changing patterns of protein synthesis reflect changing patterns of mRNA synthesis in the follicle cells.

Two recent observations made in Kafatos' laboratory make it certain

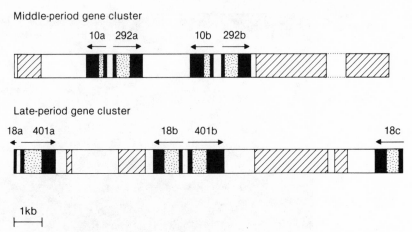

FIGURE 10–7. *Arrangement of two sets of coordinately regulated chorion protein genes. Two clones are shown that carry clustered silkmoth chorion protein genes. Genes 18 and 401 encode two different proteins expressed late in chorion (eggshell) assembly, whereas genes 10 and 292 encode two other distinct proteins expressed during the middle period of chorion assembly. The paired genes are transcribed in opposite directions with a short 5' spacer (white) between genes. Each gene has a small 5'-coding sequence (black) followed by a large intron (stippled). Genes and surrounding DNA sequences are repeated in clusters. Cross-hatched regions represent spacer regions that are variably inserted into spacers.* [From C. W. Jones and F. C. Kafatos, Structure, organization and evolution of developmentally regulated chorion genes in a silkmoth, *Cell* **22**:855–867, 1980. Copyright by the Massachusetts Institute of Technology. Redrawn from a gene map provided courtesy of C. W. Jones and F. C. Kafatos.]

that chorion genes are not only evolutionarily and functionally related; they are also genetically organized for coordinate expression. There is a mutant in the silkmoth *B. mori* that results in a defective eggshell, and it is the consequence of a deletion of a section of DNA containing about half of the chorion genes. The deleted genes are primarily late-expression genes, suggesting a clustering of genes by time of expression. Use of recombinant DNA clones of fragments of DNA containing more than a single specific chorion gene has allowed Jones and Kafatos to examine gene organization from the point of view of temporal control of expression. The arrangement of genes in two such clones are shown in Figure 10–7. The two clones contain different genes, but both contain two copies of each of two distinct genes. The genes belong to different chorion gene subfamilies, but they are coordinately expressed. Those in the upper fragment are utilized midway through choriogenesis, whereas those in the lower fragment are expressed late in the process. In both cases physically contiguous pairs of genes appear to be linked in transcription, with a common control element lying between each pair.

Eggshell structure has undergone a variety of evolutionary modifications in related species of silkmoths. This can be seen in the eggshells of *A. polyphemus* and *A. pernyi* contrasted in Figure 10–8. These species may have diverged as long ago as $10–30 \times 10^6$ years, although the not surprising dearth of a fossil record for these big moths makes this a somewhat uncertain estimate. Has the change in egg shell morphology resulted from changes in the structural genes for chorion

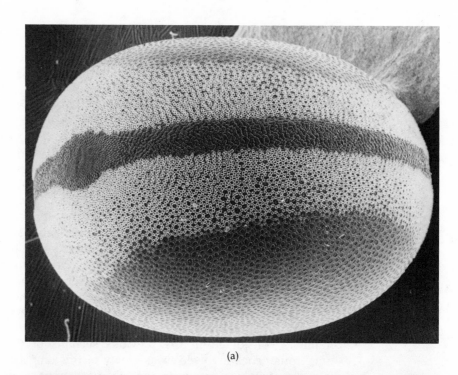

(a)

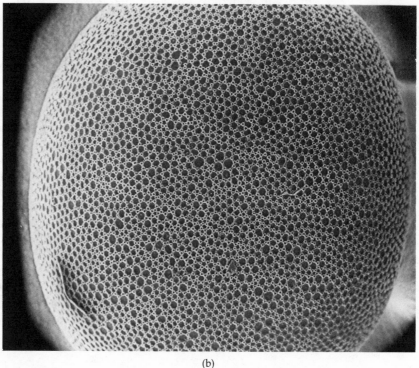

(b)

FIGURE 10–8. *The surfaces of the chorions of the eggs of two species of silkmoths. Top panel shows the surface of the eggshell of* Antheraea polyphemus. *The micropyle is at the center of the wide area in the stripe running around the perimeter of the shell. The two broad white bands consist of aeropyles. Two broad flanking regions which lack aeropyles lie on each side. The lower panel shows the surface of the shell of* A. pernyi. *The entire surface is covered by aeropyles. Detailed descriptions of chorion structure have been made by F. C. Kafatos et al., 1977, and J. C. Regier et al., 1980.* [Photographs courtesy of G. D. Mazur and F. C. Kafatos.]

311

proteins, or from changes in control of expression? Recent experiments performed in Kafatos' laboratory indicates that changes in chorion structural genes have occurred, but have been limited in extent. Morphological differences primarily reflect differences in gene expression. In light of the importance that changes in relative timing of processes have in modifying development during the course of evolution, it is interesting that the relative timing of expression of chorion genes is the same in both species of *Antheraea*. Differences in chorion gene expression are instead quantitative, with some chorion mRNA species differing greatly in amount.

There are mutations in chorion protein genes that begin to reveal the function of these switched genes in morphogenesis. *D. melanogaster* chorion proteins comprise a much smaller family than that of silkmoths, perhaps 20 genes. A mutation in one of the chorion protein genes in this species recently discussed by Digan and her collaborators produces a morphologically deranged eggshell, indicating that at least one of the chorion proteins has a role in the organization of eggshell structure.

Chorion genes bring us almost full circle to the question of gene numbers considered in the previous sections of this chapter. The numerous genes involved in assembly of this one structure and their intricate switching suggests that there may be a demand for the expression of large numbers of related genes in many of the morphogenetic processes of development. This possibility is certainly borne out by the histones, tubulins, actins, and other proteins synthesized by embryos. As these proteins, once thought to be the products of one or at most a very few genes, have been studied in more detail it has become obvious that their expression in development in reality involves whole families of structurally related and functionally coordinated genes.

The expression of large numbers of structural genes is necessary for development. Many developmentally regulated structural genes are members of evolutionary-related multigene families derived from an ancestral gene by duplication and divergence. Multiple, related genes allow fine-tuning of structural gene expression, with individual members expressed during precisely controlled periods or groups of cells in development. All of the β-globins, for example, serve the same general function, but in somewhat different ways and with different efficiencies with respect to the fetal versus the adult environment. The possession of multigene families provides organisms with evolutionary flexibility in that changes in time or location of expression of a member of a multigene family will not affect expression of other members of the family. Dissociation of developmental processes may thus be facilitated.

Eleven

The Eucaryotic Genome
and the C-Value Paradox

> We have no sufficient grounds for assuming that evolution has been
> brought about primarily by an increase in the number of genes in higher
> forms...the emphasis lies not on numbers but on kinds of new genes
> that have appeared.
>
> T. H. Morgan, The Scientific Basis of Evolution

Genome Size and Organismal Complexity

Morgan's statement on the relationship between gene numbers and
organismal complexity was written in 1932. With the recognition a
generation later that genes are composed of DNA, a much clearer idea of
the nature of genes became possible, yet this progress only served to
further confound the question of the relationship between numbers of
genes and organismal complexity. Quite simply, measurements of the
amounts of DNA present in the haploid genomes (the C-value) of a wide
spectrum of organisms indicated a general increase in DNA content with
complexity, but there was a great deal of variation in genome size such
that many morphologically primitive organisms were found to have
considerably larger genomes than morphologically more advanced
forms. This C-value paradox is illustrated in Figure 11–1.

Any such figure stands in danger of being taken as a sort of twentieth
century scale of being, and in a sense that is so. The distinct
morphological organizations make direct comparisons of relative complexity subjective. However, there are two indices of complexity that
provide rough metrics; these are the number of discernible cell types
and the number of terms used by taxonomists to describe members of a
particular group of organisms. Use of cell numbers was given theoretical
justification by S. Kauffman, who proposed that the number of
differentiated cell types in an organism is a function of the number of
stable states produced by the regulatory interactions possible within any
particular genome. Cell number estimates are relatively straightforward
for less complex organisms. Thus, bacteria produce two cell types
(vegetative cells and spores); yeasts three to four types; algae and fungi

313

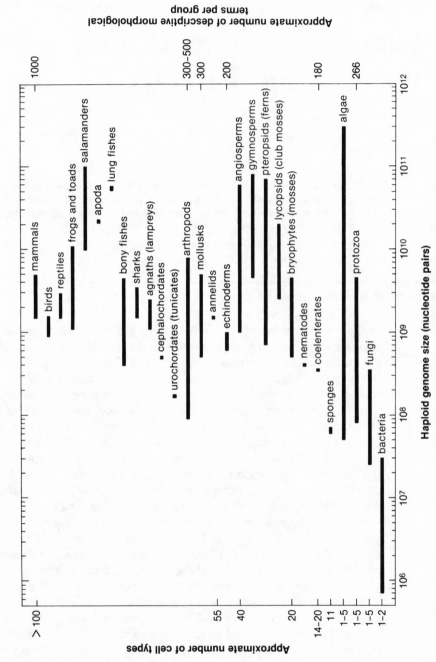

FIGURE 11-1. *The C-value paradox: noncongruence of genome size and morphological complexity. The bars show the ranges of haploid genome sizes for various higher categories of organisms. The ordering of categories is from morphologically most simple at the bottom to most complex at the top. Obviously, such an ordering is highly subjective. Two estimates of complexity are given: Approximate numbers of cell types in the body of some groups are indicated on the left vertical axis, and approximate numbers of morphological descriptive terms for certain groups are indicated on the right vertical axis.* [C-value data are from A. H. Sparrow et al, 1972.]

about five types; sponges, 11; coelenterates, 14–20; plants about 20–40; annelids about 55. Estimates for more highly differentiated animals are more difficult, and Kauffman's estimate of approximately 100 cell types for humans may be as much as one order of magnitude too low.

The use of numbers of anatomical terms used by taxonomists as an index of complexity has been suggested by Schopf et al. This may give a less objective estimate of complexity than histological differentiation, assuming that numbers of cell types have been accurately determined, which is highly problematical for complex organisms. Numbers of terms may be a function of differences in practices among taxonomists studying different groups. For example, among the mollusks, ammonites may have fewer terms applied to them than to other organisms of comparable complexity because their elaborate sutures are generally illustrated with only a minimum of verbal description. Groups under active investigation by large numbers of morphologists or taxonomists may have been more finely tagged than more obscure groups. On the whole, the use of numbers of terms appears to be a valid, if crude, expression of morphological complexity. However, it is a measure that does not necessarily correspond well with histological complexity. In fact, the meaning of correspondence between histological and morphological complexity is confounded by such organisms as the foraminifera among the protozoa for which Schopf et al. list 266 morphological terms.

The scheme of organismal complexities used in Figure 11–1 is based primarily on numbers of cell types, with secondary consideration of numbers of terms. Among related groups (for example vertebrates) phylogenetically more primitive groups are generally placed below more advanced groups. Admittedly, there is a certain artistic license to this. For instance, the lungfish is an older group than the teleost fish, but it is more closely related to the amphibians. Genome size is given as number of nucleotide pairs. A rough estimate of the information content in terms of structural genes can be obtained if one assumes that an average structural gene has a coding sequence length of 1,500 nucleotide pairs. The smallest viral genomes contain (even with overlapping genes in alternate reading frames) only a few genes. The low end of the bacterial range at about 0.7×10^6 nucleotide pairs demarks the minimum genome size required for a living cell. Similarly the low end of the fungal range defines the minimum eucaryotic genome, perhaps as few as about 10,000 average genes in yeast. The simplest multicellular animals are the sponges, which have genomes of about the same size as fungi and near the minimum for algae and protozoa. Coelenterates also have small genomes that fall in the fungal range. While coelenterates have many more cell types than fungi, some fungi are complex and fungal genomes cover a 10-fold range of C-values, so the overlap can be easily rationalized. Genome sizes for more complex organisms are scattered over a wide range of C-values. The low end of the range for any group in which a large number of species have been examined provides a reasonable estimate of the minimum genome size for the group. Individual groups commonly exhibit a wide range of values, as much as four orders of magnitude in DNA content. In some cases this range is extended even farther by polyploidy.

A Three-Part Paradox

Genome size data reveal that there is in fact not one C-value paradox, but three. The first, already alluded to, is the poor correlation between organismal complexity and genome size. This is the disquieting paradox in which fruit flies, despite their advanced histological, morphological, and developmental traits and their behavioral and musical abilities, have in common with bath sponges and bread molds a genome size of about $0.5–1.0 \times 10^8$ nucleotide pairs. Other disparities are obvious from Figure 11–1. In some cases more primitive forms greatly exceed their more advanced relatives in genome size. The lungfish have genome sizes 10–15 times those of mammals, and the formal evolutionary sequence from amphibians to reptiles to birds is accompanied by a nearly six-fold decrease in genome size. A similar relationship is seen within the insects. Bier and Müller, who measured the genome sizes of a variety of insects, observed that archaic groups have larger genomes than more recently evolved insect groups. The range encompasses two orders of magnitude. Comparisons between primitive and advanced members of a group are not a function of complexity but of ages of the taxa. It is doubtful that a lungfish is either histologically or morphologically less complex than a teleost or a frog. In fact, in many cases archaic organisms are morphologically more complex than their advanced relatives. Advanced groups have often undergone reduction in features, as for example, the number of bones in the skull and jaws of vertebrates. It is in age and rate of morphological evolution that archaic and modern groups differ. Lungfish date back nearly 400×10^6 years, whereas placental mammals are perhaps one-fourth as old. Bier and Müller concluded that the high C-values found in primitive forms reflect gene duplication rather than acquisition of new genetic information. This idea is supported by the observation of Sparrow and Nauman that genome sizes within any large group do not follow a normal distribution. Instead, the log scale distributions are found to be a series of families of peaks, each of twice the DNA content of the preceding family, which suggests that a series of genome doubling events has occurred. It is possible that all taxa may undergo processes that enlarge their nuclear DNA complements, but that the old groups are less likely to purge their genomes of excess DNA. Thus, Devonian lungfish, in the period of their rapid morphological evolution, may have had very much smaller genomes than their morphologically conservative descendants.

A second aspect of the C-value paradox is that related animals that are very similar in complexity and evolutionary advancement within their groups often have very different genome sizes. For instance, Ebeling et al. have recorded two-fold differences in genome size between species of *Bathylagus*, a genus of teleost fish; Chooi has documented a six-fold range in genome sizes among members of the plant genus *Vicia;* and finally, Laird has summarized the genome size data for several species of *Drosophila,* which show a 2.5-fold spread in C-values.

Cytological analysis makes it clear that C-value differences among related organisms usually do not result from polyploidy. In the case of *Vicia,* Chooi observed that two species are polyploid but the others were not. DNA differences affected all chromosomes, and apparently resulted from a series of localized duplications. Changes in DNA content do not appear to result from a polytenization event such that a chromosome comes to contain two or more identical strands of DNA lying side by side. Instead, the experiments of Kavenoff and Zimm have solidly established that each chromosome contains only a single DNA molecule. Changes in DNA content within a chromosome produce a proportional lengthening of that chromosome. Kavenoff and Zimm isolated chromosome-sized DNA molecules from three species of *Drosophila* with different genome sizes, and determined the length of such molecules by their viscoelastic properties. They measured DNA molecules from wild-type flies as well as from other karyotypes in which chromosomes were lengthened or shortened by translocations or deletions. The lengths of the longest chromosomes of the various species or karyotypes had a spread of about four-fold in relative length. For example, the genome sizes of *D. virilis* and *D. americana* are very similar, but the longest chromosome of *D. americana* is about twice the length of the longest chromosome of *D. virilis.* The length of the isolated DNA molecules corresponding to these species showed a similar length relationship, and the individual DNA molecules were of sufficient length to account for the DNA content of their respective chromosomes.

Organisms with similar histologies and morphologies ought to require the expression of similar numbers of genes, and this appears to be the case when appropriate comparisons are made within such groups as amphibians and insects. The salamander *Triturus* has a genome that is about seven-fold larger than that of the frog *Xenopus.* Rosbash and his collaborators found that whereas 75% of the *Xenopus* genome is composed of unique sequence DNA, the *Triturus* genome contains a broad spectrum of sequence multiplicities with only a small proportion of unique sequences. The large genome of *Triturus* seems to have evolved by means of repeated duplications of a majority of the sequences in the ancestral genome, including at least some functional genes, because Rosbash et al. found *Triturus* to contain seven times the number of ribosomal genes as *Xenopus.* However, the messenger RNAs (mRNAs) of both species are primarily transcripts of single-copy sequences of their respective genomes. It should be noted that the existence of the numerous multigene families discussed in Chapter 10 does not contradict the observation that most mRNAs are the products of unique sequences. This is so because most multigene families contain only a few members, and these, although related to each other in sequence, generally have diverged sufficiently to appear unique to the hybridization techniques used to define sequence copy number.

Despite the seven-fold larger genome of *Triturus,* the number of genes

expressed as mRNAs in the ovaries of both species is apparently the
same. The logical consequence is that the majority of the repeat
sequence DNA that constitutes most of the large *Triturus* genome seems
to be noncoding, at least in the sense of yielding mRNA.

Lengyel and Penman performed a comparable study in comparing the
mosquito *Aedes* with a more advanced dipteran, *Drosophila*, which
provided a revealing insight into an important part of the mechanistic
rationalization of the C-value paradox. The total genome of *Aedes* is six
times larger than that of *Drosophila*, although if only the single-copy
portions of the two genomes are compared the difference is reduced to
four-fold. In cultured cells of both species the majority of the mRNAs are
the products of unique-copy sequences. Further, these mRNAs are
about the same length and they contain essentially the same numbers of
distinct mRNA sequences. Thus, as in the case of *Xenopus* and *Triturus*,
two related organisms with disparate C-values express the same number
of genes as mRNAs. In addition, Lengyel and Penman observed that the
nuclear RNAs of *Aedes* were at least twice the length of those of
Drosophila. This suggests that the individual transcription units of *Aedes*
are longer than those of *Drosophila*, but that equal-sized coding regions
are excised from transcripts in processing. This idea was further
supported by the kinetics of conversion of nuclear RNAs to mRNAs in
the two species. *Drosophila* converts 20% of its transcripts into mRNA,
whereas *Aedes* converts only 3%, a six-fold difference that may reflect
transcription of noncoding sequences as well as relative sizes of
transcription units.

The question of transcription unit size will also emerge as critical in
resolving the third aspect of the C-value paradox. Organisms, even
Drosophila with its notably small genome, contain much more DNA than
can be accounted for in the numbers of genes they express. The
relationship between numbers of bands on the polytene chromosomes
of *D. melanogaster* and the number of genes expressed suggests that
there are roughly 5,000 genes in this organism. The diversity of mRNAs
measured for *Drosophila* (presented in Table 10–3) corresponds reasonably well with this estimate.

There is a third and completely independent way to estimate gene
numbers, based on mutation frequencies. Natural populations of
diploid organisms, be they *Drosophila* or humans, carry a considerable
genetic load of deleterious mutations. These include lethal alleles,
semilethals, and a variety of physiological or morphological mutations.
The extent of genetic load has been summarized by Dobzhansky, and a
few examples will make the point. Some populations of *D. melanogaster*
and *D. subobscura* exhibit up to 10% of morphologically abnormal
individuals. In a different species, *D. pseudoobscura*, approximately 30%
of second, third, and fourth chromosomes are lethal when two identical
chromosomes isolated from wild populations are present in a single
individual. Over 50% are subviable. The measurement of the rate at
which a genome acquires new mutations was pioneered by H. J. Muller
and his co-workers. Essentially, one isolates a population of flies
homozygous for a chromosome (for example, the X chromosome)

bearing no lethals. Crosses are made within the population, and in each generation progeny are counted to determine if a new lethal mutation has appeared. In the case of the X chromosome used by Muller the test is quite simple because if a lethal has appeared, the ratio of females to males in the progeny will change from a normal 1:1 to 2:1 as males carry only one X chromosome. Muller and his co-workers determined that the overall genomic rate of mutation in *D. melanogaster* is about 0.05 per gamete per generation.

Mutation rates have also been measured for individual genes in several organisms. A large number of these have been tabulated by Strickberger. In *D. melanogaster* the average rate is about 1×10^{-5} mutations in any one gene per gamete. The mutation rate per genome (U) is related to the mutation rate per locus (u) by the number of genes (N) such that

$$N = \frac{U}{u}$$

or for *D. melanogaster* $N = 5,000$. The correspondence with other estimates is remarkable. However, all of these determinations of gene numbers for *D. melanogaster* fall far below the roughly 60,000 average genes that could be encoded, based on DNA content. The problem is exacerbated in organisms with larger genomes, as for instance, humans. The human genome contains sufficient DNA to encode approximately 2×10^6 average genes. Mutation rate studies comparable to those done with *Drosophila* have been extended to humans. Crosses cannot be made as freely as is possible with flies, but data on frequencies of lethality and abnormalities are available from studies on consanguineous marriages between first cousins. These were used by Morton et al. to estimate a genomic rate of mutation for humans (0.1 per gamete per generation). In conjunction with the average rate (1×10^{-5}) of mutation of individual genes, the genomic rate estimate by Morton et al. yields a figure of only 10,000 human genes. King and Jukes have considered the genetic load to which the human population would be subjected given known rates of mutation, and have concluded that there cannot be many more than 40,000 functioning genes in humans. At 40,000 genes the total rate of mutation to lethal or nonfunctional alleles would be in the range of 0.04–0.4 per gamete per generation. The low numbers of human genes estimated by these calculations are difficult to rationalize with RNA diversities as high as 170,000 sequences in mammalian brains, unless the majority of these RNA sequences represent something other than mRNAs. The alternate possibilities are that there are large numbers of genes with mutation rates lower than 1×10^{-5}, or that a majority of mutations result in no demonstrable phenotypes.

The objection might well be raised that our "average" gene with its 1,500 nucleotide pair-long coding sequence might be an unrealistic underestimation. It is true that there are some enormous genes. For example, Daneholt and his collaborators and Lamb and Daneholt have studied a giant RNA produced in the salivary glands of the dipteran *Chironomus tentans*. This RNA enters the cytoplasm and appears to be translated to produce a very large polypeptide chain (molecular weight 850,000). This RNA is transcribed from a region of DNA of a length corresponding to approximately 37,000 nucleotide pairs. However, the vast majority of cellular proteins range around an average of 500 amino acids in length, and their mRNAs average about 2,000 nucleotides in length. The extra roughly 500 nucleotides comprise nontranslated leader and tail sequences on the 5′ and 3′ ends of the mRNA. Yet it is evident from data such as those presented by Chooi on the length of transcription units in *Drosophila* that DNA segments of 10,000–20,000 nucleotide pairs in length provide the preponderance of transcripts from which average-sized mRNAs are processed. The coding sequence evidently does not represent the entire gene.

Genome Organization in Eucaryotes

The first suspicions that eucaryotic genomes would prove to be different and more complex than those of procaryotes grew from experiments performed in the early 1960s by Hoyer, McCarthy, and Bolton. In their experiments DNA was heated to separate the strands, and the separated strands were immediately immobilized in an agar gel. Isotopically labeled DNA strands were then reacted with the immobilized DNA. Labeled strands complementary to the unlabeled strands trapped in the agar would form hybrids that could be detected as bound radioactivity. Hoyer et al. used their technique to measure the degree of evolutionary relationships between DNAs of different organisms. These experiments indicated homology between a wide range of vertebrates from salmon to human with, as expected, the greatest degree of homology between closely related species. These experiments were exciting because they presaged the study of evolution at the genome level.

DNA hybridization requires collisions between two complementary strands, and is a second-order reaction. Reassociation of single strands of concentration C is described by the expression

$$\frac{dC}{dt} = -kC^2$$

where t is time and k is the reassociation rate constant. Integration of this expression when the initial concentration at $t = 0$ is C_0 and the concentration of single-stranded DNA remaining at time t is C, gives

$$\frac{C}{C_0} = \frac{1}{1 + k\,C_0 t}$$

When the reaction is half complete

$$C_0 t_{1/2} = \frac{1}{k}$$

$C_0 t$ is a useful term in which to express the main parameter, the product of initial DNA concentration and time of reaction, which controls the

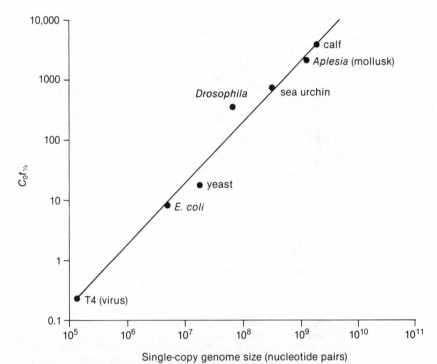

FIGURE 11–2. *Relationship of* $C_0t_{1/2}$ *value to amount of DNA in the single-copy fraction of various genomes.* [Data for T4 phage, *E. coli*, and calf from Britten and Kohne, 1968; for yeast from Hereford and Rosbash, 1977; for sea urchin from Angerer et al., 1976; for *Aplesia* from Angerer et al., 1975; for *Drosophila* from Davidson et al., 1975; and from Manning et al., 1975.]

extent of reaction. In the case of genomes, such as those of procaryotes, which consist almost solely of unique sequences, the $C_0t_{1/2}$ value is a measure of relative genome sizes. This relationship is shown in Figure 11–2.

The experiments of Hoyer and his co-workers should not have worked because the vertebrate genomes they were investigating were much larger than any procaryotic genome. If each sequence were unique, the probability of any two complementary strands colliding would be much lower than in a smaller bacterial genome, and the rate of reaction correspondingly slow. Yet the vertebrate DNA rates were higher than those observed for bacteria. The answer that emerged to this quandary was that much eucaryotic DNA consists of repeated sequences. Such sequences may be present in hundreds or even thousand of copies; thus, their concentrations were high in the preparations studied by Hoyer et al. More advanced techniques allow the study of unique as well as repeated sequences in eucaryotic DNA. The reassociation curves for eucaryotic DNAs are very different than those of procaryotes, as shown in Figure 11–3. While bacterial DNA reacts as a simple second-order reaction, calf DNA exhibits a complex reaction manifesting both fast-reacting repeat sequences (low C_0t) and slow-reacting unique sequences (high C_0t).

The majority of structural genes are unique sequences, but repeat sequences often make up a large proportion of eucaryotic genomes. For

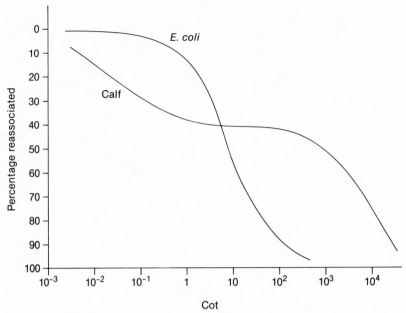

FIGURE 11–3. *Reassociation kinetics for bacterial and eucaryotic genomes. The* E. coli *genome is composed almost exclusively of single-copy sequences and exhibits a single second-order reassociation curve. The calf genome contains both repeat sequences that renature rapidly (at low Cot), and single-copy sequences that renature slowly.* [Redrawn from R. J. Britten and D. E. Kohne, Repeated sequences in DNA, *Science* **161**:529–540, 1968. Copyright 1968 by the American Association for the Advancement of Science.]

example, *Xenopus* DNA, which is fairly typical of metazoans, contains 54% of single-copy sequences, 10% of sequence families repeated about 100 times each, 31% of sequence families repeated about 2,000 times, and about 5% of sequences repeated over 10^5 times each. The number of repeat-sequence families is large, with about 18,000 different families of repeat sequences falling into the 100 times repeat class in *Xenopus*.

The members of a repeat family are related but not necessarily identical. That families differ in degree of divergence of members has been nicely demonstrated by Klein et al. Sea urchin genomes, like other eucaryotic genomes, contain several thousand repeat families. Klein and his collaborators studied recombinant DNA clones of representative members of 18 different families that ranged in degree of repetition from 3 to 12,500 members. These clones were hybridized to genomic DNA, and hybrid stabilities were determined as a test of divergence. Three clones showed very low divergence, seven were moderately divergent, whereas eight exhibited a large degree of divergence within their families. New repeat families apparently arise from sudden replication of a pre-existing sequence. Divergence of family members proceeds by nucleotide substitution in individual members, and the extent of divergence presumably reflects the age of the family. G. P. Moore et al. examined this hypothesis by using clones of repeat families from the sea urchin *Strongylocentrotus purpuratus*. These were used to measure the sizes of their respective families in *S. purpuratus*, *S. franciscanus*, and *Lytechinus pictus*. According to Durham, the two species of *Strongylocen-*

trotus diverged about 10–20 × 10⁶ years ago, whereas the two genera diverged about 150–200 × 10⁶ years ago. Closely related families were found in the three species, but their repetition frequencies differed. For example, one family was found to contain 800 members in *S. purpuratus*, 80 members in *S. franciscanus*, and only eight members in *L. pictus*. Since related repeat families are found in species that separated as long ago as 150–200 × 10⁶ years ago these families must be quite ancient. Further, these families must have undergone independent saltatory replications within the histories of the various species in a manner analogous to that diagrammed in Figure 3–2 for repeat-sequence appearances in primates.

The most highly repeated families contain millions of copies and have a rather simple sequence organization consisting of a basic unit of about 10 nucleotide pairs that is repeated in tandem. These satellite DNAs, which can comprise up to 40% of a genome, are clustered in eucaryotic chromosomes. They are generally not transcribed and apparently function in chromosome organization. The less highly repeated sequences we have been discussing are organized very differently. Davidson and his co-workers first demonstrated that members of moderately repeated sequence families are interspersed among unique sequences throughout the genome of *Xenopus*. The repeat sequences of this organism were found to average about 300 nucleotide pairs long, and to be interspersed with unique sequences ranging from 800 to over 4,000 nucleotide pairs in length. This "short-repeat" pattern was found by several investigators (see Lewin for sources) to be common among organisms as diverse as slime molds, higher plants, jelly fish, clams, sea urchins, and humans. Manning et al. discovered an altogether different "long-repeat" pattern in *Drosophila*, in which repeat sequences of 5,600 nucleotide pairs are interspersed with unique sequences with lengths of over 13,000 nucleotide pairs. A similar long-repeat pattern was discovered in bee DNA by Crain et al., but this pattern is not characteristic of all insects, because the housefly, a dipteran, exhibits the short-repeat pattern. Interspersion of repeat sequences appears to be absent in some fungi and nematodes, which suggests that the roles that have been proposed for interspersed repeats on the basis of studies of higher metazoans may not be universal.

The existence of repeat sequences interspersed among unique-sequence genes is of considerable interest because they provide a possible means of integration of single-copy gene expression. The models of Britten and Davidson and Davidson and Britten are discussed in Chapter 12, but they should be mentioned here because they have motivated much of the research on repeat sequences. These models seek to explain the observation that differentiated tissues exhibit very stable and characteristic patterns of structural gene expression. The models propose that the complex of structural genes expressed in any tissue comprise a "battery" of genes. The genes expressed in a battery are not physically linked; rather, in eucaryotes they are scattered. Integration is achieved by a network of control sequences such that an overall integrating sequence can be recognized by sequences adjacent to a set of unique sequence structural genes. The adjacent sequences must of necessity be repeated.

An integrative function for repeat sequences requires that such sequences be interspersed among structural genes, that they be transcribed, and that different tissues exhibit different patterns of repeat-sequence transcription because different patterns of structural genes are to be ultimately expressed as mRNAs. These qualifications appear to be met. The intimate interspersion of developmentally regulated structural genes and repeat sequences has been shown for β-globin genes. Shen and Maniatis examined a 44,000-nucleotide-pair-long region of the rabbit genome that contains a cluster of four β-globin genes. Twenty different repeat sequences falling into five families were found to be interspersed within this group of genes. Each globin gene was found to be flanked by at least one pair of inverted repeats of 140–400 nucleotide pairs in length, and the entire gene cluster was flanked by a pair of larger (1,400 nucleotide pairs) repeats in inverted orientation to one another.

Transcription of moderately repeated sequences has been demonstrated in sea urchin embryos. Scheller et al. used a series of clones of individual members of several repeat-sequence families to compare the expression of transcripts of these families in the nuclei of gastrula-stage embryos and adult intestine. The repeat families examined ranged in multiplicity from 20 to 1,000 copies per genome, and included both families whose members are very conserved and those whose members are widely divergent in sequence. Transcripts of all were found in the nuclei, but in distinctly different patterns, unlike unique-sequence transcripts, which are very similar in gastrula and intestine nuclei. Similarly, Costantini et al. found that members of at least 80% of the repeat-sequence families in the sea urchin genome are represented in the RNA of eggs. The most intriguing observation of these papers is that unlike structural genes, transcripts of both strands of the repeat sequences are present as RNA copies in nuclei. Repeat transcripts are also found linked to messengerlike RNAs present in the cytoplasm of eggs. Although surprising, the presence of transcripts of both strands of repeat sequences is consistent with a control hypothesis in which a regulatory transcript must bind to a complementary repeat-sequence portion of a transcript that is to be processed. It should be kept in mind, however, that consistency does not constitute a demonstration: Repeat sequences may serve entirely different functions. For example, Jelinek et al. have studied the major short-repeat family of the human genome, the *Alu* family, which is present at several hundred thousand sites in the genome. These sequences are transcribed and contribute to the nuclear RNA; however, it is not clear if they have a function in the nuclear RNA or if their transcription is a fortuitous consequence of their function or location within the DNA.

There is one other class of repeat sequences that may prove to be of considerable significance to the control of gene expression in development as well as possibly to evolutionary changes in gene organization. These are transposable elements, long sequences with short direct-repeat at each end. These sequences are inserted into the DNA at a number of sites, and they are capable of being excised and shifted in their positions within the genome. Such elements were first detected genetically in maize by McClintock in the 1950s, although the molecular nature of transposable elements has only recently been recognized. It is not clear what function these elements serve, but in their insertion adjacent to the 5' end of structural genes they can act as controlling elements to activate the adjacent gene. One of these, the *Ty1* sequence of yeast has been shown by Errede et al. and by V. M. Williamson et al. to act in this way. The study of Errede et al. is of particular interest because the genes that are activated by insertion of an adjacent *Ty1* element become responsive to control by the mating-type alleles that regulate conjugation and sporulation functions in yeast. It is possible that transposable elements provide a mechanism for the control of gene expression in development by regulated modification of the genome, although such a function remains to be demonstrated. The evolutionary possibilities for transposable elements are perhaps even more important, because these elements have the potential of moving control elements to new locations, thus adding or deleting genes from particular control networks. Similar elements have been found in *Drosophila* and in the mouse.

In contrast to hypotheses that assume a function for repeat sequences, Doolittle and Sapienza and Orgel and Crick have suggested the possibility that these elements represent "selfish" DNA. That is, there may exist DNA sequences that can insert themselves into the genome and avoid elimination. Selfish sequences hypothetically can replicate as well as or better than the bulk of the genome, and do no harm to their "host." Doolittle and Sapienza concluded their discussion with the observation that "When a given DNA or class of DNAs, of unproven phenotypic function can be shown to have evolved a strategy (such as transposition) which ensures its genomic survival, then no other explanation for its existence is necessary." This is an uncomfortable proposition.

Split Genes and Unexpected Consequences

Much of the early development of molecular genetics revolved around a series of discoveries resulting from very detailed studies of bacteriophage and bacterial genes. These studies established that the ultimate molecular products of genes, proteins, are linear polymers of amino acids, which are directly translated from mRNA composed of a linear sequence of nucleotides complementary to the genetic code of a colinear strand of DNA. This appeared to be the case in eucaryotes as well because genes could be shown to occupy discrete regions of chromosomes, and their mRNAs, like those of procaryotes, directly encode linear amino acid sequences. However, once the advent of cloning techniques made possible the isolation of individual eucaryotic genes and their detailed structural analysis, this appealingly simple

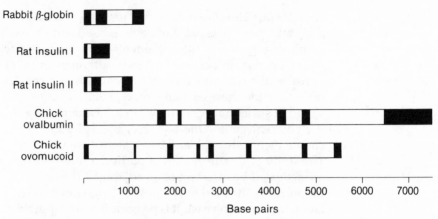

Rabbit β-globin

Rat insulin I

Rat insulin II

Chick ovalbumin

Chick ovomucoid

1000   2000   3000   4000   5000   6000   7000

Base pairs

FIGURE 11–4. *Intervening sequences (introns) in some eucaryotic structural genes. Coding sequences that give rise to the final mRNA are shown in black; introns in white.* [Redrawn from Lewin, 1980.]

view of eucaryotic genes had to be discarded. Most genes are split into several coding segments separated by noncoding intervening sequences, introns, as illustrated in Figure 11–4 for a sampling of eucaryotic genes. Numbers of introns range up to surprisingly high levels considering the precision of processing required to produce a functional mRNA: 33 introns were reported for the vitellogenin A gene by Wahli et al., and an astonishing 50 have been reported for the collagen gene by deCrombrugghe et al.

The transcription process yields a copy of the entire gene containing both intron and coding sequences in a single, large nuclear RNA. The processing of this transcript to produce a mRNA bears an uncanny resemblance to Carl Sandburg's legendary vignette:

> *Of the hook-and-eye snake unlocking itself into forty pieces, each two inches Long, then in nine seconds flat snapping itself together again.*

Intervening sequences are excised from the mRNA precursor with great precision, and then the coding sequences spliced together. An idealization of this sequence of events in the expression of a split gene is diagrammed in Figure 11–5.

The existence of introns provides much of the answer to the frustrations of the C-value paradox. Introns contain both repeat sequences and unique sequences. The total length of introns in a gene often exceeds the length of the coding sequences by as much as 10 times. The production of mRNAs from much longer transcripts is consistent with the existence of very long transcription units, such as those observed in *Drosophila*. Metazoan genomes often contain half or more of their DNA as repeat sequences, and because of the extensiveness of introns only a fraction of the remaining single-copy DNA is available to serve as mRNA coding sequences.

The reason for the existence of split genes in eucaryotes is still uncertain. The histone genes and some other eucaryotic genes have no introns; thus, introns cannot be an absolute prerequisite for gene

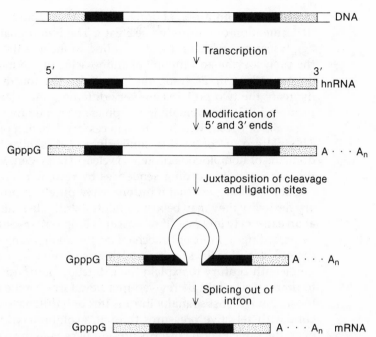

FIGURE 11–5. *Processing of a split-gene transcript to produce a mRNA. Coding sequences are shown in black, the single intron in white, and 5' and 3' nontranslated sequences are stippled. Processing involves addition of a GpppG cap to the 5' end and a poly(A) tail to the 3' end of the transcript. A processing enzyme very precisely cleaves at intron-coding sequence boundaries and splices the two coding sequences together to produce the complete mRNA coding sequence.*

expression. It is conceivable that introns are selfish sequences that are concealed by their insertion into vital genes from processes that might eliminate them. Such sequences might indeed exist, but the very prevalence of introns in all eucaryotes examined suggests that split genes have been a part of eucaryotic genomes since their origin and exist for mechanistic reasons. A purely evolutionary function has been proposed by Gilbert. Introns often separate coding sequences that encode functional domains within proteins. For example, globin genes are divided into three coding segments by two introns. The center segment encodes the heme-binding region. If, as suggested by Gilbert, introns make possible "DNA shuffling" through illegitimate recombination events, then coding sequences for individual domains can be brought into new combinations with one another. Thus, the heme-binding region of globin may have originally been part of a different split gene. Evolution of the primordial globin gene might not have required gene duplication and divergence; a simple shuffling of existing domains may have been sufficient to produce a novel protein from pre-existing parts.

The lysozymes of T4 phage and hen's egg have structures that, according to Artymiuk et al., lend themselves to an evolutionary interpretation of this kind. The two domains of hen's egg and T4 lysozyme proteins are similar. These domains are the one containing the catalytic site and a neighboring domain that is apparently involved in determination of substrate specificity. The N-terminal domain of hen's egg lysozyme, which Jung et al. found to include the signal peptide of the prelysozyme and the initial amino acids of the mature protein, has no equivalent in T4 lysozyme. Similarly, the fourth, C-terminal, domains of the two proteins are very different, and Matthews et al. have proposed that this domain in the phage enzyme functions in binding to the walls of *E. coli*, which is not a necessary function of the egg enzyme. This pattern of contrasting similarity and dissimilarity in neighboring domains of homologous proteins is clearly consistent with a shuffling of the corresponding coding sequences by recombination within introns.

Gilbert has argued that if introns serve purely in providing evolutionary flexibility they may be lost by neutral drift. He estimates this to occur at an extremely low rate. If one accepts long-term evolutionary flexibility as providing a sufficient mechanism for maintaining split genes, then the old idea of racial senility, which was so often appealed to in the late nineteenth century to explain the extinction of groups from ammonites to dinosaurs, might be revived in a new form. Senile groups would be those that had lost so many introns through drift as to become unable to cope with selective pressures through evolutionary innovations. However, the same logical fallacy is present here as in the old preformationist hypothesis discussed in Chapter 3. All introns would have been present at the beginning and would be slowly running out. It appears more likely that new introns can be generated and that they are maintained because they serve an immediate function within the cell. However, this is not to deny them an additional evolutionary role in DNA shuffling.

That introns can function in the regulation of gene expression in an important way has been shown by the elegant work of Lazowska, Jacq, and Slonimski on the *box* gene, which is located in the yeast mitochondrial genome and encodes cytochrome *b*. A physical map of the *box* gene and the clusters of mutational sites known for it is shown in Figure 11–6. The gene has six coding sequences and five introns; three distinct classes of mutants have been observed in it. Mutations in coding sequences, as expected, affect the structure of the protein, and all fall into one complementation group. The other two classes of mutants are unusual. Three clusters are located in introns. These in turn fall into three different complementation groups and block processing of the cytochrome *b* gene transcript. They also affect expression of the *oxi*-3 gene, another split mitochondrial gene that encodes the subunit I protein of cytochrome oxidase. The third class of mutants are located in intron-coding sequence boundaries.

In order to produce a functional cytochrome *b* mRNA, the intron sequences must be clipped out of the primary transcript and the coding sequences spliced together. Processing, however, has proved to be a complex multistep process. Mutations in the *box*-3 intron interfere with processing because this intron is actually translated to produce a protein required for the processing of the *box* gene transcript. The sequence of this segment of the processing protein is known because this region of the *box* gene has been sequenced.

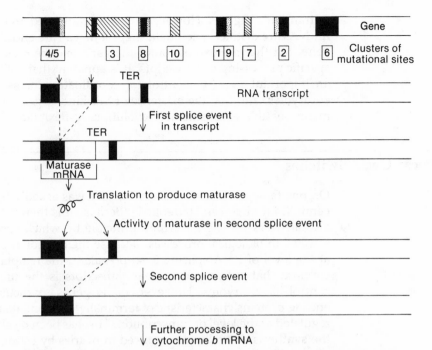

FIGURE 11–6. *Events in the processing of the yeast mitochondrial cytochrome* b *gene. The organization of the gene is diagrammed at the top of the figure. The length is approximately 7 kb. Coding sequence elements are in black; introns in white, except for two kinds of intron mutational sites. Mutations in hatched areas of introns block cytochrome* b *mRNA processing. Dotted regions exhibit mutations in intron-coding sequence boundaries. The details of processing of the left end of the RNA transcript are shown below. Splicing of the first intron produces an RNA that functions as a mRNA for a maturase responsible for the next splice step. Part of the maturase mRNA derives from the* box 3 *intron. TER marks the termination site for translation of maturase.* [Modified from J. Lazowska, C. Jacq, and P. P. Slonimski, Sequence of introns and flanking exons in wild-type and *box-3* mutants of cytochrome *b* reveals an interlaced splicing protein coded by an intron, *Cell* **22**:333–348, 1980. Copyright by the Massachusetts Institute of Technology.]

The first splicing event of the gene diagrammed in Figure 11–6 results in an RNA containing the *box-4/5* coding sequence of cytochrome *b* linked to the small coding region at the end of the *box-3* intron. This RNA, which contains both cytochrome *b* coding sequences and intron sequences, functions as a mRNA to synthesize a maturase protein required for a further processing event, which removed the *box-3* intron sequence to yield a mRNA containing only cytochrome *b* coding sequences. In a similar manner, the *box-7* intron also seems to produce a similar although distinct maturase since *box-3* and *box-7* mutants complement each other. The hypothetical *box-7* maturase seems to function not only in the processing of the cytochrome *b* gene, but also appears to be required for processing of the transcript of the *oxi-3* gene since mutations in *box-7* affect cytochrome oxidase synthesis.

A role for processing in the control of gene expression is also indicated by the case of mouse liver and salivary gland α-amylases. Hagenbüchle et al. have found that the sequences of the mRNAs of both of the liver enzymes and the salivary enzyme are identical in their coding and 3' untranslated regions. However, the 5' untranslated regions of the three mRNAs are distinct. Their data suggest that all are encoded by the same gene, but that the expression of the gene may be regulated by tissue-specific processing of transcripts. It is apparent that split genes provide an important point of control for coordinate expression of genes in eucaryotes, and that the existence of this kind of gene organization has made possible a considerable evolutionary flexibility.

Local Gene Switches

On one famous occasion George Mallory was asked why he wanted to climb Mount Everest: He replied, "Because it is there." Seemingly, this also answers for the governing principle by which control points are selected in biological processes. Because the control of gene expression at the level of RNA splicing is so prevalent, the temptation may be to suppose that splicing of nuclear transcripts is the universal level of control in eucaryotes. However, many genes, particularly those that encode proteins characteristic of terminally differentiated cell types, are regulated at the level of transcription. This has been nicely illustrated for the synthesis of ovalbumin induced in ovaries by estrogen. Roop et al. used isotopically labeled, cloned ovalbumin DNA as a probe for ovalbumin gene transcripts in ovary nuclei. They found approximately 3,000 transcripts per nucleus in tissue stimulated by estrogen, but less than two per nucleus in tissue not exposed to the hormone. The readily studied cases, such as ovalbumin, involve genes that produce massive amounts of a specialized product in response to an inducing signal. However, it is quite possible that genes responsible for major developmental decisions are also under transcriptional control. This is suggested by the transcriptional behavior of developmentally regulated puff regions in polytene chromosomes, where differentiation is clearly associated with differential patterns of transcription. The timing of activities of developmentally important genes is also suggestive. Unfertilized eggs carry a very high diversity of mRNAs in their cytoplasms, yet a large preponderance of mutants observed in experiments, such as those described in Chapter 10, are embryonic rather than maternal. Those genes must be active during the embryonic period.

Control of gene expression must involve local regulatory elements either adjacent to or within individual genes. Transcription units include not only coding sequences and introns, but also adjacent noncoding sequences on both the 5' and 3' ends of genes. The existence of controls at both transcription and processing levels indicates that local regulatory elements are included in transcription units. Some of the primary local regulatory elements include sites that must recognize external regulatory signals that specifically induce or repress transcription of a gene, a site to bind RNA polymerase, a site to initiate transcription, a site to terminate transcription, sites specifying processing, sites at which processing cuts and splices are made, and sites specifying ribosome binding and initiation of translation in the mRNA itself.

The existence of a functional local control element adjacent to a gene has been demonstrated genetically by Chovnick and his collaborators for the *rosy* locus of *Drosophila*. The *rosy* locus was originally defined as the locus of mutations producing recessive brownish eye color mutants. The mutant eye color resulted from a deficiency of the red eye-color pigment, drosopterin, resulting from a lack of the enzyme xanthine dehydrogenase. Mutations in the *rosy* locus result in structural changes in the xanthine dehydrogenase protein. Chovnick and his co-workers determined the extent of the xanthine dehydrogenase gene through intragenic recombination mapping of a very extensive collection of *rosy* mutations. There are also genic alterations that affect the level of expression of the gene. One of these has been mapped as an apparent *cis*-acting adjacent regulatory element. This variant was discovered to produce both a higher level of xanthine dehydrogenase than normal, and to produce a protein with an altered electrophoretic mobility. Analysis of intragenic recombinants showed that the regulatory site is separate from the electrophoretic site, which lies within the xanthine dehydrogenase structural gene itself. The regulatory site maps at a distance equivalent to about 3,000 base pairs from one genetically defined boundary of the structural gene. Although this mutation site appears to be very distant from the xanthine dehydrogenase structural gene, it is possible that the actual initiation site is not so distant from the regulatory mutant site. The site of initiation of transcription might well, in a manner analogous to the situation found in the chorion genes diagrammed in Figure 10–7, be close to the regulatory site and separated from the genetically defined structural gene by a large intron.

The availability of recombinant DNA-cloning and DNA-sequencing techniques has made it feasible (and fashionable) to seek local control sites in the DNA sequences adjacent to structural genes.

A number of potential control sites have been discovered. These are diagrammed in Figure 11–7, which shows an idealized mammalian transcription unit and its constituent signal sequences. This figure also shows the organization of the upstream regions of the early expressed genes of the mammalian virus SV40 and the sea urchin histone H2A gene. Both contain control sites located up to 200 nucleotide pairs upstream from the transcriptional start point. The structural gene itself starts with an initiation site at which transcription actually begins. This site marks the 5′ end of the mRNA, and in the mRNA is modified by a characteristic base, $^{7\text{-methyl}}G^{5′}$ ppp, which forms a cap important in translation. The initiation sequence shown in Figure 11–7 is a consensus sequence drawn from comparisons of initiation sequences of several genes. In reality these vary considerably. There are several internal signals, including a translation start site, splice signals at the boundaries of introns and coding sequences, and sites specifying termination of transcription and addition of polyadenylic acid tails to the 3′ end of the mRNA.

The controls of particular interest to understanding the developmental regulation of gene transcription, however, are those that lie upstream from the transcription start point. The best known of these is the

Idealized mammalian transcription unit

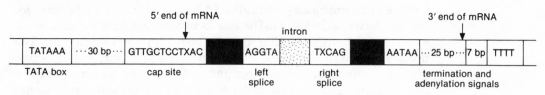

SV40 early promoter region

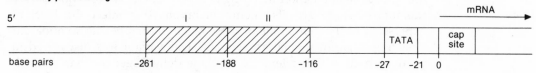

5′ prelude region of histone H2A gene

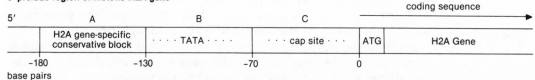

FIGURE 11–7. Signal sequences associated with eucaryotic genes. An idealized mammalian transcription unit is shown at the top of the figure. The TATA box, which may be involved in binding of RNA polymerase, lies to the 5′ end of the gene. Transcription begins about 30 base pairs away at the cap site. Coding sequences are shown in black; an intron as stippled. Internal signals include splice sites, and termination and adenylation sites. The schematic diagram of the SV40 early promoter region shows two 70-base-pair repeats (hatched) located 116 bases 5′ of the transcription start site. These sequences are required for in vivo expression of the SV40 early region. The schematic diagram at the bottom of the figure shows the 5′ prelude region for one of the members of the sea urchin histone gene cluster. Region A contains an evolutionarily conserved sequence, region B contains the TATA box, and region C the cap site. The effects of deletions of these regions on transcription are discussed in the text. [Diagrams modified from Lewin, 1980; C. Benoist and P. Chambon, In vivo sequence requirement of the SV40 early promoter region, reprinted by permission from Nature 290:304–310, copyright © 1981 Macmillan Journals Limited; D. J. Mathis and P. Chambon, The SV40 early region TATA box is required for accurate in vitro initiation of transcription, reprinted by permission from Nature 290:310–315, copyright © 1981 Macmillan Journals Limited; and R. Grosschedl and M. L. Birnstiel, 1980.]

TATAAAA sequence (TATA box) which lies about 30 nucleotide pairs from the initiation point. This sequence is very similar to the RNA polymerase recognition site first discovered in bacteria by Pribnow, which is obligatory for transcription of bacterial genes. The TATA box is required for transcription of eucaryotic genes by *in vitro* transcription systems, but recent experiments performed by Benoist and Chambon, Mathis and Chambon, and Grosschedl and Birnstiel show that the TATA box is not required for transcription *in vivo*. When cloned genes are injected into *Xenopus* oocytes, they are accurately transcribed. Deletions of specific regions can be prepared, and clones of these modified genes injected into oocytes. Both the level of transcription and the sequence of the RNA produced can be determined. By use of this experimental protocol it has been found that genes from which the TATA box has been deleted are transcribed at nearly normal levels, but that transcription is initiated at several sites not normally used. Thus, the TATA sequence acts to specify the initiation site at which the RNA polymerase begins transcription; however, this site is not obligatory for RNA polymerase binding or for initiation of transcription.

The actual modulation of transcription depends on control sites located as much as 200 nucleotide pairs upstream from the initiation site. In SV40 Benoist and Chambon have found that the structure of the upstream region is complex. There are five G-plus-C-rich sequence blocks. Two of these are included in two 72-nucleotide-pair tandem repeats located about 150 nucleotide pairs upstream from the initiation site. Experiments in which the tandem repeats have been deleted from the DNA show that these repeats are indispensable for *in vivo* transcription.

The H2A histone gene of the sea urchin *Psammechinus miliaris* also possesses control sequences distant from the initiation site. The upstream region of the H2A gene shown in Figure 11–7 can be divided into several distinct functional segments. The C segment contains the initiation site. There is a TATA box in segment B about 30 nucleotide pairs from the initiation site. About 35 nucleotide pairs further upstream from the TATA box is segment A, which contains a 30-nucleotide-long sequence that bears short inverted repeats at each end. This 30-nucleotide sequence is specific to the H2A gene and is evolutionarily conserved. Segment E begins 110 nucleotide pairs from the initiation site and extends about 340 nucleotide pairs further upstream. This segment is rich in A + T.

Grosschedl and Birnstiel tested the functional roles of each of the upstream segments by comparing the transcription of H2A clones bearing specific deletions in these segments with an unmodified histone H2A clone. Deletion of the region containing the TATA box caused a five-fold decrease in the rate of H2A gene transcription, and led to transcription starts from novel initiation sites. Deletion of the 30-nucleotide-pair H2A-conserved block in segment A led to a two-fold rise in transcription rate. Deletion of the large A + T-rich E sequence resulted in a 15–20-fold decrease in H2A transcription. The control function of this region might have been caused by its composition or by the presence of a specific sequence. To test these hypotheses, Grosschedl and Birnstiel prepared a modified clone in which the E sequence was present, but in an inverted orientation. When tested for transcription, the DNA containing the inverted segment surprisingly resulted in a five-fold increase in transcription. Both the normal H2A transcript and a transcript with an additional 90-nucleotide-long segment on its 5′ end were produced.

Upstream elements do not exhaust the possibilities for control elements. Studies by Sakonju et al. and Bogenhagen et al. have shown that the upstream end of the 5S ribosomal RNA gene of *Xenopus* can be deleted without any effect on transcription. Indeed deletion of much of the structural gene itself has no effect. The transcription control sequence lies about 50 nucleotide pairs into the structural gene. The 5S gene is transcribed by a different RNA polymerase (polymerase III) than genes that produce mRNAs (polymerase II), and this may determine the difference in location of control sites. Altogether, the existing studies of adjacent controls for gene expression are not yet well understood, but they do indicate the presence of a variety of elements located in the vicinity of genes that are involved in quantitative control of transcription and regulation of precision of initiation.

The baroque nature of eucaryotic gene organization that has emerged from contemporary molecular studies makes it necessary to add to the suggestion of Morgan, with which we began this chapter, that evolution has required not more genes but new genes. The C-value problem itself is, for the most part, easily resolved, and vanishes as a paradox. Satellite DNA, moderately repeated sequence families, and introns all greatly reduce the proportion of the genome available to serve as coding regions. These and other genomic entities can account for the majority of the DNA, and can vary considerably in amount among related organisms. In this surprising resolution of the C-value paradox a more important problem is revealed. The variety of demonstrable and potential local regulatory elements is astonishing: We are only beginning to understand their functions. Evolution among most eucaryotic groups may have involved both the acquisition of more and new genes, but the major role has been played by modifications in sophisticated control mechanisms. Evolutionary changes in gene expression most probably have occurred through changes in individual control elements, or in transposition of genes and control elements to make possible novel associations of protein domains and new associations of genes and adjacent regulators. Such changes are only effective because local regulatory elements respond to signals generated by integrative systems that govern the expression of large numbers of genes to produce integrated tissues and morphogenetic pathways.

Twelve

Regulatory Hierarchies and Evolution: A Synthesis

The geneticists are trying to make evolution fit the genes rather than to make the genes fit evolution.

Henry Fairfield Osborn

Limitations of Evolutionary Syntheses

In 1932 the famous vertebrate paleontologist H. F. Osborn wrote an article entitled "The Nine Principles of Evolution Revealed by Paleontology," which is remarkable for its grumpiness with respect to genetics and geneticists, and for its wonderfully dogmatic (and erroneous) statements of "evolutionary principles." One of these, which can serve as a showpiece, is Osborn's assertion that "All that we can say at present is that Nature does not waste time or effort with chance or fortuity or experiment, but that she proceeds directly and creatively to her marvelous adaptive ends of biomechanism." This astonishing sentence grandly sweeps away any role for natural selection operating on variant genes or genetic systems. It does provide what may be to some an agreeable picture of Nature personified acting to produce hopeful monsters and punctuated evolutionary events by design. However, except for meristic traits, which must change in a discontinuous manner, Osborn conceived of evolution as proceeding by a majestic gradualism that involved long-term continuous processes. Directedness was apparent to him in trends that, like the correlated increase in horn and body size in one group of extinct mammals, the titanotheres, were maintained throughout the evolutionary history of a lineage.

Although he was not able to clearly state what he thought the mechanism(s) to be, Osborn's perception of evolution was not ineffable nor nonmechanistic. Rather, in the nineteenth century tradition of E. D. Cope, his mentor, he continued to adhere to a variety of Lamarkism. Osborn realized that such a mechanism requires an informational feedback from the somatic body to the stable and separate germ line. In his earlier writing of a book modestly entitled *The Origin and Evolution of Life* he discussed the possibility that various environmental forces might

335

directly impinge on the "hereditary-chromatin." Osborn approached a glimmering of the functions of "chemical messengers," such as hormones in morphogenesis, and speculated at a connection between genetically determined traits and those caused by perturbation of hormonal systems.

The failure of Osborn's speculations on the causes of morphological evolution were, like those of Haeckel's syntheses, the result of a seriously flawed view of genetics and the role of genes and developmental mechanisms in morphological evolution. The directedness Osborn saw in fossil lineages is more reasonably interpreted in other ways. Changes may be continuous or discontinuous, but they will be constrained by the nature of existing developmental interactions. Trends exist because selection for certain traits can only act on available morphogenetic processes. Osborn's Lamarckian views were egregious and anachronistic, but made logical sense if evolutionary trends were to be a product of direct environmental control of changes in the hereditary material.

The object of our musings on Osborn's evolutionary speculations is cautionary. Any attempted synthesis of necessity goes beyond reasonably certain knowledge and has a high probability of looking if not preposterous, then at least naive, in the not too distant future. The efforts of Richard Goldschmidt, one of the heroes of this book, were limited by his bizarre concept of the gene, and as importantly, by the fact that developmental genetics was still in its earliest pioneering stages. Our own ability to synthesize what has gone before in this book is severely limited by a currently poor understanding of the way in which genes direct the morphogenesis of even simple metazoan structures and of the nature of high-level genetic regulatory interactions. The controls of some individual genetic loci have been described, and the organization of regulatory sequences in DNA adjacent to structural genes is beginning to be studied with methods capable of providing detailed answers. However, our ideas of integrated controls are generally extrapolations from what is known of individual structural loci. As Moses saw the Promised Land, we still see the regulatory genes that act as developmental switches from afar.

Finally, the dynamic behavior of networks involving large numbers of genes that influence each other's activities is only beginning to be a topic of theoretical studies, notably those of S. A. Kauffman. Interactions of large numbers of genes very likely provide the homeostatic stability and canalization characteristic of development, and may determine the limited numbers of stable states of differentiation (i.e., cell types) that particular genetic systems are capable of maintaining.

The Eye of the Needle

Organismal evolution exhibits a grandness and progressive elaboration of morphologies and adaptations that suggests, as it did to Haeckel and to Osborn, the existence of universal governing principles. However, the very range of evolutionary phenomena at the same time had led to the recognition of more principles than may actually exist beyond their describers' imaginations. In this book we have attempted to document the existence of characteristics of organismal evolution that reflect the crucial modifications of the developmental-genetic systems that underlie morphological change.

Rates of morphological evolution vary greatly. The bradytelic lungfish, which with their lives of alternating aestivation and activity governed by endless annual cycles of drought and torrential rain, have morphologies essentially unchanged since the Carboniferous. They contrast dramatically with the tachytelic Hawaiian drosophilids that have radiated so widely in a geologically young archipelago. The extremely slow rates tell us little except that morphologies and the processes that produce them can be maintained, as in the case of the frogs analyzed by A. C. Wilson and his collaborators, even in the face of considerable molecular evolution, which steadily changes the genes that encode the proteins from which a conservative morphology is constructed. Slow rates of organismal evolution probably reflect the effects of potent normalizing selection on maintaining a successful morphological adaptation. The modifications within a basic ontogenetic pattern exhibited by slowly evolving organisms may require only the gradual substitution of variant alleles, as visualized by classic evolutionary theory. Morphological stasis, which is extreme in living fossils, such as horseshoe crabs, lingulid brachiopods, or lungfish, is not unusual. Many, and perhaps most, species are stable in morphology during the vast majority of their histories. To borrow from Thomas Hardy, stable morphologies and the developmental systems that produce them are

> *Things mechanized*
> *By coils and pivots set to foreframed codes...*

It is the rapid, punctuational events that demand explanation, and it is the codes and consequent coils and pivots that must be changed. Morphology is reorganized, new structures appear, and yet internal integration is maintained. As in Frazzetta's metaphor, the machine must be modified while it is running.

The rapidity with which morphological evolution can occur has direct and important consequences in the other characteristics of organismal evolution. It is axiomatic that any change in morphology requires a commensurate change in the course of development. This has been evident since the writings of Müller and Haeckel, although the significance of development has not been a prevailing theme in evolutionary theory, and an adequate conception of developmental-genetic mechanisms is only now becoming possible. The demonstration of switch genes that govern the segmentation of insects or determine the differentiation of germ layers in mammalian embryos provides unequivocal evidence for the existence of genes that function specifically in the control of developmental processes. The ability to distinguish such genes is of crucial importance because genes can only be detected by the phenotypes expressed by mutant alleles. Most mutations are disruptive and pliotropic in their developmental consequences, and are thus difficult to analyze. A mutation may disrupt development because the product of the affected gene is specific to some developmental stage and the gene indeed functions in the regulation of a developmental process, or the effect may be the result of the disruption of a general metabolic pathway. The effect on development in the latter case is real enough on a phenomenological level, but the specificity is questionable. Genes whose mutant alleles always produce not disruption but a switch from one specific and distinct fate to another, as is so dramatically seen in homoeotic mutations, are unambiguously developmental control elements. Thus, we suggest that a second major characteristic of organismal evolution is that there is a genetically determined developmental program, and that evolutionary change occurs through genetic modification of this program. This conclusion should not be misconstrued as a model of total genetic determinism. We are aware of work such as that by M. J. Katz and his colleagues that indicates the existence of nongenetic flexibility in development. Ontogenetic buffer mechanisms are of great importance because they make possible the acceptance of genetic modifications with a minimum of disruption.

Not all of the genome is involved in the developmental program. In fact, much of an organism's DNA has no apparent genic function in the sense that a gene encodes information expressed through transcription. It is further clear from King and Wilson's comparison of humans with our closest evolutionary cousin, the chimpanzee, that evolution of structural genes can have little to do with morphological evolution. It is the genes that regulate the developmental program that count. That is an important conclusion because it overturns a pervasive conception of organismal evolution as an extension of processes by which point mutations are accepted in structural genes and result in amino acid substitutions in proteins. Much of evolutionary theory has been colored by this prejudice, which mistakes these clock processes or the gradual substitution in a population of an allele encoding one enzyme variant for another as evolution.

The evolutionary role of one set of regulatory switch genes, the bithorax complex of *Drosophila*, was suggested by E. B. Lewis in 1963. The transition from a multilimbed myriapodlike ancestor to a primitive six-legged insect involved the acquisition of the *bithoraxoid* (*bxd*) and *infra-abdominal* (*iab*) genes as early members of the complex. These genes suppress limb development on abdominal segments. The subsequent evolution of the diptera from winged insects required the evolution of the *bithorax* (*bx*) gene as well as other genes of the modern Bithorax Complex (*BX-C*) involved in conversion of the metathoracic wings to halteres. It is interesting that distinctly dipteran characteristics were achieved prior to the conversion of a pair of wings to halteres because a fossil four-winged dipteran dating to the Permian has been described by Riek. This frozen state in regulatory gene evolution provides strong evidence for Lewis' suggestion of a sequential origin for the BX-C.

The regulatory genes governing segmentation of the head and anterior segments of the thorax similarly evolved in a sequential manner. An overall hypothesis for the roles of both the Bithorax (BX-C) and Antennapedia (ANT-C) Complexes in the evolution of arthropod head and segmental organization leading to insects was presented in Chapter 8. However, this construct is based on an analysis of mutations which, while they reveal the function of the normal alleles of the genes constituting the complexes, do not cause the recapitulation of ancestral forms. Nor do the homoeotic mutations themselves lead to evolutionary changes, a point that has often been misunderstood. A head bearing legs in the place of antennae is not a hopeful monster: It is a statement of gene function couched in bizarre and challenging terms. It has no evolutionary future.

Can mutations in regulatory genes acting as switches actually be shown to provide a basis for morphological evolution? We have already suggested that very strong evidence exists in genes that control the symmetry of coiling of snails, or that control the counting events underlying ambulacral number in echinoderms, or toe number in guinea pigs, but the relation of these genes to evolutionary events is (as is so often the case when dealing with evolution) inferential. Fortunately, an elegant study has been carried out by Sternberg and Horvitz to answer explicitly the question of the evolutionary role of switches.

Sternberg and Horvitz explored the developmental-genetic basis for morphological differences between two species of small nematode worms. The nematode *Caenohrabdites elegans* has been catapulted to popularity as a research organism in recent years because, although it is a complex metazoan, it has approximately only 2,000 genes (about the same number as *E. coli*), a limited and constant number of somatic cells, and is highly amenable to studies of the genetic control of development. The sterling qualities of *C. elegans* are nicely presented in a short review by Edgar. Three characteristics of this organism provided Sternberg and Horvitz with the prerequisites for a detailed evolutionary comparison of gonadal cell lineages between *C. elegans* and the quaintly named "sour paste nematode" *Panagrellus redivivus* that belongs to the same order but to a different family. Adult nematodes are constructed of a small number of somatic cells (808 in the case of *C. elegans*), cell fates are rigidly determined, and it is possible to precisely trace comparable cell lineages in the two species.

The ovaries of *C. elegans* and *P. redivivus* are shown in Figure 12–1. *P. redivivus* has only one ovary directed anteriorly from the vulva, whereas *C. elegans* has two ovaries opening into a common vulva. The basic cell lineage program for ovary development is essentially the same in the two species, with evolutionary modifications superimposed in later stages. The gonad primordium consists of four cells lying in an anterior-posterior orientation. These four cells, called Z1, Z2, Z3, Z4, are shown in Figure 12–1. The two center cells, Z2 and Z3, give rise to germ cells, whereas the terminal cells, Z1 and Z4, follow an invariant pattern of cell lineages to produce the somatic structures of the gonad. The early lineages of the Z1 and Z4 cells of the two species are compared in Figure 12–1. The patterns of division are identical, as are the fates of the cells finally produced. There is, however, one crucial difference. The Z1 and Z4 cells of the *C. elegans* ovary primordium each produce a cell called a

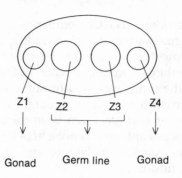

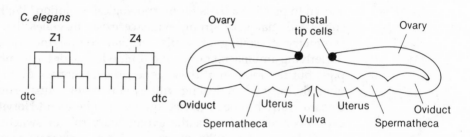

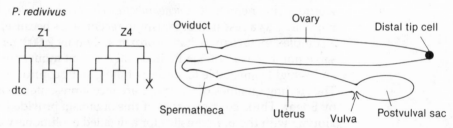

FIGURE 12–1. Evolutionary modification of cell lineage in development of the gonads of nematode worms. The top diagram shows the gonad primordium consisting of four cells, Z1, Z2, Z3, Z4, in both Caenorhabdites elegans *and* Panagrellus redivivus. *The double-armed gonad of* C. elegans *is shown (center, right) with a diagram (center, left) of the cell lineages of the Z1 and Z4 cells that produce the somatic structures of the gonad. The single-armed gonad of* P. redivivus *and a diagram of the cell lineages of the Z1 and Z4 cells of this species are shown at the bottom. Note the programmed cell death of the Z4-derived distal tip cell (DTC) indicated by X, with consequent lack of development of the posterior arm of the gonad. [Drawn from photographs and diagrams provided courtesy of P. W. Sternberg and H. R. Horvitz.]*

distal tip cell (DTC), whereas in *P. redivivus* the corresponding product of the Z4 cell undergoes programmed cell death. The morphological consequences of this single change in cell fate are striking and instructive.

The double-armed ovary of *C. elegans* apparently represents the primitive situation in nematodes, whereas single-armed ovaries have evolved independently several times. Kimble and White have shown by microbeam laser deletion of DTCs in *C. elegans* that this cell is required for growth of a gonadal arm. Destruction of the DTC results in cessation of mitosis in the germ cells. Because growth of the ovary is driven by germ-cell mitosis, an arm missing the DTC ceases growth. A *C. elegans* ovary in which the posterior distal tip cell product of the Z4 cell has been ablated grows only an anterior ovary arm, and comes to resemble the ovary of *P. redivivus*. The evolution of a single-armed ovary may require only a single mutational change in cell fate to the programmed cell death of the cell otherwise programmed to become the DTC. Single-gene mutations that cause switches in cell fate are well documented in *C. elegans*. Mutants exhibiting specific switches to programmed cell death in normally viable cells are known, and one such mutant, acting like an internal ablation of the DTC, causes *C. elegans* to develop a single-armed ovary. Sternberg and Horvitz proposed that if switches occur in a regulatory cell, which they define as a cell that exerts control over other cells, they provide the potential for discontinuous evolutionary transitions. Thus, among the genetic changes that have occurred in the phylogeny of *P. redivivus* the alterations in the fate of the Z4 cell lineage have almost certainly resulted from a single gene change, and yet have had a drastic morphological consequence.

Sternberg and Horvitz observed three classes of cell lineage transformations in their detailed comparisons of *C. elegans* and *P. redivivus* for which there are mutational equivalents in *C. elegans*. The first includes switches in cell fates, as exemplified in the Z4 lineage. The others include alterations in the number of cell divisions characteristic of a lineage and altered segregation such that the developmental potential normally associated with one cell is transferred to its sister cell. The existence of mutations in *C. elegans*, which exhibit equivalent transformations to those seen in evolution, provides a system capable of revealing a great deal about the genetics of the switch genes that govern these events. In this, nematodes allow a genetic approach to the problem of cell determination so well documented in the classic spiralian and tunicate systems, which have not proven amenable to genetic studies. It remains to be seen if nematode switch genes are equivalent to the homoeotic switch genes of *Drosophila*.

Of all the modes by which ontogeny is modified in the course of evolution, changes in timing have received the most attention. The whole of embryonic development presents a sweep of changes of movements and structural elaboration in time. The process has an air of inevitability, the blossoming of an orchestrated program in which all events occur in precise temporal sequence. To a large extent this is true, but numerous cases of dissociation of timing of ontogenetic processes from one another exist, and a vast array of evolutionary examples show that heterochrony is indeed a very common agent in evolution. There is a sound mechanistic reason for this in view of the need to maintain an integrated developmental program. Heterochrony often results in non-disruptive modifications in a developmental path. Existing integrated processes are shifted with respect to each other, but overall functional integrity is maintained: A reproductively mature organism with a larval morphology retains a set of environmental adaptations and a working body structure. Other, more subtle heterochronies are also reinforced by existing systems of developmental homeostasis. A process, such as growth, that is initiated somewhat earlier or later than normal may ultimately lead to a change in proportion between two structures, but unless the change interferes with a necessary interactive event with some other tissue, the already established morphogenetic processes will occur. A modification in proportions imposed relatively late in development should be easily accommodated if selectively advantageous in the animal's external environment.

Surprisingly, although a few mutations affecting timing have been identified, little research has been directed at understanding the genetic basis for temporal controls in development. This is part of a larger sphere of ignorance in biology spanning a range of temporal phenomena from the control of timing of DNA synthesis in cells to the control of the circadian rhythms of animals. The small numbers of mutations so far detected that change timing of developmental events suggest that there are individual genes that specifically regulate timing. Some, such as the gene that controls the decision of salamanders of the genus *Ambystoma* to undergo metamorphosis or to become neotenous, appear to be analogous to the disruptive mutations discussed earlier. In the case of salamanders neoteny results from an underproduction of thyroxine and consequent effects on all tissues. Other genes are highly intriguing in having the properties of specific switches. The *anemic* mutation of the axolotl, which delays the onset of adult globin synthesis, may belong to this class. Genetic elements that control the timing of expression of enzymes in the ontogeny of several organisms have been identified. These elements, called "temporal genes" by K. Paigen, are closely linked to the structural genes they control, and they are *cis*-acting.

The heterochronic genes that at present are most readily available for study are those that affect lineage switches in development of the nematode *C. elegans*. The first of these, the *lin-4* mutation analyzed by Chalfie and his co-workers, causes reiterations in cell lineages such that lineages fail to progress to normal final adult progeny. Instead, cells characteristic of the first larval instar are produced over and over. Thus, although the wild-type worm proceeds through a set of ontogenetic stages,

$$\text{embryo} \rightarrow \text{L1} \rightarrow \text{L2} \rightarrow \text{L3} \rightarrow \text{L4} \rightarrow \text{adult}$$

lin-4 mutants follow a highly abnormal pattern of larval cell lineages so that development effectively follows the path

$$\text{embryo} \rightarrow \text{L1} \rightarrow \text{L1} \rightarrow \text{L1} \rightarrow \text{L1} \rightarrow \text{L1}$$

Instead of progressing through a normal sequence of larval stages, stage L1 is reiterated. Only ectodermal lineages are affected by the *lin-4* allele, so that many somatic lineages and gonadal development are unaffected, and the final molt possesses a sexually mature gonad. At each larval stage the cuticle has a characteristic and stage-specific morphology and protein composition. According to Cox, Staprans, and Edgar there may be as many as 30 different collagenlike proteins expressed in a stage-specific manner through larval development. The cuticle of the reiterative *lin-4* mutant has a morphology like that of normal L2-stage larvae and the corresponding cuticular protein composition. This is as expected because one of the functions of the normal L1 larva is to make L2 cuticle at the first molt.

A second similar, but distinct, gene is currently being studied by Ambros and Horvitz. Some mutant alleles at this locus skip the cell divisions characteristic of cell lineages during the L1 stage. The result is an acceleration of morphological development. Larval stage transitions of *C. elegans* may represent metamorphoses, major stage transitions in gene regulation. Horvitz, Sternberg, and Ambros as well as Edgar propose that these stages may represent "temporal compartments" analogous to the spatial compartments of *Drosophila* that are revealed in part by the functions of the *Antennapedia* and *bithorax* switch genes. Thus, there may be temporal switches in *C. elegans*.

A final and highly significant characteristic of organismal evolution is that morphogenesis appears to be governed by a relatively small number of regulatory genes. The best-understood metazoan genetic system, *D. melanogaster*, possesses a total of about 5,000 genes. A significant proportion is required for development, but only a minority can be shown to be required to make developmental decisions. This theme emerges repeatedly from the specific examples examined in this book. Only 10 genes must be changed to produce the considerable differences in head morphology that have evolved between the Hawaiian drosophilids *D. heteronura* and *D. silvestris* shown in Figure 3–8. It is not clear how these particular morphogenetic genes act, and indeed they may affect a variety of processes, including positional information, timing, and inductive events. A detailed developmental analysis would be necessary to dissect out the actual modifications in the developmental program. The 10 genes may represent only a fraction of the genes actually involved in head morphogenesis, the fraction involved in a particular evolutionary change. Nevertheless, the small number is consistent with other studies from which numbers of control genes have been extracted.

The best examples of morphogenetic regulatory genes are provided by the genes that control the location, number, and identity of *Drosophila* head and body segments. Segmental control genes can be divided into two general classes: those active during oogenesis in the establishment of the positional information of the egg, and those active during embryogenesis in the interpretation of positional information. Only two genes involved in establishment of positional information have been well studied. The existence of these genes was revealed by two maternal effect mutations, *bicaudal* and *dorsal*, which have been investigated by Nüsslein-Volhard. Other maternal effect genes with analogous functions have also been discovered. According to Wieschaus, the frequency with which such mutations are detected indicates that it is unlikely that there are many more than 20 of these genes altogether.

There are two subclasses of genes active during embryogenesis that can be detected by mutations that cause abnormalities in segmentation. Members of the first subclass presumably encode products that function to interpret maternally established positional information for the determination of the location and number of segments, whereas members of the second subclass interpret positional information to determine segmental identities. A systematic search by Nüsslein-Volhard and Wieschaus revealed 15 mutant loci scattered throughout the genome that produced abnormalities in segment numbers and polarities: An additional seven genes were subsequently found. The mutant alleles fall into three discrete categories: those that affect a large region of the embryo, those that affect pairs of segments, and those that affect single segments. These categories may define a stepwise establishment of segments, and it appears likely from the number of mutants screened that the 22 genes constitute the preponderance of genes of this type.

Interpretation of positional information to determine segmental identities involves the functioning of the genes *Polycomb* (*Pc*) and *extra sex combs* (*esc*), which, as discussed by Lewis and by Struhl, are regulators of the Antennapedia and Bithorax Complexes. The approximately 15 genes of these two complexes are expressed during embryonic development and determine segmental identities and subsequent morphological differentiation of each segment. Thus, about 50–60 genes acting during oogenesis and embryonic development are sufficient to provide the basic developmental program for segmentation. Without question, a number of other genes function in subsequent morphogenetic steps. However, it appears probable that whereas a large number of genes may be required for the aggregate of all of the subprograms for morphogenesis of structures within any segment, each structure may require only a few major commands. This conclusion still needs confirmation, but the requirement for only four genes to determine the number of digits appearing in the feet of guinea pigs lends credence to the need for limited numbers of switches in any particular subprogram.

The consequences for evolution are profound. Ontogeny involves the activity of many genes expressed in a whole set of very stable processes. In *Drosophila* about one-third of the total number of detectable genes are expressed in a developmentally specific manner, and are needed for successful completion of specific developmental stages. Nevertheless, the number of switches is small, and changes in switch functions may have correspondingly great effects in morphogenesis. It is important to note, however, that evolution is not a single-step affair. The chief significance of alterations in genes with regulatory functions may be to produce changes in ontogeny that provide the raw material for further changes in a new direction. Further change and consolidation of the novel direction occur through mutational events in genes modifying the principal regulatory gene. Canalization and integration can be retained in the midst of evolutionary transitions in morphogenesis.

Integration

In his analysis of scientific revolutions, Thomas Kuhn suggested that the chief distinguishing characteristic of revolutions is a change in world view, or what Kuhn called a paradigm. Observations that were previously difficult to interpret would now fit into a coherent theoretical framework, a framework with predictive as well as explanatory powers. Biology has experienced its share of revolutions, with its most profound, Darwinian evolution, extending its influence in various guises beyond biology to society at large. One of the major recent revolutions took place in 1961 when François Jacob and Jacques Monod published their own operon model for the control of gene expression in bacteria. This model linked the expression of structural genes to control by a protein encoded by a regulatory gene located elsewhere in the genome, and explained how small molecules entering the cell could interact with the regulatory protein to mediate the coordinate expression of structural genes involved in particular metabolic pathways. The operon model opened an exciting period of prodigious investigation of gene regulation in bacteria. Not surprisingly, the predictive success of the model and its tremendous heuristic value made it inevitable that it would come to permeate ideas about the organization of metazoan genomes. The optimistic early application of the operon model directly to metazoans has been recognized as naive as metazoan genomes have become better understood. Nevertheless, a legacy remains of a concept of control of structural gene expression by products of distant regulatory elements that interact with a control site adjacent to the structural gene.

As in procaryotes, much of gene regulation must involve control of structural genes required for cell differentiation, maintenance, and metabolism. Unlike procaryotes, metazoans produce a large number of cell types organized into discrete and stable tissue types. The control of expression of structural genes in these cell or tissue types requires a coordinate expression of a set of genes specific to each cell type, as well as expression of a larger set of genes active in many or all cell types. A widely known model for this kind of cell-specific coordinate control of gene sets was proposed by Britten and Davidson in 1969. A schematic diagram of their model is presented in Figure 12–2. Integration depends on the activation of sensor sequences that respond to external influences to which the cell is exposed (e.g., hormones or inductive signals from neighboring cells). These sequences in turn activate specific integrator genes each of which produces a specific activator molecule. The activators in turn interact with corresponding regulatory sequences adjacent to structural genes and switch them on by allowing their transcription. Britten and Davidson proposed that the activators are transcripts of moderately repetitive DNA sequences, and that these interact directly by sequence complementarity with corresponding repeat elements acting as adjacent control elements. Obviously, protein activators would serve as well in the formal model.

In a more recent formulation of their model Davidson and Britten have suggested that in cases such as sea urchin embryos, in which all cells or stages appear to contain the same nuclear transcripts but different sets of mRNAs, that the integrative function might occur via repeat sequence transcripts acting to influence the processing of transcripts bearing complementary copies of the repeat sequence. However, the work of Derman et al. shows that in the mammalian tissues they examined, mRNA diversity is a function of differential

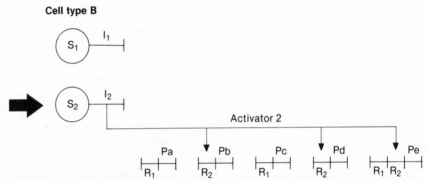

FIGURE 12-2. *Function of a tissue integrative system in controlling patterns of expression of gene batteries in two different cell types. Heavy arrows indicate inductive or hormonal signals from outside the cell. These interact with an appropriate sensor to activate an integrator gene. The integrator gene product interacts with specific control elements (R for receptor) adjacent to structural genes (P for producer).* [Based on the model of Britten and Davidson.]

transcription. Despite the potential importance of RNA processing in control of gene repression, the original Britten and Davidson model for control of transcription may serve as a reasonable approximation of the manner in which tissue-specific integration of gene action may occur.

There are several features of the model that should be noted. First, genes can be used in different combinations to yield characteristic tissue-specific batteries of active genes. Second, batteries can include vast numbers of genes, and as pointed out by Kauffman, large-scale interactive networks are stable. Stability can be increased by feedback loops, in which the product of an activated structural gene member of a battery may in turn act to keep the appropriate integrator genes active, thus forming a forcing loop that maintains tissue differentiation, even in the absence of the initial inducing signal. Third, a simple set of external signals is all that is necessary to activate a complex battery of genes defining the biochemical differentiation of a tissue. This is consistent with the molecular simplicity of hormones and inducers.

Control elements with some of the characteristics postulated for regulators of tissue integration have been demonstrated. Abraham and Doane, in a very graphic example showed that localized expression of the structural gene for α-amylase in the posterior midgut of *Drosophila* is controlled by a *trans*-acting regulatory gene. The expression of aldehyde oxidase in *Drosophila* has been found by W. J. Dickinson to be subject to control by both distant *trans*-acting and apparently adjacent *cis*-acting regulatory elements. Control is tissue-specific. For instance, one interesting variant has been found to cause elevated levels of aldehyde oxidase expression in male accessory sex glands, but to have no effect on enzyme levels in other tissues. The variant allele is nearby the structural gene, and acts in *cis*. *Cis*-acting control elements have also been found to operate in the developmental regulation of alcohol dehydrogenase in species of Hawaiian drosophilids by Dickinson and Carson. It is interesting to note that according to Rabinow and Dickinson, the *cis*-acting regulatory element for alcohol dehydrogenase controls gene expression at the level of synthesis of mRNA.

In accordance with our previous generalizations about the control of development, the Britten and Davidson model exhibits switch functions. However, the integrative controls provided are not sufficient to govern morphogenesis. For instance, the forelimbs and hindlimbs of vertebrates contain identical tissues, with striated muscle, skin, nerve, connective tissue, and so on. If arm and leg tissues were compared by some high-resolution analysis, they would exhibit the same pattern of structural gene expression, the same tissue-specific gene batteries, and the same tissue integrative systems. Tissue integration is real and important in development, but morphogenesis requires a different kind of integration involving hierarchical systems capable of making binary decisions in response to spatially determined patterns of information. This we call organismal integration.

The nature of organismal integration and its genic regulation is most clearly apparent in the control of segmental identity in *Drosophila*. Meso- and metathorax are composed of much the same tissues, but they are arranged significantly differently in the two segments. The subprograms for the shaping of individual segmental structures require genetic direction of cellular morphogenetic processes, cell movements, cell shape-changes, cell division patterns, and cell-cell affinities. These subprograms for the translation of genetic information into shape are still not well understood, although Garcia-Bellido and Ripoll have discussed some mutants affecting these processes. These give promise for future genetic dissection. It is not yet possible to define the actual machinery of morphogenesis, but it is possible to outline the system of gene switches by which segmental identities are initially established.

The essential elements of the model proposed by Garcia-Bellido are that the blastoderm-stage embryo contains an anterior-posterior system of positional information (as well as a dorsoventral system). Activator genes, which probably correspond to the regulatory genes *extra sex combs* and *Polycomb*, are expressed to an extent determined by the local level of positional information in a small region of the blastoderm surface. The size of the band on the blastoderm giving rise to a segment has been shown by Lohs-Schardin et al. to be on the order to three- or four-cell diameters in width. The level of activator gene expression in each narrow band of blastoderm cell determines which of the selector genes (members of the Bithorax or Antennapedia Complexes) are in turn switched on or off. The combination of selector genes active in each protosegment controls the expression of sets of genes responsible for the realization of individual segmental morphological subprograms. The activator gene substances act as repressors. When the selector genes in the Bithorax Complex (BX-C) are almost fully repressed, a set of subsequent genes specifying a pattern of morphogenetic events leading to the structures of the mesothorax is activated. With the switching on of additional members of the BX-C, progressively more divergent sets of realization-level genes are activated. The result is a discontinuous morphological progression from the "ground state" morphology, which for the segments governed by the BX-C is the mesothorax. Thus, metathorax has many features in common with the mesothorax, whereas abdominal segments differ quite radically. The most extreme segmental divergence is in the eighth abdominal segment, in which the activator genes are inactive and all members of the BX-C active.

This regulatory model defines a hierarchical control system operating through a cascade of switch genes. The activator genes act as variable controllers whose activity levels are determined by the characteristics of positional information in a segment. The selector genes are differentially activated as a function of position, and themselves act as combinatorial switches for a final set of morphological subprograms. This system is not concerned with tissue-level integration. The structural genes of the tissue batteries are certainly expressd in coordination with the genes that control morphogenesis, but because the same or very similar batteries are expressed in the tissues of different compartments they are probably regulated by tissue integrative systems distinct from the organismal integrative system. The cuticles of mesothorax and metathorax are identical in composition but differ in morphology. In a very general sense tissue integration systems may provide the basic cell types, whereas the organismal integration system provides the directions for shaping. Numerous examples of dissociation between cytodifferentiation and morphogenesis support the existence of these coordinated but separate integration systems.

A Brief Anecdotal and Undocumented History

If, as we suggest, there are distinct tissue and organismal integrative systems, then their evolutionary histories in the origin and diversification of metazoans have followed considerably different pathways. The origins of animals occurred late in the Precambrian, and our first metazoan ancestors have left no record of their passage. Like Haeckel, we must construct our historical fantasies from living organisms and what we perceive (a risky business) to be general principles.

The earliest multicellular animals faced two considerable problems: generating and maintaining stable tissues and the invention of ontogeny. As in all evolutionary transitions there was no *de novo* invention. The genetic and cellular systems necessary for rudimentary developmental processes already existed in the antecedent protistans. Some protista, such as ciliates, have extremely complex cellular morphologies, whereas others, such as *Volvox*, approach multicellular organisms in exhibiting a differentiation into somatic and germ cells and a set of morphogenetic movements in development that resemble animal gastrulation.

Protists have thrived in extremely diverse ways, but they are restricted to relatively small sizes imposed by physical constraints and by limitations to the mass of cytoplasm that can be maintained by a single nucleus. Those limits are pushed by large ciliates on the order of a millimeter or so in length that are sizeable enough to prey on the smallest metazoans. Ciliates have increased their sizes and complexities while remaining single-celled organisms by deploying polyploid macronuclei to control somatic functions. These nuclei can sustain the transcription rates needed by big cells, but processes of intracellular transport are sluggish enough to limit the range of control of even a macronucleus. For all the successes of protistans in environments ranging from tidepools to the rumens of cows there are, nevertheless, distinct advantages to being large—particularly when the question of who is to be on the menu is being decided.

The cellular preadaptations of protistans would have been the heritage of the first multicellular forms. These preadaptations would have been sufficient to provide for the differentiation required in the most simple imaginable metazoan, one not far removed in complexity from some living protistans. Such an animal would have been small by metazoan standards, and in some respects would have resembled the large protistan *Opalina,* a large flagellated organism possessing several hundred nuclei in a common cytoplasm. Only a segregation of nuclei from each other by membrane boundaries would be necessary to convert an *Opalina*-like protistan to a multicellular condition. Differentiation of internal digestive cells from external ciliated epithelial cells requires only the introduction of a single switch to control differential expression of programs in two spatially distinct cell populations of functions previously common to a single cell type. Both tissue and organismal integrative controls thus originated in a rudimentary fashion to produce an organism formally similar to the planula larva of coelenterates or to the acoel flatworms. These represent grades of morphological organization that, for obvious reasons, have been dear to people engaged in phylogenetic speculations about the origins of metazoans.

Our hypothetical animal has one additional feature, distinct germ cells. The somatic cell-germ cell dichotomy is fundamental to metazoan organization. Metazoans are sexual organisms. Production of gametes requires cells capable of undergoing meiosis and then subsequent fusion with another haploid cell to initiate development. If an organism is to have any degree of cellular differentiation, not all cells can be competent to give rise to gametes. Thus, unlike most protistans, where the entire organism gives rise to gametes, a separate cell population must be established. The precursors of this basic functional segregation are seen in organisms as primitive as the cellular slime mold *Dictyostelium*, which in fruiting produces mortal somatic cells, the stalk cells, and immortal germ cells, the spores. By the account of the Old Testament book of Genesis, death was the price of knowledge: More prosaically, it was the price of multicellularity.

With the evolution of the necessary systems of induction combined with the evolution of new tissue interaction systems, novel gene batteries could be generated to produce novel tissue types. Because even very primitive metazoan groups possess several cell types, this level of evolution may have produced early on many of the basic metazoan tissues. Muscle, nerve, gut, and epithelium among others appear to be very ancient cell types and to have remained extraordinarily stable through long periods of evolutionary time while organismal integration systems have been modified. The resultant dramatic changes in form seen within the mammals or insects have entailed old tissues being molded into new shapes. This brief discussion of metazoan origins is so oversimplified as to suggest that accounting for the origins of metazoan animals and their ontogenetic processes is a trivial matter. It is not, but we are so ignorant of the actual course of events that we can only speculate. The point of this speculation is to show that many tissue integration systems are extremely old. So are the basic organismal integration systems of the phyla so clearly established by the Cambrian and Ordovician. However, changes in organismal integration systems appear to provide most of the impetus to morphological evolution.

Novelties

In 1860 Louis Agassiz, who was to lead the scientific opposition to Darwinian evolution in America, expressed his initial outrage at the publication of the *Origin of Species* in a passionate review. Among Agassiz's arguments against Darwin's evidence for evolution is the following:

> Would the supporters of the fanciful theories lately propounded only extend their studies a little beyond the range of domesticated animals, would they investigate the alternate generations of the Acalephs, the extraordinary modes of development of the Helminth, the reproduction of the Salpae, etc., etc., they would soon learn that there are, in the world, far more astonishing phenomena, strictly circumscribed between the natural limits of unvarying species, than the slight differences produced by the intervention of men, among domesticated animals, and, perhaps, cease to be so confident as they seem to be, that these differences are trustworthy indications of the variability of species.

Agassiz's argument can be stood on its head. The complexities of developmental programs, discussed in Chapter 9, provide the raw material for evolutionary change. Heterochronies allow juvenile characteristics to be retained in reproductively mature adults. Modifications of temporal relationships between subprograms or uncoupling between inductive events involved in morphogenesis provide evolutionary mechanisms to alter the course of development, and thus provide new larval life history strategies, such as direct terrestrial development in some tropical frogs, or novel adult tissues or morphologies. Anecdotal examples of such developmental transformations are numerous, and it is their genetic and developmental mechanisms that we wish to understand. However, there is perhaps an important point in Agassiz's comment that has not lost its force. There is a certain danger in attempting to pare the complexity of evolutionary phenomena down to manageable paradigms. In our own analysis there may be a risk of leaving an oversimplification that only regulatory genes and not structural genes are important in morphological evolution, and that of regulatory genes, only organismal integrators and not tissue integrators are really significant. This is a far more one-dimensional conclusion than we would like to draw. In fact, many evolutionary events have involved a subtle interweaving of several kinds of genetic changes, and have resulted in novel structures and behaviors that have opened new adaptive possibilities. One such event was the origin of the mammary gland.

The significance of the mammary gland is that it is a relatively new organ with its origins lying in the transition between the mammal-like reptiles and the true mammals. The possession of mammary glands increased the efficiency of mammalian reproduction and, through the bond between mother and offspring, initiated a suite of behavioral changes that may have contributed to the progressive evolution of the mammalian brain. The most likely progenitor for mammary glands lies in sweat or other cutaneous glands. Evaporative cooling by means of sweat glands may have been an early adaptation of a group of animals evolving homeothermy and sophisticated mechanisms of thermoregulation. As suggested by Chadwick, ancestral mammals were small animals and their young in their first few days of life were very likely at risk from dehydration. Sweat glands may have been especially well developed in brooding mammal-like reptiles as a way of providing their young with water and salts.

The morphogenesis of the mammary gland in the three living subclasses of mammals, as reviewed by Raynaud, provides a view of stages in the increasing evolutionary sophistication of the mammary gland. In monotremes, the egg-laying mammals, there is no defined nipple. Rather, about 100 tubular glands open to the surface at each side of the midline. At the opening of each gland there is a stiff hair. The secretion of the gland runs along the hair to be licked up by the nursing young. In marsupials the mammary precursors differentiate to produce three types of buds. These in turn give rise to mammary hairs, or to mammary or sebaceous glands. The mammary hairs of marsupials are transitory, and in placental mammals the mammary anlagen are no longer associated with those for hairs or sebaceous glands.

As was discussed in Chapter 5, the morphogenesis of the basic monopodial mammary gland is induced in mammary epithelium by mammary mesenchyme. The paramount change in organizational integration patterns required establishing a link between the control of gland morphogenesis and the hormones that stimulate proliferation of glandular elements during maturation and pregnancy. Proliferation in these phases of ontogeny involve replication of basic glandular components. The evolutionary changes in mammary gland structures have entailed concentration of glandular elements into discrete aggregates of glands connected to a nipple. As in all evolutionary changes in morphology, changes occurred in genetic systems controlling morphogenesis of the cutaneous glands from which mammary glands arose; however, we have chosen this example for discussion precisely because such a large part of the evolution of the mammary gland has been so obviously played by changes in tissue integration and in the evolution of structural genes.

Descriptions of mammary development such as provided by Forsyth and Hayden indicate that hormones are not required for the initial stages of mammary induction and development, but that a group of hormones, estrogen, growth hormone, and adrenal steroids, are necessary for maturation in adolescence. The vigorous proliferation of ducts and alveoli during pregnancy requires several hormones, notably estrogen, progesterone, and prolactin. Membrane receptor sites for these hormones are present on the surfaces of mammary gland cells. Evidently, differentiation and function of this tissue occurs in response to signals that would, by the model of Britten and Davidson, interact with sensor elements.

Although remote in time, the connection between prolactin and the evolution of the mammary gland is reasonable in view of descriptions by Bern, by Dent, and by Nicoll of the diverse roles played by prolactin. Prolactin is a protein hormone related in sequence to growth hormone, from which, as suggested by Niall, it probably diverged after a gene duplication event early in the history of the vertebrates. Unlike most hormones, prolactin did not become restricted in early vertebrates to a particular and specialized set of functions. Instead, it has remained available to serve in a variety of processes. In all classes of vertebrates prolactin is important in osmotic regulation, and it has significant effects in the differentiation of epithelial structures related to reproduction. These include secretion by the skin of discus fish of mucus upon which the young feed, the development of nuptial pads by male frogs, the development of brood patches by nesting birds, and lactation in female mammals. Prolactin may well be important in the control of ion pumping in the mammary gland. Although prolactin stimulates activity of the sebaceous glands of mammals, sweat glands do not appear to be responsive.

If mammary glands evolved from sweat glands, as appears probable, the regulation of the former by prolactin and other hormones was acquired in the evolution of a new tissue integration system. This is not an unreasonable proposition because the ancestral mammary gland would have had to be hormonally linked to the control of reproductive activities, whereas its evolutionary precursor would not.

In its mechanism of action prolactin appears to function as a signal for the activation of a specialized battery of genes in the mammary gland. According to a review by J. M. Rosen of recent studies in his own and other laboratories, prolactin interacts with mammary cells to increase the levels of milk protein mRNAs. During lactation the mRNAs for the caseins and α-lactalbumin come to account for over 80% of total cellular mRNA. Rosen and his collaborators found by use of specific cloned DNA probes that casein mRNA rises 300-fold from a very low level found in the virgin rat mammary. Nakhasi and Qasba observed the same effect with α-lactalbumin mRNA. Prolactin causes both an increase in transcription rate and a decrease in the rate of messenger degradation.

The evolutionary integration of a new battery of genes for the specialized cytodifferentiation of mammary glands involved two processes. The first was the establishment of a hormonal control linkage so that secretory activity of the ancestral gland would be elicited in concert with reproduction. The initial battery of structural genes activated in this way were those typical of the glands that served as evolutionary precursors to the mammary gland. The cellular machinery for secretion in the mammary gland, which has been described by Mepham, is similar to that of other secretory cells, and most likely existed in the ancestral gland. Thus, an already existing battery of genes may have been captured by a new control exerted over a subset of epidermal glands. Evolution of a true mammary gland would subsequently require the evolutionary integration of specialized structural genes to provide an optimal nutrient secretion. New members of a mammary gene battery could have arisen in two ways: by recruitment of pre-existing genes, and by evolution of novel genes.

The integration of pre-existing structural genes into a battery may be achieved by placement of a new cis-acting regulatory sequence adjacent to the structural gene. As has been explicitly pointed out by Dickinson, modification of cis-acting regulatory elements would make possible evolutionary changes in the program of expression of individual genes without disrupting the overall developmental program. The appearance of a new cis-acting regulatory element could result from point mutations in a pre-existing cis-acting element, but more rapid changes could occur by a radically different mechanism of transposition of appropriate pre-existing regulatory elements from elsewhere in the genome. Eucaryotic cells contain transposable elements capable of stable insertion into the genome. It has been suggested by Echols that the repression system that maintains stable integration can be overcome by certain stresses. One particularly interesting case has been observed in D. melanogaster. When some wild-caught strains are mated with strains that have been long maintained in the laboratory, a high frequency of apparently spontaneous mutagenesis results. This hybrid dysgenesis results from an increase in the mobility of once-stable integrated transposable elements.

Echols visualizes an analogous induction of novel genotypes under environmental conditions in which a population is poorly adapted. These proposals of rapid regulatory evolution are supported by Dickinson's observations that in tissues of closely related species of Hawaiian drosophilids homologous structural genes for alcohol dehydrogenase and aldehyde oxidase are regulated at strikingly different levels. Use of cloned probes for these genes should allow a direct experimental test to be made of the proposition that these changes have resulted from transposition of cis-acting regulatory elements.

The final component in the evolution of the mammary gland is the origin of novel structural genes for gland-specific functions. Milk contains several proteins unique to the mammary gland. These, reviewed by Jenness, include several caseins, β-lactoglobulin, and of most significance here, α-lactalbumin. It is this protein, whose function is the synthesis of lactose, as lucidly reviewed by E. A. Jones, that provides the nicest example of the origin of a novel structural gene as part of the evolution of a new organ. The critical papers on the function and evolution of α-lactalbumin come from the work of Brodbeck and Ebner and Brew and his colleagues. The enzyme that carries out the synthesis of lactose from uridine 5'-diphosphate (UDP)-galactose and UDP-glucose is galactosyltransferase. This enzyme normally has a poor affinity for glucose, except when the enzyme is complexed with α-lactalbumin. The complex has a high affinity for glucose and results in the unique ability of the mammary gland to synthesize lactose. The α-lactalbumin molecule has the ability to modify the synthetic properties of galactosyltransferase from a wide evolutionary spectrum of organisms, including the astonishing demonstration by Powell and Brew that it can cause the galactosyltransferase from onion to synthesize lactose.

The evolutionary origin of α-lactalbumin is clear. This protein has significant amino acid homology with lysozyme, which hydrolizes bacterial cell-wall mucopolysaccharide and is present in a variety of mammalian body fluids. Genes for both α-lactalbumin and lysozyme exist in the same animal. Brew et al. have invoked the likely mechanism of duplication of an ancestral lysozyme gene followed by divergence.

The evolution of the mammary gland involved a suite of genetic changes. Modifications in tissue integration allowed a set of cutaneous glands, probably sweat glands, to become linked to the hormonal system controlling reproduction. A novel tissue integration system was thus established, and evolution of novel structural protein genes ensued. Hopper and McKenzie have reported that the milk of the echidna, a monotreme, contains not a typical α-lactalbumin, but a lysozymelike protein with α-lactalbumin activity, a possible living fossil protein. The evolution of this and other novel milk proteins would have proceeded in conjunction with the integration of a new set of structural genes into a battery expressed in a gland coming under a new integrative control. Concomitant changes in integrative systems involved in morphogenesis would also have occurred to produce both the specialized glands themselves and their integration into an organized mammary structure connected to a nipple. It is significant that the early, hormone-independent steps in mammary development require induction of glandular epithelium by mesenchyme. The experiments of Sakakura et al. discussed in Chapter 5 suggest that changes in the genes controlling induction in this system were required, but that the precursors were already present, and rather few genetic changes may have been needed. The overall impression is that although the evolution of the present-day structure may have been the result of numerous gene changes at several levels of control as well as in the origin new structrual genes, the initial steps required a rather small number of modifications of already existing morphogenetic, hormonal, and tissue integrative processes.

If the notion of developmental constraints limiting evolutionary directions has any meaning, it is in the sense that modifications of already existing developmental processes provide the most readily available route for evolutionary change. Once a modification becomes established, it in turn makes acceptance of changes in certain directions more feasible than others. But if existing developmental patterns constrain, they also provide opportunities for rapid evolutionary departures when selection pressures on morphology change because of their dissociability and apparently simple genetic controls.

A century has passed since Haeckel's views of the relationship of ontogeny to phylogeny were at the peak of their influence. Since then, embryology and evolution have largely followed their own separate paths. Evolutionary theory has been highly integrated with one aspect of genetics, although developmental biology followed Roux's program for experimental embryology and by and large ignored genetics. Richard Goldschmidt recognized that the common basis for understanding evolution lay in the application of genetics to the study of development. His vision has been delayed and modified, but it remains. But the central and still unsolved problem is, how do genes direct the making of an organism? The solution will allow us to answer the still very real question posed by Charles Bonnet over 200 years ago: "Now, I beg you to tell me what mechanics will preside over the formation of a brain, a heart, a lung, and so many other organs?"

Bibliography

Abraham, I., and W. W. Doane, 1978. Genetic regulation of tissue-specific expression of *Amylase* structural genes in *Drosophila melanogaster*. *Proc. Natl. Acad. Sci. USA* **75**:4446–4450.

Agassiz, L., 1860. Contributions to the Natural History of the United States. *Am. J. Sci. Arts Ser.* 2. **30**:142–154.

Alberch, P., 1980. Ontogenesis and morphological diversification. *Am. Zool.* **20**:653–667.

Alberch, P., 1981. Convergence and parallelism in foot morphology in the neotropical salamander *Bolitoglossa*. I. Function. *Evolution* **35**:84–100.

Alberch, P., and J. Alberch, 1981. Heterochronic mechanisms of morphological diversification and evolutionary change in the neotropical salamander *Bolitoglossa occidentalis* (Amphibia: Plethodontidae). *J. Morphol.* **167**:249–264.

Allen, G., 1978. *Thomas Hunt Morgan: The Man and His Science*. Princeton University Press, Princeton, N.J.

Anderson, D. T., 1969. On the embryology of the cirripede crustaceans *Tetraclita rosea* (Krauss), *T. purpurascens* (Wood), *Chthamalus antennatus* (Darwin) and *Chamaesipho columna* (Spengler) and some considerations of crustacean phylogenetic relationships. *Phil. Trans. R. Soc. B.* **256**:183–235.

Anderson, D. T., 1973. *Embryology and Phylogeny in Annelids and Arthropods*. Pergamon Press, Oxford.

Anderson, S., and J. K. Jones, Jr., 1967. *Recent Mammals of the World: A Synopsis of Families*. Ronald Press Company, New York.

Angerer, R. C., E. H. Davidson, and R. J. Britten, 1975. DNA sequence organization in the mollusc *Aplesia californica*. *Cell* **6**:29–39.

Angerer, R. C., E. H. Davidson, and R. J. Britten, 1976. Single copy DNA and structural gene sequence relationships among four sea urchin species. *Chromosoma* **56**:213–226.

Arnold, J. M., 1971. Cephalopods. In *Experimental Embryology of Marine and Freshwater Invertebrates* (G. Reverberi, ed.), North Holland Publishing Company, Amsterdam, pp. 265–311.

Artymiuk, P. J., C. C. F. Blake, and A. E. Sippel, 1981. Genes pieced together—Exons delineate homologous structures of diverged lysozymes. *Nature* **290**:287–288.

Avise, J. C., J. C. Patton, and C. F. Aquadro, 1980. Evolutionary genetics of birds. Comparative molecular evolution in new world warblers and rodents. *J. Heredity* **71**:303–310.

Axel, R., P. Feigelson, and G. Schutz, 1976. Analysis of the complexity and diversity of mRNA from chicken liver and oviduct. *Cell* **7**:247–254.

Baker, R. J., and J. W. Bickham, 1980. Karyotypic evolution in bats: Evidence of extensive and conservative chromosomal evolution in closely related taxa. *Syst. Zool.* **29**:239–253.

Barnes, R. D., 1974. *Invertebrate Zoology*, 3rd ed. W. B. Saunders Company, Philadelphia.

357

Bassindale, R., 1936. The developmental stages of three English barnacles, *Balanus balanoides* (Linn.), *Chthamalus stellatus* (Poli), and *Verruca stroemia* (O. F. Müller). *Proc. Zool. Soc. Lond.* **106**:57–74.

Bateson, W., 1894. *Materials for the Study of Variation.* Macmillan and Co., London.

Bauchot, R., and H. Stephan, 1964. Le poids encéphalique chez les insectivores Malgaches. *Acta Zoologica* **45**:63–75.

Baxter, J. D., N. L. Eberhard, J. W. Apriletti, L. K. Johnson, R. D. Ivarie, B. S. Schacter, J. A. Morris, P. H. Seeburg, H. M. Goodman, K. R. Latham, J. R. Polansky, and J. A. Martial, 1979. Thyroid hormone receptors and responses. *Rec. Progr. Horm. Res.* **35**:97–147.

Beckingham-Smith, K., and J. R. Tata, 1976. Cell death—Are new proteins synthesized during hormone-induced tadpole tail regression? *Exp. Cell Res.* **100**:129–146.

Beebe, D. C., and J. Piatigorsky, 1977. The control of δ-crystallin gene expression during lens cell development: Dissociation of cell elongation, cell division, δ-crystallin synthesis, and δ-crystallin mRNA accumulation. *Dev. Biol.* **59**:174–182.

Bell, B. M., 1976. Phylogenetic implications of ontogenetic development in the class Edrioasteroidea (Echinodermata). *J. Paleontol.* **50**:1001–1019.

Bell, B. M., 1976. *A Study of North American Edrioasteroidea.* New York State Museum and Science Services, Albany, N. Y.

Benbow, R. M., and C. C. Ford, 1975. Cytoplasmic control of nuclear DNA synthesis during development of *Xenopus laevis:* A cell free assay. *Proc. Natl. Acad. Sci. USA* **72**:2437–2441.

Bennett, D., 1975. The T-locus of the mouse. *Cell* **6**:441–454.

Benoist, C., and P. Chambon, 1981. In vivo sequence requirements of the SV40 early promoter region. *Nature* **290**:304–310.

Berg, D. E., 1980. Control of gene expression by a mobile recombinational switch. *Proc. Natl. Acad. Sci. USA* **77**:4880–4884.

Berkner, L. V., and L. C. Marshall, 1964. The history of oxygenic concentration in the earth's atmosphere. *Dis. Faraday Soc.* **37**:122–141.

Bern, H. A., 1975. Prolactin and osmoregulation. *Am. Zool.* **15**:937–949.

Berrill, N. J., 1955. *The Origin of Vertebrates.* Clarendon Press, Oxford.

Bier, K., and W. Müller, 1969. DNS-Messungen bei Insekten und eine Hypothese über retardierte Evolution und besonderen DNS-Reichtum im Tierreich. *Biol. Zentralblatt* **88**:425–449.

Birnstiel, M. L., K. Gross, W. Schaffner, and J. Telford, 1975. Biochemical dissection of the histone gene cluster of sea urchin. *Fed. Eur. Biochem. Soc. Meet. (Proc.)* **38**:3–24.

Bishop, J. O., J. S. Beckmann, M. S. Campo, N. D. Hastie, M. Izquierdo, and S. Perlman, 1975. DNA-RNA hybridization. *Phil. Trans. R. Soc. Lond. B* **272**:147–157.

Bloemendal, H., 1977. The vertebrate eye lens. *Science* **197**:127–138.

Bogenhagen, D. F., S. Sakonju, and D. D. Brown, 1980. A control region in the center of the 5S RNA gene directs specific initiation of transcription. II. The 3' border of the region. *Cell* **19**:27–35.

Bonner, J. T., 1958. *The Evolution of Development.* Cambridge University Press, Cambridge.

Bonnet, C., 1764. *Contemplation de la Nature,* Amsterdam. Quoted by C. O. Whitman, 1894, *The palingenesia and the germ doctrine of Bonnet.* In *Biological Lectures of The Marine Biological Laboratory of Woods Hole, Mass.* Ginn and Company, Boston, Mass., pp. 241–272.

Bookstein, F. L., P. D. Gingerich, and A. G. Kluge, 1978. Hierarchical linear modeling of the tempo and mode of evolution. *Paleobiology* **4**:120–134.

Boveri, T., 1899. Die Entwicklung von *Ascaris megalocephala* mit besonderer Rücksicht auf die Kernverhältnisse. *Festschr. f. C. v. Kupffer*. G. Fischer, Jena, pp. 383–430.

Boveri, T., 1902. Über mehrpolige Mitosen als Mittel zur Analyse der Zellkerns. *Verh. phys. med. Gesellsch. (Würzburg)*, **35**:67–90.

Brandhorst, B. P., and K. M. Newrock, 1981. Post-transcriptional regulation of protein synthesis in *Ilyanassa* embryos and isolated polar lobes. *Dev. Biol.* **83**:250–254.

Brew, K., and R. L. Hill, 1975. Lactose biosynthesis. *Rev. Physiol. Biochem. Pharmacol.* **72**:105–158.

Brew, K., and R. L. Hill, 1975. Lactose synthetase. *Adv. Enzymol.* **43**:411–490.

Brew, K., H. M. Steinman, and R. L. Hill, 1973. A partial amino acid sequence of α-Lactalbumin-I of the grey kangaroo. *J. Biol. Chem.* **248**:4739–4742.

Briggs, R., 1972. Further studies on the maternal effect of the *o* gene in the Mexican axolotl. *J. Exp. Zool.* **181**:271–280.

Briggs, R., 1973. Development genetics of the axolotl. In *Genetic Mechanisms of Development* (F. H. Ruddle, ed). Academic Press, New York, pp.169–199.

Briggs, R., and G. Cassens, 1966. Accumulation in the oocyte nucleus of a gene product essential for embryonic development beyond gastrulation. *Proc. Natl. Acad. Sci. USA* **55**:1103–1109.

Briggs, R. and J. T. Justus, 1968. Partial characterization of the component from normal eggs which corrects the maternal effect of gene *o* in the Mexican axolotl (*Ambystoma mexicanum*). *J. Exp. Zool.* **167**:105–115.

Britten, R. J. and E. H. Davison, 1969. Gene regulation for higher cells: A theory. *Science* **165**:349–357.

Britten, R. J., and D. E. Kohne, 1968. Repeated sequences in DNA. *Science* **161**:529–540.

Britten, R. J. and D. E. Kohne, 1970. Repeated segments of DNA. Sci. Am. **222**:24–31.

Brodbeck, U., and K. E. Ebner, 1966. Resolution of a soluble lactose synthetase into two protein components and solubilization of microsomal lactose synthetase. *J. Biol. Chem.* **241**:762–764.

Brown, D. D., and I. B. Dawid, 1968. Specific gene amplification in oocytes. *Science* **160**:272–279.

Brown, D. D., D. Carroll, and R. D. Brown, 1977. The isolation and characterization of a second oocyte 5S DNA from *Xenopus laevis*. *Cell* **11**:1045–1056.

Brown, G. W., Jr., and P. P. Cohen, 1958. Biosynthesis of urea in metamorphosing tadpoles. In *A Symposium on the Chemical Basis of Development* (W. D. McElroy and B. Glass, eds.) John Hopkins University Press, Baltimore, Md., pp. 495–513.

Brown, G. W., Jr., and P. P. Cohen, 1960. Comparative biochemistry of urea synthesis. *Biochem. J.* **75**:82–91.

Brown, M. R. and H. C. Bold, 1964. Comparative studies of the algal genera *Tetracystis* and *Chlorococcum*. *Phycol. Stud.*, University of Texas publication no. 6417, 213 pp.

Bryant, P. J., 1978. Pattern formation, growth control and cell interactions in *Drosophila* imaginal discs. In *Determinants of Spatial Organization* (S. Subtelny and I. R. Konigsberg, eds). Academic Press, New York, pp. 295–316.

Bull, A., 1966. *Bicaudal*, a genetic factor which affects the polarity of the embryo in *Drosophila melanogaster*. *J. Exp. Zool.* **161**:221–241.

Bullough, W. S., 1975. Mitotic control in adult mammalian tissues. *Biol. Rev. Cambridge Phil. Soc.* **50**:99–127.

Bush, G. L., 1975. Modes of animal speciation. *Ann. Rev. Ecol. Systemat.* **6**:339–364.

Bush, G. L., S. M. Case, A. C. Wilson, and J. L. Patton, 1977. Rapid speciation and chromosomal evolution in mammals. *Proc. Natl. Acad. Sci. USA* **74**:3942–3946.

Busslinger, M., S. Rusconi, and M. L. Birnstiel, 1982. An unusual evolutionary behavior of a sea urchin histone gene cluster. Manuscript in press.

Cahn, P. H., 1959. Comparative optic development in *Astyanax mexicanus* and in two of its blind cave derivatives. *Bull. Am. Mus. Nat. Hist.* **115**:72–112.

Calman, W. T., 1909. Appendiculata: Part 3, Crustacea. *Treatise in Zoology*, Part VII (E. R. Lankester, ed.). Adam and Charles Black, London, 364 pages.

Cardellini, P., J. Gabrion, and M. Sala, 1978. Electrophoretic patterns of larval, neotenic and adult haemoglobin of *Triturus helveticus* Raz. *Acta Embryol. Exp.* **2**:151–161.

Carlson, S. S., G. A. Mross, A. C. Wilson, R. T. Mead, L. D. Wolin, S. F. Bowers, N. T. Foley, A. O. Muijsers, and E. Margoliash, 1977. Primary structure of mouse, rat, and guinea pig cytochrome c. *Biochemistry* **16**:1437–1442.

Carson, H. L. and K. Y. Kaneshiro, 1976. *Drosophila* of Hawaii: Systematics and ecological genetics. *Ann. Rev. Ecol. Syst.* **7**:311–345.

Carson, H. L. and K. Y. Kaneshiro, 1976. *Drosophila* of Hawaii: Systematics and Ecological Genetics. *Ann. Rev. Ecol. Syst.* **7**:311–345.

Castle, W. E. and J. C. Phillips, 1909. A successful ovarian transplantation in the guinea pig and its bearing on problems of genetics. *Science* **30**:312–313.

Chadwick, A., 1977. Comparison of milk-like secretions found in non-mammals. *Symp. Zool. Soc. Lond.* **41**:341–358.

Chalfie, M., H. R. Horovitz, and J. E. Sulston, 1981. Mutations that lead to reiterations in the cell lineages of *C. elegans*. *Cell* **24**:59–69.

Chambon, P., F. Perrin, K. O'Hare, J. L. Mandel, J. P. Le Pennec, M. Le Meur, A. Krust, R. Heilig, P. Gerlinger, F. Gannon, M. Cochet, R. Breathnach, and C. Benoist, 1979. Structure and expresion of ovalbumin and closely related chicken genes. In *Eucaryotic Gene Regulation* (R. Axel, T. Maniatis, and C. F. Fox, ed.). Academic Press, New York, pp. 259–279.

Chan, L. N., and W. Gehring, 1971. Determination of blastoderm cells in *Drosophila melanogaster*. *Proc. Natl. Acad. Sci USA*. **68**:2217–2221.

Cherry, L. M., S. M. Case, and A. C. Wilson, 1978. Frog perspective on the morphological difference between humans and chimpanzees. *Science* **200**:209–211.

Childs, G., R. Maxson, and L. H. Kedes, 1979. Histone gene expression during sea urchin embryogenesis: Isolation and characterization of early and late messenger RNAs of *Strongylocentrotus purpuratus* by gene-specific hybridization and template activity. *Dev. Biol.* **73**:153–173.

Chooi, W. Y., 1971. Variation in nuclear DNA content in the genus *Vicia*. *Genetics* **68**:195–211.

Chooi, W. Y., 1976. RNA transcription and ribosomal protein assembly in *Drosophila melanogaster*. In *Handbook of Genetics*, Vol. 5 (R. C. King, ed.), Plenum Press, New York, pp. 219–265.

Chovnick, A., W. Gelbart, M. McCarron, B. Osmond, E. P. M. Candido, and D. L. Baillie, 1976. Organization of the *rosy* locus in *Drosophila melanogaster*: Evidence for a control element adjacent to the xanthine dehydrogenase structural element. *Genetics* **84**:233–255.

Chovnick, A., W. Gelbart, and M. McCarron, 1977. Organization of the *rosy* locus in *Drosophila melanogaster*. *Cell* **11**:1–10.

Cisne, J. L., 1974. Trilobites and the origin of arthropods. *Science* **186**:13–18.

Clark, H., 1953. Metabolism of the black snake embryo. I. Nitrogen excretion. *J. Exp. Biol.* **30**:492–501.

Clark, H., and B. F. Sisken, 1956. Nitrogenous excretion by embryos of the viviparous snake *Thamnophis s. sirtalis* (L.) *J. Exp. Biol.* **33**:384–393.

Clausen, R. E. and T. H. Goodspeed, 1925. Interspecific hybridization in *Nicotiana*. II. A tetraploid *glutinosa-tabacum* hybrid, an experimental verification of Winge's hypothesis. *Genetics* **10**:279–284.

Cleveland, D. W., M. A. Lopata, N. J. Cowan, R. J. MacDonald, W. J. Rutter, and M. W. Kirschner, 1980. A study of the number and evolutionary conservation of genes coding for α- and β-tubulin and β- and γ-cytoplasmic actin using specific cloned cDNA probes. *Cell* **20**:95–105.

Cloud, P., 1976. Beginnings of biospheric evolution and their biogeochemical consequences. *Paleobiology* **2**:351–387.

Cobban, W. A., 1951. Scaphitid cephalopods of the Colorado Group. *U.S. Geol. Surv. Prof. Paper. no. 239.*

Cohen, P. P., 1969. Biochemical aspects of metamorphosis: Transition from ammonotelism to ureotelism. *Harvey Lect.* **60**:119–154.

Cohen, P. P., and G. W. Brown, Jr. 1963. Evolution of nitrogen metabolism. In *Evolutionary Biochemistry: Proceedings of the 5th International Congress of Biochemistry* (A. I. Oparin, ed.). Macmillan Publishing Co., Inc., New York, pp. 129–138.

Cohn, R. H., J. C. Lowry, and L. H. Kedes, 1976. Histone genes of the sea urchin *(S. purpuratus)* cloned in *E. coli:* Order, polarity, and strandedness of the five histone-coding and spacer regions. *Cell* **9**:147–161.

Conklin, E. G., 1905. Organization and cell-lineage of the ascidian egg. *J. Acad. Natl. Sci. (Phila.)* **13**:1–119.

Conklin, E. G., 1905. Mosaic development in ascidian eggs. *J. Exp. Zool.* **2**:145–223.

Conklin, E. G., 1911. The organization of the egg and the development of single blastomeres of *Phallusia mamillata. J. Exp. Zool.* **10**:393–407.

Conrad, G. W., D. C. Williams, F. R. Turner, K. M. Newrock, and R. A. Raff, 1973. Microfilaments in the polar lobe constriction of fertilized eggs of *Ilyanassa obsoleta. J. Cell Biol.* **59**:228–233.

Conway-Morris, S., and H. B. Wittington, 1979. The animals of the Burgess shale. *Sci. Am.* **241(1)**:122–133.

Corruccini, R. S., M. Baba, M. Goodman, R. L. Ciochon, and J. E. Cronin, 1980. Non-linear macromolecular evolution and the molecular clock. *Evolution* **34**:1216–1219.

Costantini, F. D., R. J. Britten, and E. H. Davidson, 1980. Message sequences and short repetitive sequences are interspersed in sea urchin egg poly(A)$^+$ RNAs. *Nature* **287**:111–117.

Costantini, F. D., R. H. Scheller, R. J. Britten, and E. H. Davidson, 1978. Repetitive sequence transcripts in the mature sea urchin oocyte. *Cell* **15**:173–187.

Cox, G. N., S. Staprans, and R. S. Edgar, 1981. The cuticle of *Caenorhabditis elegans*. II. Stage-specific changes in ultrastructure and protein composition during post-embryonic development. *Dev. Biol.* **86**:458–470.

Crain, W. R., E. H. Davidson, and R. J. Britten, 1976. Contrasting patterns of DNA sequence arrangement in *Apis mellifer* (honey bee) and *Musca domestica* (housefly). *Chromosoma* **59**:1–12.

Cronin, J. E., N. T. Boaz, C. B. Stringer, and Y. Rak, 1981. Tempo and mode in hominid evolution. *Nature* **292**:113–122.

Czihak, G., 1971. Echinoids. In *Experimental Embryology of Marine and Fresh-Water Invertebrates*. (G. Reverberi, ed.). North Holland Publishing Company, Amsterdam, pp. 383–506.

Daly, H. V., and A. Sokoloff, 1965. *Labiopedia*, a sex-linked mutant in *Tribolium confusum* Duval (Coleoptera: Tenebrionidae). *J. Morphol.* **117**:251–270.

Daly, H. V., J. T. Doyen, and P. R. Ehrlich, 1978. *Introduction to Insect Biology and Diversity*. McGraw-Hill Book Company, New York.

Dan, K., and M. Ikeda, 1971. On the system controlling the time of micromere formation in sea urchin embryos. *Dev. Growth Differen.* **13**:285–301.

Daneholt, B., 1974. Transfer of genetic information in polytene cells. *Int. Rev. Cytol. (Suppl.)* **4**:417–462.

Daneholt, B., 1975. Transcription in polytene chromosomes. *Cell* **4**:1–9.

Daneholt, B., K. Andersson, and M. Fagerlind, 1977. Large-sized polysomes in *Chironomus tentans* salivary glands and their relation to Balbiani ring 75S RNA. *J. Cell. Biol.* **73**:149–160.

Darwin, C., 1859. *The Origin of Species*. John Murray, London.

Darwin, C., 1872. *The Origin of Species*, 6th ed. John Wanamaker, Philadelphia.

Davidson, E. H., 1976. *Gene Activity in Early Development*, 2nd ed. Academic Press, New York.

Davidson, E. H., and R. J. Britten, 1979. Regulation of gene expression: Possible role of repetitive sequences. *Science* **204**:1052–1059.

Davidson, E. H., G. A. Galau, R. C. Angerer, and R. J. Britten, 1975. Comparative aspects of DNA organization in metazoa. *Chromosoma (Berl.)* **51**:253–259.

Davidson, E. H., B. R. Hough, C. S. Amenson, and R. J. Britten, 1973. General interspersion of repetitive with non-repetitive sequence elements in the DNA of *Xenopus*. *J. Mol. Biol.* **77**:1–24.

Day, A., 1965. The evolution of a pair of sibling allotetraploid species of cobwebby gilias (Polemoniaceae). *Aliso* **6**:25–75.

Dayhoff, M O., 1976, *Atlas of Protein Sequence and Structure*, Vol. 5, Suppl. 2, and Vol. 5, Suppl. 3, 1978. National Biomedical Research Foundation, Washington, D.C.

Dayhoff, M. O. and R. V. Eck, 1969. Inferences from protein sequence studies. In *Atlas of Protein Sequence and Structure*, Vol. 4 (M. O. Dayhoff, ed.). National Biomedical Research Foundation, Washington, D.C., pp. 1–5.

DeBeer, G. R., 1958. *Embryos and Ancestors*, 3rd ed. Clarendon Press, Oxford.

DeCrombrugghe, B., G. Vogeli, H. Ohkubo, Y. Yamada, E. Avvedimento, M. Sobel, M. Mudryi, and I. Pasten, 1981. The collagen gene. *J. Supramol. Struc. Cell Biochem.* Suppl. **5**:380.

DeRobertis, E. M., and J. B. Gurdon, 1977. Gene activation in somatic nuclei after injection into amphibian oocytes. *Proc. Natl. Acad. Sci. USA* **74**:2470–2474.

DeWet, J. M. J., 1979. Origins of polyploids. In *Polyploidy: Biological Relevance* (Walter H. Lewis, ed.). Plenum Press, New York, pp. 3–16.

Denis, H., and J. Brachet, 1969. Gene expression in interspecific hybrids. I. DNA synthesis in the lethal cross *Arbacia lixula* ♂ × *Paracentrotus lividus* ♀. *Proc. Natl. Acad. Sci. USA.* **62**:194–201.

Denis, H., and J. Brachet, 1969. Gene expression in interspecific hybrids. II. RNA synthesis in the lethal cross *Arbacia lixula* ♂ × *Paracentrotus lividus* ♀. *Proc. Natl. Acad. Sci. USA.* **62**:438–445.

Dent, J. N., 1968. Survey of amphibian metamorphosis. In *Metamorphosis* (W. Etkin and L. I. Gilbert, eds.). Appleton-Century-Crofts, New York, pp. 271–311.

Dent, J. N., 1975. Integumentary effects of prolactin in the lower vertebrates. *Am. Zool.* **15**:923–935.

Derman, E., K. Krauter, L. Walling, C. Weinberger, M. Ray, and J. E. Darnell, Jr., 1981. Transcriptional control in the production of liver-specific mRNAs. *Cell* **23**:731–739.

Deuchar, E. M., 1975. *Cellular Interactions in Animal Development.* Chapman and Hall, London.

Dhouailly, D., 1973. Dermo-epidermal interactions between birds and mammals: Differentiation of cutaneous appendages. *J. Embryol. Exp. Morphol.* **30**:587–603.

Dhouailly, D., 1975. Formation of cutaneous appendages in dermo-epidermal recombinations between reptiles, birds and mammals. *Roux Arch.* **177**:323–340.

Dhouailly, D., and P. Sengel, 1973. Interactions morphogenes entre l'épiderme de reptile et de derme d'oiseau ou de mammifere. *C. r. Acad. Sci. Ser. D.* **277**:1221–1224.

Dhouailly, D., G. E. Rogers, and P. Sengel, 1978. The specification of feather and scale protein synthesis in epidermal-dermal recombinations. *Dev. Biol.* **65**:58–68.

Dickerson, R. E., 1971. The structure of cytochrome *c* and the rates of molecular evolution. *J. Mol. Evol.* **1**:26–45.

Dickinson, W. J., 1975. A genetic locus affecting the developmental expression of an enzyme in *Drosophila melanogaster. Dev. Biol.* **42**:131–140.

Dickinson, W. J., 1978. Genetic control of enzyme expression in *Drosophila:* A locus influencing tissue specificity of aldehyde oxidase. *J. Exp. Zool.* **206**:333–342.

Dickinson, W. J., 1980. Evolution of patterns of gene expression in Hawaiian picture-winged *Drosophila. J. Mol. Evol.* **16**:73–94.

Dickinson, W. J., 1980. Complex *cis*-acting regulatory genes demonstrated in *Drosophila* hybrids. *Dev. Genet.* **1**:229–240.

Dickinson, W. J., 1980. Tissue specificity of enzyme expression regulated by diffusible factors: Evidence in *Drosophila* hybrids. *Science* **207**:995–997.

Dickinson, W. J., and H. L. Carson, 1979. Regulation of the tissue specificity of an enzyme by a *cis*-acting genetic element: Evidence from interspecific *Drosophila* hybrids. *Proc. Natl. Acad. Sci. USA* **76**:4559–4562.

Digan, M. E., A. C. Spradling, G. L. Waring, and A. P. Mahowald, 1979. The genetic analysis of chorion morphogenesis in *Drosophila melanogaster.* In *Eucaryotic Gene Regulation* (R. Axel, T. Maniatis, and C. F. Fox, eds.). Academic Press, New York, pp. 171–181.

Dilcher, D. L., 1974. Approaches to the identification of angiosperm leaf remains. *Bot. Rev.* **40**:1–157.

Dobkin, S., 1961. Early stages of pink shrimp from Florida waters. *Fish. Bull. Fish Wildlife Ser.* **61**:321–354.

Dobzhansky, T., 1970. *Genetics of the Evolutionary Process.* Columbia University Press, New York.

Dodd, M. H. I., and J. M. Dodd, 1976. The biology of metamorphosis. In *Physiology of the Amphibia,* Vol. III (B. Lofts, ed.). Academic Press, New York, pp. 467–599.

Dohmen, M. R., and N. H. Verdonk, 1979. The ultrastructure and role of the polar lobe in development of molluscs. In *Determinants of Spatial Organization* (S. Subtelny and I. R. Konigsberg, eds.). Academic Press, New York, pp. 3–27.

Doolittle, W. F., and C. Sapienza, 1980. Selfish genes, the phenotype paradigm and genome evolution. *Nature* **284**:601–603.

Driesch, H., 1894. *Analytische Theorie der Organischen Entwicklung.* Engelmann, Leipzig.

Ducibella, T., 1974. The occurrence of biochemical metamorphic events without anatomical metamorphosis in the axolotl. *Dev. Biol.* **38**:175–186.

Dunn, L. C., 1964. Abnormalities associated with a chromosome region in the mouse. I. Transmission and population genetics of the t-region. *Science* **144**:260–263.

Durcia, D. S., and H. M. Krider, 1977. Studies on the ribosomal cistrons in interspecific *Drosophila* hybrids. I. Nucleolar dominance. *Dev. Biol.* **59**:62–74.

Durham, J. W., 1966. Echinoids, classification. In *Treatise on Invertebrate Paleontology, Part U: Echinodermata 3, 1* (R. C. Moore, ed.). The Geological Society of America, New York, and the University of Kansas Press, Lawrence, Kansas, pp. 270–295.

Durham, J. W., and K. E. Caster, 1963. Helicoplacoidea: A new class of echinoderms. *Science* **140**:820–822.

Dutta, S. K., and R. K. Chaudhuri, 1975. Differential transcription of non-repeated DNA during development of *Neurospora crassa*. *Dev. Biol.* **43**:35–41.

Ebeling, A. W., N. B. Atkin, and P. Setzer, 1971. Genome sizes of teleostean fishes: Increases in some deep-sea species. *Am. Natural.* **105**:549–561.

Echols, H., 1981. SOS functions, cancer and inducible evolution. *Cell* **25**:1–2.

Ede, D. A., 1978. *An Introduction to Developmental Biology.* Halsted Press, New York.

Edgar, R. S., 1980. The genetics of development in the nematode *Caenorhabdites elegans*. In *The Molecular Genetics of Development* (T. Leighton and W. Loomis ed.). Academic Press, New York, pp. 213–235.

Edgar, R. S., 1981. Personal communication.

Efstratiadis, A., P. Lomedico, N. Rosenthal, R. Kolodner, R. Tizard, F. Perler, L. Villa-Komaroff, S. Naber, W. Chick, S. Broome, and W. Gilbert, 1979. The structure and transcription of rat preproinsulin genes. In *Eucaryotic Gene Regulation* (R. Axel, T. Maniatis, and C. F. Fox, eds.). Academic Press, New York, pp. 301–315.

Efstratiadis, A., J. W. Posakony, T. Maniatis, R. M. Lawn, C. O'Connell, R. A. Spritz, J. K. De Riel, B. G. Forget, S. M. Weissman, J. L. Slightom, A. E. Blechl, O. Smithies, F. E. Baralle, C. C. Shoulders, and N. J. Proudfoot, 1980. The structure and evolution of the human β-globin gene family. *Cell* **21**:653–668.

Eldredge, N., 1971. The allopatric model and phylogeny in Paleozoic invertebrates. *Evolution* **25**: 156–167.

Eldredge, N., and S. J. Gould, 1972. Punctuated equilibria: An alternative to phyletic gradualism. In *Models in Paleobiology* (T. J. M. Schopf, ed.). Freeman, Cooper and Company, San Francisco, pp. 82–115.

Epp, L. G., 1978. A review of the eyeless mutant in the Mexican axolotl. Developmental Genetics of the Mexican Axolotl. *Am. Zoologist*, **18**:267–272.

Erdmann, K., 1933. Zur Entwicklung des knöcheren Skelets von *Triton* und *Rana* unter besonder Berücksichtigung der Zeitfolge der Ossifikationen. *Ztscher. f. anat. u. Entwicklungsgesch* **101**:566–651.

Ernst, S. G., B. R. Hough-Evans, R. J. Britten, and E. H. Davidson, 1980. Limited complexity of the RNA in micromeres of sixteen-cell sea urchin embryos. *Dev. Biol.* **79**:119–127.

Errede, B., T. S. Cardillo, F. Sherman, E. Dubois, J. Deschamps, and J-M Wiame, 1980. Mating signals control expression of mutations resulting from insertion of a transposable repetitive element adjacent to diverse yeast genes. *Cell* **25**:427–436.

Etkin, W., 1970. The endocrine mechanism of amphibian metamorphosis, an evolutionary achievement. *Mem. Soc. Endocrinol.* **18**:137–155.

Fallon, J. F., and G. M. Crosby, 1977. Polarizing zone activity in limb buds of amniotes. In *Verterbrate Limb and Somite Morphogenesis*, (D. A. Ede, J. R.

Hinchliffe, and M. Balls, eds.) Cambridge University Press, Cambridge, pp. 55–69.

Feduccia, A., 1980. *The Age of Birds.* Harvard University Press, Cambridge, Mass.

Firtel, R. A., 1972. Changes in the expression of single-copy DNA during development of the cellular slime mold *Dictyostelium discoideum. J. Mol. Biol.* **66**:363–377.

Firtel, R. A., 1980. Personal communication on the complexity of RNA in polysomes of *Dictyostelium.*

Fisher, J. R., and R. E. Eakin, 1957. Nitrogen excretion in developing chick embryos. *J. Embryol. Exp. Morphol.* **5**:215–224.

Fitch, W. M., and C. H. Langley, 1976. Protein evolution and the molecular clock. *Fed. Proc.* **35**:2092–2097.

Fletcher, B. N., G. M. Phillip, and C. W. Wright, 1967. Echinodermata: Eleutherozoa. In *The Fossil Record* (W. B. Harland et al., eds.). Geological Society of London, pp. 583–599.

Ford, E. B., and J. S. Huxley, 1929. Genetic rate-factors in *Gammarus. Roux Arch.* **117**:67–79.

Forget, B. G., C. A. Marotta, S. M. Weissman, I. M. Verma, R. P. McCaffrey, and D. Baltimore, 1974. Nucleotide sequences of human globin messenger RNA. *Ann. N. Y. Acad. Sci.* **241**:290–309.

Forsyth, I. A., and T. J. Hayden, 1977. Comparative endocrinology of mammary growth and lactation. *Symp. Zool. Soc. Lond.* **41**:135–163.

Frazzetta, T. H., 1975. *Complex Adaptations in Evolving Populations.* Sinauer Associates, Sunderland, Mass.

Freeman, G., 1976. The role of cleavage in the localization of developmental potential in the Ctenophore *Mnemiopsis leidyi. Devel. Biol.* **49**:143–177.

Freeman, G., 1977. The transformation of the sinistral form of the snail *Lymnaea peregra* into its dextral form (abst.). *Am. Zool.* **17**:946.

Freeman, G., 1978. The role of asters in the localization of the factors that specify the apical tuft and the gut of the nemertine *Cerebratulus lacteus. J. Exp. Zool.* **206**:81–108.

Freeman, G., 1979. The multiple roles which cell division can play in the localization of developmental potential. In *Determinants of Spatial Organization* (S. Subtelny and I. R. Konigsberg, eds.). Academic Press, New York, pp. 53–76.

French, V., P. Bryant, and S. Bryant, 1976. A theory of pattern regulation in epimorphic fields. *Science* **193**:969–981.

Fuchs, E., and H. Green, 1978. The expression of keratin genes in epidermis and cultured epidermis cells. *Cell* **15**:887–897.

Galau, G. A., M. E. Chamberlin, B. R. Hough, R. J. Britten, and E. H. Davidson, 1976. Evolution of repetitive and nonrepetitive DNA. In *Molecular Evolution* (F. J. Ayala, ed.). Sinauer Associates, Sunderland, Mass. pp. 200–224.

Galau, G. A., W. H. Klein, M. M. Davis, B. J. Wold, R. J. Britten, and E. H. Davidson, 1976. Structural gene sets active in embryos and adult tissues of the sea urchin. *Cell* **7**:487–505.

Gans, M., C. Audit, and M. Masson, 1975. Isolation and characterization of sex linked female-sterile mutants in *Drosophila melanogaster. Genetics* **81**:683–704.

Garcia-Bellido, A., 1977. Homoeotic and atavic mutations in insects. *Am. Zool.* **17**:613–629.

Garcia-Bellido, A., and P. Ripoll, 1978. Cell lineage and differentiation in *Drosophila.* In *Results and Problems in Cell Differentiation,* Vol. 9 (W. Gehring, ed.). Springer-Verlag Publishing Co., Inc., Berlin, pp. 119–156.

Garen, A., and W. J. Gehring, 1972. Repair of the lethal developmental defect in deep orange embryos of *Drosophila* by injection of normal cytoplasm. *Proc. Natl. Acad. Sci. USA* **69**:2982–2985.

Garstang, W., 1922. The theory of recapitulation: A critical restatement of the biogenetic law. *J. Linn. Soc. Zool.* **35**:81–101.

Gehring, W. J., 1973. Genetic control of determination in the *Drosophila* embryo. In *Genetic Mechanisms of Development* (F. H. Ruddle, ed.). Academic Press, New York, pp. 103–128.

Gemmill, J. F., 1912. The development of the starfish, *Solaster endica* (Forbes). *Trans. Zool. Soc. Lond.* **20**:1–72.

Gilbert, W., 1979. Introns and exons: Playgrounds of evolution. In *Eucaryotic Gene Regulation* (R. Axel, T. Maniatis, and C. F. Fox, eds.). Academic Press, New York, pp. 1–12.

Gill, J. R., and W. A. Cobban, 1966. The Red Bird Section of the Upper Cretaceous Pierre Shale in Wyoming. *U.S. Geol. Surv. Prof. Paper. no. 393-A.*

Gill, J. R., and W. A. Cobban, 1973. Stratigraphy and geologic history of the Montana Group and equivalent rocks, Montana, Wyoming, and North and South Dakota. *U.S. Geol. Surv. Prof. Paper no. 776.*

Gillespie, D., 1977. Newly evolved repeated DNA sequences in primates. *Science* **196**:889–891.

Gingerich, P. D., 1976. Paleontology and phylogeny: Patterns of evolution at the species level in early Tertiary mammals. *Am. J. Sci.* **276**:1–28.

Girton, J. R. and P. J. Bryant, 1980. The use of cell lethal mutations in the study of *Drosophila* development. *Dev. Biol.* **77**:233–243.

Gitlin, D., 1944. The development of *Eleutherodactylus portoricensis*. *Copeia* (no. 2):91–98.

Glaessner, M. F., 1971. Geographic distribution and time range of the Ediacara Precambrian fauna. *Geol. Soc. Am. Bull.* **82**:509–514.

Glaessner, M. F., and M. Wade, 1966. The Late Precambrian fossils from Ediacara, South Australia. *Paleontology* **9**:599–628.

Gluecksohn-Waelsch, S., 1963. Lethal genes and analysis of differentiation. *Science* **142**:1269–1276.

Gold, J. R., 1980. Chromosomal change and rectangular evolution in North American cyprinid fishes. *Genet. Res.* **35**:157–164.

Goldschmidt, R., 1938. *Physiological Genetics*. McGraw-Hill Book Company, New York.

Goldschmidt, R., 1940. *The Material Basis of Evolution*. Yale University Press, New Haven, Conn.

Goldschmidt, R., 1952. Evolution as viewed by one geneticist. *Am. Scientist* **40**:84–98.

Goldsmith, M. R., and G. Basehoar, 1978. Organization of the chorion genes of *Bombyx mori*, a multigene family. I. Evidence for linkage to chromosome 2. *Genetics* **90**:291–310.

Goldsmith, M. R. and E. Clermont-Rattner, 1979. Organization of the chorion genes of *Bombyx mori*, a multigene family. II. Partial localization of three gene clusters. *Genetics* **92**:1173–1185.

Golosow, N., and C. Grobstein, 1962. Epithelio-mesenchymal interactions in pancreatic morphogenesis. *Dev. Biol.* **4**:242–255.

Goodman, M., G. W. Moore, and G. Matsuda, 1975. Darwinian evolution in the geneology of haemoglobin. *Nature* **253**:603–608.

Goss, R. J., 1964. *Adaptive Growth*. Logos Press and Academic Press, New York and London.

Gould, S. J., 1969. An evolutionary microcosm: Pleistocene and recent history of the land snail *P. (Poecilozonites)* in Bermuda. *Bull. Mus. Comp. Zool.* **138**:407–532.

Gould, S. J., 1971. Geometric similarity in allomeric growth: A contribution to the problem of scaling in the evolution of size. *Am. Natur.* **105**:113–136.

Gould, S. J., 1974. The origin and function of "bizarre" structures: Antler size and skull size in the "Irish Elk"; *Megaloceros giganteus. Evolution* **28**:191–220.

Gould, S. J., 1977. *Ontogeny and Phylogeny.* The Belknap Press of Harvard University Press, Cambridge, Mass.

Gould, S. J., and N. Eldredge, 1977. Punctuated equilibria: The tempo and mode of evolution reconsidered. *Paleobiology* **3**:115–151.

Graham, C. F., 1973. The necessary conditions for gene expression during early mammalian development. In *Genetic Mechanisms of Development* (F. H. Ruddle, ed.). Academic Press, New York, pp. 202–224.

Graham, C. F., K. Arms, and J. B. Gurdon, 1966. The induction of DNA synthesis by frog egg cytoplasm. *Dev. Biol.* **14**:349–381.

Grant, U., 1971. *Plant Speciation.* Columbia University Press, New York.

Greenleaf, A. L., J. R. Weebs, R. A. Voelker, S. Ohnishi, and B. Kickson, 1980. Genetic and biochemical characterization of mutants at an RNA polymerase II locus in *Drosophila melanogaster. Cell* **21**:785–792.

Grigliatti, T., and D. T. Suzuki, 1970. Temperature-sensitive mutations in *Drosophila.* V. Mutation affecting concentrations of pteridines. *Proc. Natl. Acad. Sci. USA* **67**:1101–1108.

Grossbach, U., 1973. Chromosome puffs and gene expression in polytene cells. *Cold Spring Harbor Symp. Quant. Biol.* **38**:619–627.

Grosschedl, R., and M. L. Birnstiel, 1980. Identification of regulatory sequences in the prelude sequences of an H2A histone gene by the study of specific deletion mutants *in vivo. Proc. Natl. Acad. Sci. USA* **77**:1432–1436.

Grosschedl, R., and M. L. Birnstiel, 1980. Spacer DNA sequences upstream of the TATAAATA sequence are essential for promotion of H2A histone gene transcription *in vivo. Proc. Natl. Acad. Sci. USA* **77**:7102–7106.

Gruenwald, P., 1952. Development of the excretory system. *Ann. N.Y. Acad. Sci.* **55**:142–146.

Grüneberg, H., 1963. *The Pathology of Development, a Study of Inherited Skeletal Disorders in Animals.* John Wiley and Sons, Inc., New York.

Gudernatsch, J. F., 1912. Feeding experiments on tadpoles. I. The influence of specific organs given as food on growth and differentiation: A contribution to the knowledge of organs with internal secretion. *Roux Arch.* **35**:457–483.

Guerrier, P., 1970. Les caractères de la segmentation et la détermination de la polarité dorsoventrale dans le développement de quelques Spiralia. II. *Sabellaria alveolata* (Annélide polychète). *J. Embryol. Exp. Morphol.* **23**:639–665.

Guerrier, P., and J. A. M. Van den Biggelaar, 1979. Intracellular activation and cell interactions in so-called mosaic embryos. In *Cell Lineage, Stem Cells and Cell Determination* (N. LeDouarin, ed.,). Elsevier/North-Holland Biomedical Press, Amsterdam, pp. 29–36.

Gurdon, J. B., 1968. Changes in somatic cell nuclei inserted into growing and maturing amphibian oocytes. *J. Embryol. Exp. Morphol.* **20**:401–414.

Gurdon, J. B., 1974. *The Control of Gene Expression in Animal Development.* Harvard University Press, Cambridge, Mass.

Hadorn, E., 1961. *Developmental Genetics and Lethal Factors* (English trans.). John Wiley and Sons, New York.

Hadorn, E., 1974. *Experimental Studies of Amphibian Development.* Springer-Verlag Publishing Co., Inc., New York.

Hadorn, E., 1978. Transdetermination. In *The Genetics and Biology of Drosophila,* Vol. 2c (M. Ashburner and T. Wright, eds.). Academic Press, London.

Haeckel, E., 1879. *The Evolution of Man: A Popular Exposition of the Principal Points of Human Ontogeny and Phylogeny.* D. Appleton and Company, New York.

Hagenbüchle, O., M. Tosi, U. Schibler, R. Bovey, P. K. Wellauer, and R. A. Young, 1981. Mouse liver and salivary gland alpha-amylase mRNAs differ only in 5′ non-translatable sequences. *Nature* **289**:643–646.

Hahn, W. E., D. E. Pettijohn, and J. Van Ness, 1977. One strand equivalent of *Escherichia coli* genome is transcribed: Complexity and abundance classes of mRNA. *Science* **197**:582–585.

Hake, S. C., 1980. *The Genome of Zea mays: Its Organization and Homology to Related Grasses.* Ph. D. dissertation, Washington University, St. Louis, Mo.

Haldane, J. B. S., 1932. The time of action of genes and its bearing on some evolutionary problems. *Am. Natural.* **66**:5–24.

Haldane, J. B. S., 1949. Suggestions as to quantitative measurement of rates of evolution. *Evolution* **3**:51–56.

Hallam, A., 1975. Evolutionary size increase, and longevity in Jurassic bivalves and ammonites. *Nature* **258**:439–446.

Hamburger, V., 1960. *A Manual of Experimental Embryology.* University of Chicago Press, Chicago.

Hampé, A., 1959. Contribution à l'étude du développement et de la régulation des déficiences et des excédents dans la patte de l'embryon de poulet. *Arch. D'Anat. Microscop. Morphol. Exp. Suppl.* **48**:347–478.

Hampé, A., 1960. Le compétition entre les éléments osseux du zeugopode de poulet. *J. Embryol. Exp. Morphol.* **8**:241–245.

Hardison, R. C., E. T. Butler, III, E. Lacy, T. Maniatis, N. Rosenthal, and A. Efstratiadis, 1979. The structure and transcription of four linked rabbit β-like globin genes. *Cell* **18**:1285–1297.

Hardy, T., 1903. *The Dynasts,* Part I. Quoted from the 1978 Macmillan (London) edition, p. 147.

Harper, C. W., Jr., 1975. Origin of species in geologic time: Alternatives to the Eldredge-Gould model. *Science* **190**:47–48.

Harris, P., M. Osborn, and K. Weber, 1980. A spiral array of microtubules in the fertilized sea urchin egg cortex examined by indirect immunofluorescence and electron microscopy. *Exp. Cell Res.* **128**:227–236.

Harris, P., M. Osborn, and K. Weber, 1980. Distribution of tubulin containing structures in the egg of the sea urchin *Strongylocentrotus purpuratus* from fertilization through first cleavage. *J. Cell Biol.* **84**:668–679.

Harrison, R. G., 1918. Experiments on the development of the forelimb of *Amblystoma,* a self-differentiating, equi-potential system. *J. Exp. Zool.* **25**:413–462.

Hastie, N. D., and J. O. Bishop, 1976. The expression of three abundance classes of messenger RNA in mouse tissues. *Cell* **9**:761–774.

Haugh, B. N., and B. M. Bell, 1980. Fossilized viscera in primitive echinoderms. *Science* **209**:653–657.

Hayashi, Y., 1965. Differentiation of the beak epithelium as studied by a xenoplastic induction system. *Jap. J. Exp. Morphol.* **19**:116–123.

Hennig, B., 1975. Change of cytochrome *c* structure during development of the mouse. *Eur. J. Biochem.* **55**:167–183.

Hereford, L. M., and M. Rosbash, 1977. Number and distribution of polyadenylated RNA sequences in yeast. *Cell* **10**:453–462.

Hersh, A. H., 1934. Evolutionary relative growth in the titanotheres. *Am. Natur.* **58**:537–561.

Herskowitz, I. H., 1949. *Hexaptera*, a homoeotic mutant in *Drosophila melanogaster*. *Genetics* **34**:10–25.

Herth, W., and K. Sander, 1973. Mode and timing of body pattern formation (regionalization) in the early embryonic development of cyclorrhaphic dipertans. (Protophormia, *Drosophila*). *Wilhelm Roux Arch.* **172**:1–27.

Heyer, W. R., 1969. The adaptive ecology of the species groups of the genus *Leptodactylus* (Amphibia, Leptodactylidae). *Evolution* **23**:421–428.

Hillman, N., M. I. Sherman, and C. Graham, 1972. The effect of spatial arrangement on cell determination during mouse development. *J. Embryol. Exp. Morphol.* **28**:263–278.

Hinchliffe, J. R., and P. V. Thorogood, 1974. Genetic inhibition of mesenchymal cell death and the development of form and skeletal pattern in the limbs of *talpid*^{3} (*ta*^{3}) mutant chick embryos. *J. Embryol. Exp. Morphol.* **31**:747–760.

Hinegardner, R. T., 1975. Morphology and genetics of sea urchin development. *Am. Zool.* **15**:679–689.

His, W., 1888. On the principles of animal morphology. *Proc. Roy. Soc. Edinburgh* **15**:287–298.

Hochman, B., 1973. Analysis of a whole chromosome in *Drosophila*. *Cold Spring Harbor Symp. Quant. Biol.* **38**:581–589.

Holmquist, R., T. H. Jukes, and S. Pangburm, 1973. Evolution of transfer RNA. *J. Mol. Biol.* **78**:91–116.

Hood, L., J. H. Campbell, and S. C. R. Elgin, 1975. The organization and evolution of antibody genes and other multigene families. *Ann. Rev. Genet.* **9**:305–353.

Hopper, K. E., and H. A. McKenzie, 1974. Comparative studies of α-lactalbumin and lysozyme: Echidna lysozyme. *Mol. Cell Biochem.* **3**:93–108.

Hori, H., 1975. Evolution of 5S RNA. *J. Mol. Evol.* **7**:75–86.

Horvitz, H. R., 1981. Personal communication.

Hough-Evans, B. R., B. J. Wold, S. G. Ernst, R. J. Britten, and E. H. Davidson. 1977. Appearance and persistence of maternal RNA sequences in sea urchin development. *Dev. Biol.* **60**:258–277.

Hough-Evans, B. R., M. Jacobs-Lorena, M. R. Cummings, R. J. Britten, and E. H. Davidson, 1980. Complexity of RNA in eggs of *Drosophila melanogaster* and *Musca domestica*. *Genetics* **95**:81–94.

Houke, M. S. and R. T. Hinegardner, 1980. Personal communication.

Hoyer, B. H., B. J. McCarthy, and E. T. Bolton, 1964. A molecular approach in the systematics of higher organisms. *Science* **144**:959–967.

Humphrey, R. R., 1967. Albino axolotls from an albino tiger salamander through hybridization. *J. Heredity* **58**:95–101.

Humphrey, R. R., 1972. Genetic and experimental studies on a mutant gene (*c*) determining absence of heart action in embryos of the Mexican axolotl (*Ambystoma mexicanum*). *Dev. Biol.* **27**:365–375.

Hunt, L. T., S. Hurst-Calderone, and M. O. Dayhoff, 1978. Globins, In *Atlas of Protein Sequence and Structure*, Vol. 5, Suppl. 3 (M. O. Dayhoff, ed.). National Biomedical Research Foundation, Washington D.C. pp. 229–249.

Huxley, J. S., 1932. *Problems of Relative Growth*. Methuen and Company, London. Second edition published in 1972, by Dover Publications Inc., New York.

Huxley, T. H., 1869. Geological reform. *Quart. J. Geol. Soc. Lond.* **25**:xxxviii–liii.

Hyman, L. H., 1940. *The Invertebrates*, Vol. I: *Protozoa Through Ctenophora*. McGraw-Hill Book Company, New York.

Hyman, L. H., 1955. *The Invertebrates*, Vol. IV: *Echinodermata*. McGraw-Hill Book Company, New York.

Ikeda, K., S. Ozawa, and S. Hagiwara, 1976. Synaptic transmission reversibly conditioned by single gene mutation in *Drosophila melanogaster. Nature* **259:**489–491.

Illmensee, K., 1978. *Drosophila* chimeras and the problem of determination. In *Results and Problems in Cell Differentiation*, Vol. 9 (W. Gehring, ed.). Springer-Verlag Publishing Co., Inc., Berlin, pp. 51–69.

Illmensee, K., and A. P. Mahowald, 1974. Transplantation of posterior polar plasm in *Drosophila*. Induction of germ cells at the anterior pole of the egg. *Proc. Natl. Acad. Sci. USA* **71:**1016–1020.

Illmensee, K., A. P. Mahowald, and M. R. Loomis, 1976. The ontogeny of germ plasm during oogenesis in *Drosophila. Dev. Biol.* **49:**40–65.

Jacob, F., 1977. Evolution and tinkering. *Science* **196:**1161–1166.

Jacob, F., and J. Monod, 1961. On the regulation of gene activity. *Cold Spring Harbor Symp. Quant. Biol.* **26:**193–211.

Jacobs, L. L., and D. Pilbeam, 1980. Of mice and men: Fossil based divergence dates and molecular "clocks". *J. Hum. Evol.* **9:**551–555.

Jacobson, A. G., 1963. The determination and positioning of the nose, lens and ear. III. Effects of reversing the antero-posterior axis of epidermis, neural plate and neural fold. *J. Exp. Zool.* **154:**293–303.

Jacobson, A. G., 1966. Inductive processes in embryonic development. *Science* **152:**25–34.

Jacobson, A. G., and J. T. Duncan, 1968. Heart induction in salamanders. *J. Exp. Zool.* **167:**79–103.

Jarry, B., and D. Falk, 1974. Functional diversity within the *rudimentary* locus of *Drosophila melanogaster. Mol. Gen. Genet.* **135:**113–122.

Jefferies, R. P. S., 1975. Fossil evidence concerning the origin of the chordates. *Symp. Zool. Soc. Lond.* **36:**253–318.

Jefferies, R. P. S., K. A. Joysey, C. R. C. Paul, and W. H. C. Ramsbottom, 1967. Echinodermata: Pelmatozoa. In *The Fossil Record* (W. B. Harland et al., eds.). The Geological Society of London, pp. 565–581.

Jelinek, W. R., T. P. Toomey, L. Leinwand, C. H. Duncan, P. A. Biro, P. V. Choudary, S. M. Weissman, C. M. Rubin, C. M. Houck, P. L. Deininger, and C. W. Schmid, 1980. Ubiquitous, interspersed repeated sequences in mammalian genomes. *Proc. Natl. Acad. Sci. USA* **77:**1398–1402.

Jenkin, P. M., 1970. *Control of Growth and Metamorphosis*. Pergamon Press, Oxford.

Jenness, R., 1974. The composition of milk. In *Lactation*, Vol. 3 (B. L. Larson and V. R. Smith, eds.). Academic Press, New York, pp. 3–107.

Johanson, D., and M. Edey, 1981. *Lucy. The Beginnings of Humankind*. Simon and Schuster, New York.

Johnson, W. E., H. L. Carson, K. Y. Kaneshiro, W. W. M. Steiner, and M. M. Cooper, 1975. Genetic variation in Hawaiian *Drosophila*. II. Allozymic differentiation in the *D. planitibia* subgroup. In *Isozymes*, Vol. IV: *Genetics and Evolution* (C. L. Markert, ed.). Academic Press, New York, pp. 563–584.

Jones, C. W., and F. C. Kafatos, 1980. Structure, organization and evolution of developmentally regulated chorion genes in a silkmoth. *Cell* **22:**855–867.

Jones, C. W., and F. C. Kafatos, 1981. Structure and organization of developmentally-regulated chorion genes from *Antheraea*. In *Levels of Genetic Control in Development* (S. Subtelny and U. K. Abbott; eds.). Alan R. Liss, New York, pp. 69–81.

Jones, C. W., N. Rosenthal, G. C. Rodakis, and F. C. Kafatos, 1979. Evolution of two major chorion multigene families as inferred from cloned cDNA and protein sequences. *Cell* **18:**1317–1332.

Jones, E. A., 1977. Synthesis and secretion of milk sugars. *Symp. Zool. Soc. Lond.* **41:**77–94.

Jones, J. S., 1981. An uncensored page of fossil history. *Nature* **293**:427–428.

Juberthie-Jupeau, L., 1974. Action de la temperature sur le développement embryonnaire de *Glomeris marginata* (Villers). *Symp. Zool. Soc. Lond.* **32**:289–300.

Judd, B. H., M. W. Shen, and T. C. Kaufman, 1972. The anatomy of a segment of the X chromosome of *Drosophila melanogaster*. *Genetics* **71**:139–156.

Juenger, E., 1960. *The Glass Bees*. The Noonday Press, New York.

Jung, A., A. E. Sippel, M. Grez, and G. Schutz, 1980. Exons encode functional and strucutal units of chicken lysozyme. *Proc. Natl. Acad. Sci. USA* **77**:5759–5763.

Kafatos, F. C., A. Efstratiadis, B. G. Forget, and S. M. Weissman, 1977. Molecular evolution of human and rabbit β-globin mRNA. *Proc. Natl. Acad. Sci. USA* **74**:5618–5622.

Kafatos, F. C., J. C. Regier, G. D. Mazur, M. R. Nadel, H. M. Blau, W. H. Petri, A. R. Wyman, R. E. Gelinas, P. B. Moore, M. Paul, A. Efstratiadis, J. N. Vournakis, M. R. Goldsmith, J. R. Hunsley, B. Baker, J. Nardi, and M. Koehler, 1977. The eggshell of insects: Differentiation-specific proteins and the control of their synthesis and accumulation during development. In *Results and Problems in Cell Differentiation*, (W. Beerman, ed.) Vol. 8. Springer-Verlag, Berlin, pp. 45–145.

Kalthoff, K., 1969. Der Einfluss verschiedener Versuchsparameter auf die Häufigkeit der Missbildung "Doppelabdomen" in UV-bestrahlten Eiern von *Smittia* spec. (Diptera, Chironomidae). *Zool. Anz. (Suppl. BD.)* **33**:59–65, *Verh. Zool. Ges.*

Kalthoff, K., 1979. Analysis of a morphogenetic determinant in an insect embryo (*Smittia* spec., Chironomidae, Diptera). In *Determinants of Spatial Organization* (S. Subtelny and I. R. Konigsberg, eds.). Academic Press, New York, pp. 97–126.

Kalthoff, K., and K. Sander, 1968. Der Entwicklungsgang der Missbildung "Doppelabdomen" im partiell UV-bestrahlten Ei von *Smittia parthenogenetica* (Dipt. Chironomidae). *Wilhelm Roux Arch.* **161**:129–146.

Kamalay, J. C., and R. B. Goldberg, 1980. Regulation of structural gene expression in tobacco. *Cell* **19**:935–946.

Kandler-Singer, I., and K. Kalthoff, 1976. RNase sensitivity of an anterior morphogenetic determinant in an insect egg (*Smittia* sp., Chironomidae, Diptera). *Proc. Natl. Acad. Sci. USA* **73**:3739–3743.

Katz, M. J., and R. J. Lasek, 1978. Evolution of the nervous system: Role of ontogenetic mechanisms in the evolution of matching populations. *Proc. Natl. Acad. Sci. USA* **75**:1349–1352.

Katz, M. J., R. J. Lasek, and I. R. Kaiserman-Abramof, 1981. Ontophyletics of the nervous system: Eyeless mutants illustrate how ontogenetic buffer mechanisms channel evolution. *Proc. Natl. Acad. Sci. USA* **78**:397–401.

Katz, M. J., 1982. Ontophyletics: Studying evolution beyond the genome. *Perspect. Biol. Med.*, in press.

Kauffman, E. G., 1978. Evolutionary rates and patterns among Cretaceous Bivalvia. *Phil. Trans. R. Soc. Lond. B* **284**:277–304.

Kauffman, S., 1971. Gene regulation networks: A theory for their global structure and behaviors. *Curr. Topics Dev. Biol.* **6**:145–182.

Kauffman, S. A., 1975. Control circuits for determination and transdetermination: Interpreting positional information in a binary epigenetic code. In *Cell Patterning: Ciba Foundation Symposium No. 29*. Elsevier Publishing Co., Amsterdam, pp. 201–214.

Kauffman, S. A., 1977. Characteristic waves, compartments and binary decisions in *Drosophila* development. *Am. Zool.* **17**:631–648.

Kauffman, S. A., 1981. A theory for the evolution of metazoan gene regulation. Unpublished manuscript.

Kauffman, S., R. Shymko, and K. Trabert, 1978. Control of sequential compartment formation in *Drosophila*. *Science* **199**:259–270.

Kaufman, T. C., M. P. Shannon, M. W. Shen, and B. H. Judd, 1975. A revision of the cytology and ontogeny of several deficiencies in the 3A1-3C6 region of the X chromosome of *Drosophila melanogaster*. *Genetics* **79**:265–282.

Kavenoff, R., and B. H. Zimm, 1973. Chromosome-sized DNA molecules from *Drosophila*. *Chromosoma* **41**:1–27.

Kedes, L. H., P. R. Gross, G. Cognetti, and A. L. Hunter, 1969. Synthesis of nuclear and chromosomal proteins on light polyribosomes during cleavage in the sea urchin embryo. *J. Mol. Biol.* **45**:337–351.

Keller, R. E., 1975. Vital dye mapping of the gastrula and neurula of *Xenopus laevis*. I. Prospective areas and morphological movements of the superficial layer. *Dev. Biol.* **42**:222–241.

Keller, R. E., 1976. Vital dye mapping of the gastrula and neurula of *Xenopus laevis*. II. Prospective areas and morphogenetic movements of the deep layer. *Dev. Biol.* **51**:118–137.

Kemphues, K. J., R. A. Raff, T. C. Kaufman, and E. C. Raff, 1979. Mutation in a structural gene for a β-tubulin specific to testis in *Drosophila melanogaster*. *Proc. Natl. Acad. Sci. USA* **76(8)**:3991–3995.

Kemphues, K. J., T. C. Kaufman, R. A. Raff, and E. C. Raff, 1979. Mutation in a structural gene for a β-tubulin specific to testis in *Drosophila melanogaster*. *Proc. Natl. Acad. Sci. USA* **76**:3991–3995.

Kimble, J., and J. White, 1981. On the control of germ cell development in *Caenorhabdites elegans*. *Dev. Biol.* **81**:208–219.

Kindle, K. L., and R. A. Firtel, 1978. Identification and analysis of *Dictyostelium* actin genes, a family of moderately repeated genes. *Cell* **15**:763–778.

King, J. L, and T. H. Jukes, 1969. Non-Darwinian evolution. *Science* **164**:788–798.

King, M. -C., and A. C. Wilson, 1975. Evolution at two levels in humans and chimpanzees. *Science* **188**:107–116.

Kirschner, M., J. C. Gerhart, K. Hara, and G. A. Ubbels, 1980. Initiation of the cell cycle and establishment of bilateral symmetry in *Xenopus* eggs. In *The Cell Surface: Mediator of Developmental Processes* (S. Subtelny and N. K. Wessells eds.). Academic Press, New York, pp. 187–215.

Kiss, I., G. Bencze, E. Fekete, A. Fodor, J. Gausz, P. Maróy, J. Szabad, and J. Szidonya, 1976. Isolation and characterization of X-linked lethal mutants affecting differentiation of the imaginal discs in *Drosophila melanogaster*. *Theoret. Appl. Genet.* **48**:217–226.

Kitchen, H., 1974. Animal hemoglobin heterogeneity. *Ann. N.Y. Acad. Sci.* **241**:12–24.

Kitchen, H., and I. Brett, 1974. Embryonic and fetal hemoglobins in animals. *Ann. N.Y. Acad. Sci.* **241**:653–671.

Kleene, K. C., and T. Humphreys, 1977. Similarity of hnRNA sequences in blastula and pluteus stage sea urchin embryos. *Cell* **12**:143–155.

Klein, W. H., T. L. Thomas, C. Lai, R. H. Scheller, R. J. Britten, and E. H. Davidson, 1978. Characteristics of individual repetitive sequence families in the sea urchin genome studied with cloned repeats. *Cell* **14**:889–900.

Kleinenberg, N., 1886. Die Entstehung des Annelids aus der Larve von Lopadorhynchus. Nebst Bemarkungen über die Entwicklung anderer Polychäten. *Ztschr. Wiss. Zool.* **44**:1–227.

Knoll, A. H., and E. S. Barghoorn, 1975. Precambrian eucaryotic organisms: A reassessment of the evidence. *Science* **190**:52–54.

Kohne, D. E., 1970. Evolution of higher-organism DNA. *Quart. Rev. Biophys.* **33**:327–375.

Kohne, D. E., J. A. Chiscon, and B. H. Hoyer, 1972. Evolution of primate DNA sequences. *J. Hum. Evol.* **1**:627–644.

Kollar, E. J., and G. R. Baird, 1970. Tissue interactions in embryonic mouse tooth germs. I. Reorganization of the dental epithelium during tooth germ reconstruction. *J. Embryol. Exp. Morphol.* **24**:159–171.

Kollar, E. J., and G. R. Baird, 1970. Tissue interactions in embryonic mouse tooth germs. II. The inductive role of the dental papilla. *J. Embryol. Exp. Morphol.* **24**:173–186.

Kollar, E. J., and C. Fisher, 1980. Tooth induction in chick epithelium: Expression of quiescent genes for enamel synthesis. *Science* **207**:993–995.

Korschelt, E., and K. Heider, 1900. *Textbook of the Embryology of the Invertebrates.* The Macmillan Publishing Co., Inc., New York.

Kuhn, T. S., 1970. *The Structure of Scientific Revolutions.* University of Chicago Press, Chicago.

Kulikowski, R. R., and F. J. Manasek, 1978. The cardiac lethal mutant of *Ambystoma mexicanum:* A re-examination. *Am. Zool.* **18**:349–358.

Kunkel, N. S., and E. S. Weinberg, 1978. Histone gene transcripts in the cleavage and mesenchyme blastula embryo of the sea urchin, *S. purpuratus. Cell* **14**:313–326.

Kuroda, Y., 1977. Studies on *Drosophila* embryonic cells *in vitro.* II. Tissue- and time-specificity of a lethal gene, deep orange. *Dev. Growth Different.* **19**:57–66.

Kurtén, B., 1958. A differentiation index, and a new measure of evolutionary rates. *Evolution* **12**:146–157.

Lacy, E., R. C. Hardison, D. Quon, and T. Maniatis, 1979. The linkage arrangement of four β-like globin genes. *Cell* **18**:1273–1283.

Laird, C. D., 1973. DNA of *Drosophila* chromosomes. *Ann. Rev. Genet.* **7**:177–204.

Laird, C. D., B. L. McConaughy, and B. J. McCarthy, 1969. Rate of fixation of nucleotide substitution in evolution. *Nature* **224**:149–154.

Lamb, M. M., and B. Daneholt, 1979. Characterization of active transcription units in Balbiani rings of *Chironomus tentans. Cell* **17**:835–848.

Lamb, M. M., and C. D. Laird, 1976. Increase in nuclear Poly (A)-containing RNA at syncytial blastoderm in *Drosophila melanogaster* embryos. *Dev. Biol.* **52**:31–42.

Lankester, E. R., 1877. Notes on the embryology and classification of the animal kingdom: Comprising a revision of speculations relative to the origin and significance of the germ-layers. *Quart. J. Microsc. Sci.* **17**:399–454.

Lazowska, J., C. Jacq, and P. P. Slonimski, 1980. Sequence of introns and flanking exons in wild-type and *box 3* mutants of cytochrome *b* reveals an interlaced splicing protein coded by an intron. *Cell* **22**:333–348.

Leakey, R. E., and R. Lewin, 1977. *Origins.* E. P. Dutton and Co., Inc., New York.

Lemanski, L. F., D. J. Paulson, and G. S. Hill, 1979. Normal anterior endoderm corrects the heart defect in cardiac mutant salamander (*Ambystoma mexicanum*). *Science* **204**:860–862.

Lengyel, J., and S. Penman, 1975. hnRNA size and processing as related to different DNA content in two Dipterans: *Drosophila* and *Aedes. Cell* **5**:281–290.

Levi-Setti, R., 1975. *Trilobites*. University of Chicago Press, Chicago.

Levy, W. B., and B. J. McCarthy, 1975. Messenger RNA complexity in *Drosophila melanogaster*. *Biochemistry* **14**:2440–2446.

Lewin, B., 1980. *Gene Expression*, Vol. 2: *Eucaryotic Chromosomes*, 2nd ed. John Wiley and Sons, Inc., New York.

Lewis, E. B., 1963. Genes and developmental pathways. *Am. Zool.* **3**:33–56.

Lewis, E. B., 1978. A gene complex controlling segmentation in *Drosophila*. *Nature* **276**:565–570.

Lewis, R., B. Wakimoto, R. Denell, and T. Kaufman, 1980. Genetic analysis of the Antennapedia gene complex (ANT-C) and adjacent chromosomal regions of *Drosophila melanogaster*. II. Polytene chromosome segments 84A-84B1, 2. *Genetics* **95**:383–397.

Lillie, F. R., 1895. The embryology of the Unionidae. *J. Morphol.* **10**:1–100.

Lillie, F. R., 1898. Adaptation in cleavage. In *Biological Lectures of the Marine Biological Laboratory of Woods Hole, Mass.* Ginn and Company, Boston, pp. 43–67.

Lillie, F. R., 1927. The gene and the ontogenetic process. *Science* **66**:361–368.

Lohs-Schardin, M., C. Cremer, and C. Nüsslein-Volhard, 1979. A fate map for the larval epidermis of *Drosophila melanogaster*: Localized cuticle defects following irradiation of the blastoderm with an ultraviolet laser microbeam. *Dev. Biol.* **73**:239–255.

Lüscher, M., 1960. Hormonal control of caste differentiation in termites. *Ann. N.Y. Acad. Sci.* **89**:549–563.

Lüscher, M., 1963. Functions of the corpora allata in the development of termites. *Proc. 16th Int. Congr. Zool.* **4**:244–250.

Lüscher, M., 1969. Die Bedeutung des Juvenilhormones für die Differenzierung der Soldaten bei der Termite *Kolotermes flavicollis*. In *Proceedings of the Sixth Congress of the International Union for the Study of Social Insects*, Zoological Institute, University of Bern, Bern, pp. 165–170.

Lutz, B., 1947. Trends towards non-aquatic and direct development in frogs. *Copeia* (No. 4): 242–252.

Lutz, B., 1948. Ontogenetic evolution in frogs. *Evolution* **2**:29–39.

Lynn, W. G., 1942. The embryology of *Eleutherodactylus nubicola*, an anuran which has no tadpole stage. *Contr. Embryol.* **30**(190):27–62.

Lynn, W. G., 1947. The effects of thiourea and phenylthriourea upon the development of *Plethodon cinereus*. *Biol. Bull.* **93**:199.

Lynn, W. G., and A. M. Peadon, 1955. The role of the thyroid gland in direct development in the anuran, *Eleutherodactylus martinicenis*. *Growth* **19**:263–286.

Lyon, M. F., P. H. Glenister and M. L. Lamoreux, 1975. Normal spermatozoa from androgen resistant germ cells of chimaeric mice and the role of androgen in spermatogenesis. *Nature* **258**:620–622.

MacBride, E. W., and W. K. Spencer, 1938. Two new Echinoides, *Aulechinus* and *Ectinechinus* and an adult plated Holothuran, *Eothuria*, from the Upper Ordovician of Girvan, Scotland. *Phil. Trans. Roy. Soc. Lond. B* **229**:91–136.

MacDonald, R. J., M. M. Crerar, W. F. Swain, R. L. Pictet, G. Thomas, and W. J. Rutter, 1980. Structure of a family of rat amylase genes. *Nature* **207**:117–122.

Macurda, D. B., Jr., 1980. Abnormalities of the Carboniferous blastoid *Pentremites*. *J. Paleontol.* **54**:1155–1162.

Maglio, V. J., 1973. Origin and evolution of the Elephantidae. *Trans. Am. Phil. Soc. NS* **63**(3):1–149.

Mahowald, A. P., J. H. Caulton, and W. J. Gehring, 1979. Ultrastructural studies of oocytes and embryos derived from female flies carrying the *grandchildless* mutation in *Drosophila subobscura*. *Dev. Biol.* **69**:118–132.

Manglesdorf, P. C., 1958. The mutagenic effect of hybridizing maize and teosinte. *Cold Spring Harbor Symp. Quant. Biol.* **23**:409–421.

Mangold, O., 1961. Grundzüge der Entwicklungsphysiologie der Wirbeltiere mit besonderer Berücksichtigung der Missbildungen auf grund experimenteller Arbeiten an Urodelen. *Acta Genet. Med. et. Gmel.* **10**:1–49.

Mangold, O., and Seidel, F., 1927. Homoplastische und heteroplastiche Verschmelzung ganzer Tritonkeime. *Roux Arch.* **111**:494–665.

Manning, J. E., C. W. Schmid, and N. Davidson, 1975. Interspersion of repetitive and nonrepetitive DNA sequences in the *D. melanogaster* genome. *Cell* **4**:141–156.

Manton, S. M., 1972. The evolution of arthropod locomotory mechanisms, part 10. *J. Linn. Soc. Zool.* **51**:203–400.

Margoliash, E., 1963. Primary structure and evolution of cytochrome *c*. *Proc. Natl. Acad. Sci. USA* **50**:672–679.

Marshall, L. G., and R. S. Corruccini, 1978. Variability, evolutionary rates, and allometry in dwarfing lineages. *Paleobiology* **4**:101–119.

Mathis, D. J., and P. Chambon, 1981. The SV40 early region TATA box is required for accurate in vitro initiation of transcription. *Nature* **290**:310–315.

Matthews, B. W., M. G. Grütter, W. F. Anderson, and S. J. Remington, 1981. Common precursor of lysozymes of hen egg-white and bacteriophage T4. *Nature* **290**:334–335.

Maxson, L. E. R., and A. C. Wilson, 1979. Rates of molecular and chromosomal evolution in salamanders. *Evolution* **33**:734–740.

McClintock, B., 1956. Controlling elements and the gene. *Cold Spring Harbor Symp. Quant. Biol.* **21**:197–216.

McKusick, V. A., 1978. *Mendelian Inheritance in Man*, 5th ed. The Johns Hopkins University Press, Baltimore and London.

Mepham, T. B., 1977. Synthesis and secretion of milk proteins. *Symp. Zool. Soc. Lond.* **41**:57–75.

Merrell, D. J., 1975. In defense of frogs. *Science* **189**:838.

Meyer, W. J., B. R. Migeon and C. J. Migeon, 1975. A locus on the human X-chromosome for dihydrotestosterone receptor and androgen insensitivity. *Proc. Natl. Acad. Sci. USA* **72**:1469–1472.

Millard, N., 1945. The development of the arterial system of *Xenopus laevis*, including experiments on the destruction of the larval aortic arches. *Trans. Roy. Soc. S. Afr.* **30**:217–234.

Minganti, A., 1959. Androgenetic hybrids in ascidians. I. *Ascidia malaca* (♀) x *Phallusia mamillata* (♂). *Acta Embryol. Morphol. Exp.* **2**:244–256.

Mintz, B., 1965. Genetic mosaicism in adult mice of quadriparental lineage. *Science* **148**:1232–1233.

Mintz, B., 1967. Gene control of mammalian pigmentary differentiation. I. Clonal origin of melanocytes. *Proc. Natl. Acad. Sci. USA* **58**:344–351.

Mintz, B., and K. Illmensee, 1975. Normal genetically mosaic mice produced from malignant teratocarcinoma cells. *Proc. Natl. Acad. Sci. USA* **72**:3585–3589.

Mizuno. S., Y. R. Lee, A. H. Whiteley, and H. R. Whiteley, 1974. Cellular distribution of RNA populations in 16-cell stage embryos of the sand dollar, *Dendraster excentricus*. *Dev. Biol.* **37**:18–27.

Mohler, J. D., 1977. Developmental genetics of the *Drosophila* egg. I. Identification of 59 sex-linked cistrons with maternal effects on embryonic development. *Genetics* **85**:259–272.

Monod, J., 1971. *Chance and Necessity*. Alfred A. Knopf, New York.

Moore, G. P., R. H. Scheller, E. H. Davidson, and R. J. Britten, 1978.

Evolutionary change in the repetition frequency of sea urchin DNA sequences. *Cell* **15**:649–660.

Moore, J. A., 1941. Developmental rate of hybrid frogs. *J. Exp. Zool.* **86**:405–422.

Moore, N. W., C. E. Adams, and L. E. A. Rowson, 1968. Developmental potential of single blastomeres of the rabbit egg. *J. Reprod. Fertil.* **17**:527–531.

Morgan, T. H., 1932. *The Scientific Basis of Evolution.* W. W. Norton & Company, Inc., New York.

Morgan, T. H., 1934. *Embryology and Genetics.* Columbia University Press, New York.

Mortin, M., and G. Lefevre, 1981. An RNA polymerase II mutation in *Drosophila melanogaster* that mimics *Ultrabithorax*. *Chromosoma* **82**:237–247.

Morton, N. E., J. F. Crow, and H. J. Muller, 1956. An estimate of the mutational damage in man from data on consanguineous marriages. *Proc. Natl. Acad. Sci. USA* **40**:855–863.

Müller, F., 1864. *Für Darwin.* Translated into English by W. S. Dallas as *Facts and Arguments for Darwin*, 1869. J. Murray, London.

Muller, H. J., 1962. *Studies in Genetics.* Indiana University Press, Bloomington.

Muller, H. J., and G. Pontecorvo, 1942. Recessive genes causing interspecific sterility and other disharmonies between *Drosophila melanogaster* and *D. simulans*. *Genetics.* **27**:157.

Nachtsheim, H., 1950. The Pelger-anomaly in man and rabbit. *J. Hered.* **41**:131–137.

Nakhasi, H. L., and P. K. Qasba, 1979. Quantification of milk proteins and their mRNAs in rat mammary gland at various stages of gestation and lactation. *J. Biol. Chem.* **254**:6016–6025.

Nash, G., and G. Fankhauser, 1959. Changes in the pattern of nitrogen excretion during the life cycle of the newt. *Science* **130**:714–716.

Needham, J., 1931. *Chemical Embryology*, Vol. 2. Cambridge University Press, London.

Needham, J., 1933. On the dissociability of the fundamental process in ontogenesis. *Biol. Rev.* **8**:180–223.

Newell, N. D., 1942. Late Paleozoic pelecypods: Mytilacea, part 2. *Kansas Geol. Surv.* **10**:1–115.

Newell, N. D., 1949. Phyletic size increase, an important trend illustrated by fossil invertebrates. *Evolution* **3**:103–124.

Newrock, K. M., and R. A. Raff, 1975. Polar lobe specific regulation of translation in embryos of *Ilyanassa obsoleta*. *Dev. Biol.* **42**:242–261.

Newrock, K. M., C. R. Alfageme, R. V. Nardi, and L. H. Cohen, 1977. Histone changes during chromatin remodeling in embryogenesis. *Cold Spring Harbor Symp. Quant. Biol.* **42**:421–431.

Newrock, K. M., L. H. Cohen, M. B. Hendricks, R. J. Donnelly, and E. S. Weinberg, 1978. Stage-specific mRNAs coding for subtypes of H2A and H2B histones in the sea urchin embryo. *Cell* **14**:327–336.

Niall, H. D., 1981. The chemistry of prolactin. In *Prolactin* (R. B. Jaffe, ed.). Elsevier, New York, pp. 1–17.

Nicoll, C. S., 1980. Ontogeny and evolution of prolactin's functions. *Fed. Proc.* **39**:2563–2566.

Nicoll, C. S., 1981. Role of prolactin in water and electrolyte balance in vertebrates. In *Prolactin* (R. B. Jaffe, ed.). Elsevier, New York, pp. 127–166.

Nieuwkoop, P. D., and J. Faber, 1956. *Normal Table of Xenopus laevis (Daudin).* North-Holland Publishing Company, Amsterdam.

Noble, G. K., 1925. An outline of the relation of ontogeny to phylogeny within the amphibia. *Am. Mus. Nov.* **165**:1–17; **166**:8–10.

Noble, G. K., 1927. The value of life history data in the study of the evolution of amphibia. *Ann. N.Y. Acad. Sci.* **30**:31–128.

Noble, G. K., 1931. *The Biology of the Amphibia.* McGraw-Hill Book Company, New York.

Norby, S., 1973. The biochemical genetics of *rudimentary* mutants of *Drosophila melanogaster.* I. Aspartate carbamyl transferase levels in complementing and non-complementing strains. *Hereditas* **73**:11–16.

Norris, D. O., and W. A. Gern, 1976. Thyroxine-induced activation of hypothalamo-hypophysial axis in neotenic salamander larvae. *Science* **194**:525–527.

Nüsslein-Volhard, C., 1977. Genetic analysis of pattern-formation in the embryo of *Drosophila melanogaster. Roux Arch.* **183**:249–268.

Nüsslein-Volhard, C., 1979. Maternal effect mutations that alter the spatial coordinates of the embryo of *Drosophila melanogaster.* In *Determinants of Spatial Organization* (S. Subtelny and I. R. Konigsberg, eds.). Academic Press, New York, pp. 185–211.

Nüsslein-Volhard, C., and E. Wieschaus, 1980. Mutations affecting segment number and polarity in *Drosophila. Nature* **287**:795–801.

Oehler, J. H., D. Z. Oehler, and M. D. Muir, 1976. On the significance of tetrahedral tetrads of Precambrian algal cells. *Orig. Life* **7**:259–267.

Ohno, S., 1979. Major Sex-Determining Genes. *Monographs on Endocrinology,* Volume II. Springer-Verlag, Berlin.

Olson, S. L., 1973. Evolution of the rails of the South Atlantic islands (Aves: Rallidae). *Smithson. Contr. Zool.* **152**:1–53.

Oppenheimer, C., and C. A. Mitchell, 1901. *Ferments and Their Actions.* C. Griffin and Company, London.

Orgel, L. E., and F. H. C. Crick, 1980. Selfish DNA: The ultimate parasite. *Nature* **284**:604–607.

Ortolani, G., 1954. Resultati definitivi sulla distribuzione dei territori presumtivi degli organi, nel germe di Ascidie allo stadio. VIII. Determinati, con le marche al carbone. *Pubbl. Staz. Zool. Napoli* **25**:161–187.

Osborn, H. F., 1917. *The Origin and Evolution of Life.* Charles Scribner's Sons, New York.

Osborn, H. F., 1932. The nine principles of evolution revealed by paleontology. *Am. Natural.* **66**:52–60.

Osborn, M., and K. Weber, 1976. Cytoplasmic microtubules in tissue culture cells appear to grow from an organizing structure towards the plasma membrane. *Proc. Natl. Acad. Sci. USA* **73**:867–871.

Ospovat, D., 1976. The influence of Karl Ernst von Baer's embryology, 1829–1859: A reappraisal in light of Richard Owen's and William B. Carpenter's "Palaeontological Application of 'von Baer's Law'". *J. Hist. Biol.* **9**:1–28.

Osterud, H. L., 1918. Preliminary observations on the development of *Leptasterias hexactis. Pub. Puget Sound Biol. Sta.* **2**:1–15.

Ouweneel, W., 1976. Developmental genetics of homoeosis. *Adv. Genet.* **16**:179–248.

Ovcharenko, V. N., 1969. Transitional forms and species differentiation of brachiopods. *Paleontol. J.* **1**:67–73.

Ozawa, T., 1975. Evolution of *Lepidolina multiseptata* (Permian foraminifer) in East Asia. *Mem. Fac. Sci. Kyushu Univ., Ser. D. Geol.* **23**:117–164.

Paigen, K., 1980. Temporal genes and other developmental regulators in mammals. In *The Molecular Genetics of Development* (T. Leighton and W. Loomis, eds.). Academic Press, New York, pp. 419–470.

Parsley, R. L., and L. W. Mintz, 1975. North American Paracrinoidea. *Bull. Am. Paleontol.* **68**:5–115.

Paul, C. R. C., 1977. Evolution of primitive echinoderms. In *Patterns of Evolution as Illustrated by the Fossil Record* (A. Hallam, ed.). Elsevier Publishing Co., Inc., Amsterdam, pp. 123–158.

Penners, A., 1926. Experimentelle Untersuchungen zum Determinationsproblem am Keim von *Tubifex revulorum* Lam. II. Die Entwicklung teilweise abgetöteter keime. *Z. Wiss. Zool.* **127**:1–140.

Perlman, S. M., P. J. Ford, and M. M. Rosbash, 1977. Presence of tadpole and adult globin RNA sequences in oocytes of *Xenopus laevis. Proc. Natl. Acad. Sci. USA* **74**:3835–3839.

Petersen, J. L., J. R. Larsen, and C. B. Craig, 1976. *Palpantenna,* a homoeotic mutant in *Aedes aegypti. J. Hered.* **67**:71–78.

Phillips, T. L., and G. A. Leisman, 1966. *Paurodendron,* a rizomorphic Lycopod. *Am. J. Bot.* **53**:1086–1100.

Pilbeam, D., and S. J. Gould, 1974. Size and scaling in human evolution. *Science* **186**:892–901.

Poodry, C. A., L. Hall, and D. T. Suzuki, 1973. Developmental properties of *shibire^{ts}*: A pleiotropic mutation affecting larval and adult locomotion and development. *Dev. Biol.* **32**:378–388.

Poulson, D., 1940. The effects of certain X-chromosome deficiencies on the embryonic development of *Drosophila melanogaster. J. Exp. Zool.* **83**:271–325.

Powell, J. T., and K. Brew, 1974. Glycosyltransferases in the Golgi membranes of onion stem. *Biochem. J.* **142**:203–209.

Prager, E. M., and A. C. Wilson, 1975. Slow evolutionary loss of the potential for interspecific hybridization in birds: A manifestation of slow regulatory evolution. *Proc. Natl. Acad. Sci. USA* **72**:200–204.

Pribnow, D., 1975. Phage T-7 early promoters: Nucleotide sequences of two RNA polymerase binding sites. *J. Mol. Biol.* **99**:419–444.

Quinn, T. C., and G. B. Craig, 1971. Phenogenetics of homoeotic mutant *proboscipedia* in *Aedes albopictus. J. Hered.* **62**:3–12.

Rabinow, L., and W. J. Dickinson, 1981. A cis-acting regulator of enzyme tissue specificity in *Drosophila* is expressed at the RNA level. *Mol. Gen. Genet.,* **183**:264–269.

Raff, E. C., 1979. The control of microtubule assembly *in vivo. Int. Rev. Cytol.* **59**:1–96.

Raff, E. C., M. Fuller, T. C. Kaufman, K. M. Kemphues, J. Rudolph, and R. A. Raff, 1982. Regulation of tubulin gene expression during embryogenesis in *Drosophila melanogaster. Cell* **28**:33–40.

Raff, R. A., 1972. Polar lobe formation by embryos of *Ilyanassa obsoleta:* Effects of inhibitors of microtubule and microfilament function. *Exp. Cell Res.* **71**:455–459.

Raff, R. A., 1977. The molecular determination of morphogenesis. *Bioscience* **27**:394–401.

Raff, R. A., and H. R. Mahler, 1972. The non-symbiotic origin of mitochondria. *Science* **177**:575–582.

Raff, R. A., and E. C. Raff, 1970. Respiratory mechanisms and the metazoan fossil record. *Nature* **228**:1003–1005.

Raup, D. M., and A. Michelson, 1965. Theoretical morphology of the coiled shell. *Science* **147**:1294–1295.

Rawls, J. M., and J. W. Fristrom, 1975. A complex genetic locus that controls the first three steps of pyrimidine biosynthesis in *Drosophila. Nature* **233**:738–740.

Raynaud, A., 1961. Morphogenesis of the mammary gland. In *Milk: The Mammary Gland and its Secretion,* Vol. I (S. K. Kon and A. T. Cowie, eds.). Academic Press, New York, pp. 3–46.

Rees, L. J., 1970. Studies on the larval structure and metamorphosis of *Balanus balanoides* (L.). *Phil Trans. Roy. Soc. Lond. B* **256**:237–280.

Reeves, R., 1977. Hormonal regulation of epidermis-specific protein and messenger RNA synthesis in amphibian metamorphosis. *Dev. Biol.* **60**:163–179.

Regier, J. C., G. D. Mazur, and F. C. Kafatos, 1980. The silkmoth chorion: Morphological and biochemical characterization of four surface regions. *Dev. Biol.* **76**:286–304.

Reverberi, G. and A. Minganti, 1946. Fenomeni di evocazione nello sviluppo dell'uove di Ascidia. Risultati dell'indagine sperimentale sull'uovo di *Ascidiella aspersa* e di *Ascidia malaca* allo stadio di otto blastomeri. *Pubbl. Staz. Zool. Napoli* **20**:199–252.

Rice, T. B., and A. Garen, 1975. Localized defects of blastoderm formation in maternal effect mutants of *Drosophila. Dev. Biol.* **43**:277–286.

Riek, E. F., 1977. Four-winged Diptera from the Upper Permian of Australia. *Proc. Linn. Soc. N.S. Wales* **101**:250–255.

Rodgers, W. H., and P. R. Gross, 1978. Inhomogeneous distribution of egg RNA sequences in the early embryo. *Cell* **14**:279–288.

Roop, D. R., J. L. Nordstrom, S. Y. Tsai, M. J. Tsai, and B. W. O'Malley, 1978. Transcription of structural and intervening sequences in the ovalbumin gene and identification of potential mRNA precursors. *Cell* **15**:671–685.

Rosbash, M., M. S. Campo, and K. S. Gummerson, 1975. Conversation of cytoplasmic poly (A)-containing RNA in mouse and rat. *Nature* **258**:682–686.

Rosbash, M., P. J. Ford, and J. O. Bishop, 1974. Analysis of the C-value paradox by molecular hybridization. *Proc. Natl. Acad. Sci. USA* **71**:3746–3750.

Rosen, J. M., 1981. Mechanism of action of prolactin in the mammary gland. In *Prolactin* (R. B. Jaffe, ed.). Elsevier, New York, pp. 127–166.

Ross, M. H., 1964. Pronotal wings in *Blatella germanica* and their possible evolutionary significance. *Am. Midland Natur.* **71**:161–180.

Roux, W., 1894. The problems, methods and scope of developmental mechanics. An introduction to the "Archiv für Entwicklungsmechanik der Organismen," translated by W. M. Wheeler. In *Biological Lectures of the Marine Biological Laboratory of Woods Hole, Mass.*, 1895. Ginn and Company, Boston, pp. 149–190.

Ruderman, J. V., and P. R. Gross, 1974. Histones and histone synthesis in sea urchin development. *Dev. Biol.* **36**:286–298.

Russell, E. S., 1916. *Form and Function: A Contribution to the History of Animal Morphology.* John Murray, London.

Rutter, W. J., J. D. Kemp, W. S. Bradshaw, W. R. Clark, R. A. Ronzio, and T. G. Sanders, 1968. Regulation of specific protein synthesis in cytodifferentiation. *J. Cell Physiol.* **72**(Suppl. 1):1–18.

Ruud, G., 1925. Die Entwicklung isolierter Keimfragmente frühester Stadien von *Triton taeniatus. Roux Arch.* **105**:209–293.

Sadoglu, P., 1967. The selective value of eye and pigment loss in Mexican cave fish. *Evolution* **21**:541–549.

Sakakura, T., Y. Nishizuka, and C. J. Dawe, 1976. Mesenchyme-dependent morphogenesis and epithelium-specific cytodifferentiation in mouse mammary gland. *Science* **194**:1439–1441.

Sakonju, S., D. F. Bogenhagen, and D. D. Brown, 1980. A control region in the center of the 5S RNA gene directs specific initiation of transcription. I. The 5' border of the region. *Cell* **19**:13–25.

Salser, W., S. Bowen, D. Browne, F. El Adli, N. Federoff, K. Fry, H. Heindell, G. Paddock, R. Poon, B. Wallace, and P. Whitcome, 1976. Investigation of the organization of mammalian chromosomes at the DNA sequence level. *Fed. Proc.* **35**:23–35.

Salthe, S. N., and W. E. Duellman, 1973. Quantitative constraints associated with reproductive mode in anurans. In *Evolutionary Biology of the Anurans* (J. V. Vial, ed.). University of Missouri Press, Columbia, pp. 229–249.

Sandburg, C., 1936. They have yarns. In *The People, Yes*. Harcourt, Brace, Jovanovich, New York.

Sander, K., M. Lohs-Schardin, and M. Baumann, 1980. Embryogenesis in a *Drosophila* mutant expressing half the normal segment number. *Nature* **287**:841–843.

Sarich, V. M., 1972. Generation time and albumin evolution. *Biochem. Genet.* **7**:205–212.

Sarich, V. M., and J. E. Cronin, 1976. Molecular systematics of the primates. In *Molecular Anthropology* (M. Goodman and R. E. Tashian, eds.). Plenum Press, New York, pp. 141–170.

Saunders, J. W., 1948. The proximo-distal sequence of origin of the parts of the chick wing and the role of the ectoderm. *J. Exp. Zool.* **108**:363–403.

Saunders, J. W., and J. F. Fallon, 1966. Cell death in morphogenesis. In *Major Problems in Developmental Biology* (M. Locke, ed.) Academic Press, London, pp. 289–314.

Schaffner, W., G. Kunz, H. Daetwyler, J. Telford, H. O. Smith, and M. L. Birnstiel, 1978. Genes and spacers of cloned sea urchin histone DNA analyzed by sequencing. *Cell* **14**:655–671.

Scheller, R. H., F. O. Costantini, M. R. Kozlowski, R. J. Britten, and E. H. Davidson, 1978. Specific representation of cloned repetitive DNA sequences in sea urchin RNAs. *Cell* **15**:189–203.

Scheller, R. H., L. B. McAllister, W. R. Crain, Jr., D. S. Durica, J. W. Posakony, T. L. Thomas, R. J. Britten, and E. H. Davidson, 1981. Organization and expression of multiple actin genes in the sea urchin. *Mol. Cell. Biol.* **1**:609–628.

Schlampp, K. W., 1892. Das Auge des Grottenolmes, *(Proteus anguineus)*. *Zeitschr. f. Wiss. Zool.* **53**:537–557.

Schopf, J. W., 1978. The evolution of the earliest cells. *Sci. Am.* **239**(3):110–138.

Schopf, T. J. M., D. M. Raup, S. J. Gould, and D. S. Simberloff, 1975. Genomic versus morphologic rates of evolution: Influence of morphologic complexity. *Paleobiology* **1**:63–70.

Schubiger, G., and W. Wood, 1977. Determination during early embryogenesis in *Drosophila melanogaster*. *Am. Zool.* **17**:565–576.

Schuler, M. A., and E. B. Keller, 1981. The chromosomal arrangement of two linked actin genes in the sea urchin *S. purpuratus*. *Nucleic Acids Res.* **9**:591–604.

Schwartz, R. M., and M. O. Dayhoff, 1978. Cytochromes. In *Atlas of Protein Sequence and Structure*, Vol. 5, Suppl. 3 (M. O. Dayhoff, ed.). National Biomedical Research Foundation, Washington, D.C. pp. 29–44.

Sedgwick, A., 1888. The development of the cape species of *Peripatus*. *Quart. J. Microscop. Sci.* **28**:373–398.

Selander, R. K., S. Y. Yang, R. C. Lewontin, and W. E. Johnson, 1970. Genetic variation in the horseshoe crab *(Limulus polyphemus)*, a phylogenetic "relic". *Evolution* **24**:402–414.

Sengel, P., 1971. The organogenesis and arrangement of cutaneous appendages in birds. *Adv. Morphogen.* **9**:181–230.

Sengel, P., 1976. *Morphogenesis of Skin*. Cambridge University Press, Cambridge.

Shannon, M. P., T. C. Kaufman, M. W. Shen, and B. H. Judd, 1972. Lethality patterns and morphology of selected lethal and semi-lethal mutations in the zeste-white region of *Drosophila melanogaster*. *Genetics* **72**:615–638.

Shearn, A., 1978. Mutational dissection of imaginal disc development. In *The Genetics and Biology of Drosophila*, Vol. 2C (M. Ashburner and T. Wright, eds.). Academic Press, London, pp. 443–510.

Shearn, A., and A. Garen, 1974. Genetic control of imaginal disc development in *Drosophila*. *Proc. Natl. Acad. Sci. USA* **71**:1393–1397.

Sheldon, L., 1889. On the development of *Peripatus novae-zelandiae*, *Quart. J. Microscop. Sci.* **29**:283–294.

Shellenbarger, D. L., and J. D. Mohler, 1975. Temperature-sensitive mutations of the *Notch* locus in *Drosophila melanogaster*. *Genetics* **81**:143–162.

Shellenbarger, D. L., and J. D. Mohler, 1978. Temperature-sensitive periods and autonomy of pleitropic effects of $l(1)N^{tsl}$, a conditional *Notch* lethal in *Drosophila*. *Dev. Biol.* **62**:432–446.

Shen, C.-K., and T. Maniatis, 1980. The organization of repetitive sequences in a cluster of rabbit β-like globin genes. *Cell* **19**:379–391.

Sim, G. K., F. C. Kafatos, C. W. Jones, M. D. Koehler, A. Efstratiadis, and T. Maniatis, 1979. Use of a cDNA library for studies on evolution and developmental expression of the chorion multigene families. *Cell* **18**:1303–1316.

Simpson, G. G., 1953. *The Major Features of Evolution*. Columbia University Press, New York.

Simpson, G. G., 1960. The history of life. In *Evolution After Darwin*, Vol. 1: *The Evolution of Life* (S. Tax, ed). University of Chicago Press, Chicago, pp. 117–180.

Sipfle, D. A., 1969. On the intelligibility of the epochal theory of time. *The Monist* **53**:505–518.

Skaife, S. H., 1955. *Dwellers in Darkness: An Introduction to the Study of Termites*. Longman Group Limited, London.

Snodgrass, R. E., 1935. *Principles of Insect Morphology*. McGraw-Hill Book Company, New York.

Snodgrass, R. E., 1952. *A Textbook of Arthropod Anatomy*. Cornell University Press, Ithaca, N.Y.

Sparrow, A. H., and A. F. Nauman, 1976. Evolution of genomic size by DNA doublings. *Science* **192**:524–529.

Sparrow, A. H., H. J. Price, and A. G. Underbrink, 1972. A survey of DNA content per cell and per chromosome of prokaryotic and eucaryotic organisms: Some evolutionary considerations. *Brookhaven Symp. Biol.* **23**:451–494.

Spemann, H., 1938. *Embryonic Development and Induction*. Yale University Press, New Haven, Conn.

Spemann, H., and H. Mangold, 1924. Über Induktion von Embryonalanlagen durch Implantation artfremder Organisatoren. *Roux Arch.* **100**:599–638.

Spooner, B. S., H. I. Cohen, and J. Faubion, 1977. Development of the embryonic mammalian pancreas: The relationship between morphogenesis and cytodifferentiation. *Dev. Biol.* **61**:119–130.

Stanley, S. M., 1973. An ecological theory for the sudden origin of multicellular life in the late Precambrian. *Proc. Natl. Acad. Sci. USA* **70**:1486–1489.

Stanley, S. M., 1973. An explanation for Cope's rule. *Evolution* **27**:1–26.

Stanley, S. M., 1976. Fossil data and the Precambrian-Cambrian evolutionary transition. *Am. J. Sci.* **276**:56–76.

Stanley, S. M., 1979. *Macroevolution: Pattern and Process*. W. H. Freeman and Company, San Francisco.

Sternberg, P. W. and H. R. Horvitz, 1981. Gonadal cell lineages of the nematode

Panagrellus redivivus and implications for evolution by modification of cell lineage. *Dev. Biol.* **88**:147–166.

Stevens. N. M., 1905. Studies in spermatogenesis with especial reference to the "accessory chromosome." *Carn. Inst. Wash.* publ. **36**.

Stevens, N. M., 1908. A study of the germ cells of certain Diptera. *J. Exp. Zool.* **5**:359–379.

Strathmann, R. R., 1975. Limitations on diversity of forms: Branching of ambulacral systems of echinoderms. *Am. Natur.* **109**:177–190.

Strickberger, M. W., 1976. *Genetics,* 2nd ed. Macmillan Publishing Co., Inc., New York.

Strub, S., 1977. Localization of cells capable of transdetermination in a specific region on the male forleg disk of *Drosophila. Roux Arch.* **182**:69–74.

Struhl, G., 1981. A gene product required for the correct initiation of segmental determination in *Drosophila. Nature* **293**:36–41.

Sturtevant, A. H., 1920. Genetic studies on *Drosophila simulans* I. Introduction. Hybrids with *Drosophila melanogaster. Genetics* **5**:488–500.

Sutton, W. S., 1902. On the morphology of the chromosome group in *Brachystola magna. Biol. Bull.* **4**:24–39.

Sutton, W. S., 1903. The chromosomes in heredity. *Biol. Bull.* **4**:231–251.

Suzuki, D. T., 1970. Temperature-sensitive mutations in *Drosophila melanogaster. Science* **170**:695–706.

Suzuki, D. T., 1974. Behavior in *Drosophila melanogaster:* A geneticist's view. *Can. J. Genet. Cytol.* **16**:713–735.

Takamura, T., and T. K. Watanabe, 1979. Characterization of the *Lethal hybrid rescue (Lhr)* gene of *Drosophila simulans. Nat. Inst. Genet., Japan, Ann. Rep.* **30**:50–51.

Tarkowski, A. J., and J. Wroblewska, 1967. Development of blastomeres of mouse eggs isolated at the 4- and 8-cell stage. *J. Embryol. Exp. Morphol.* **18**:155–180.

Tauber, C. A., and M. J. Tauber, 1977. Sympatric speciation based on allelic changes at three loci: Evidence from natural populations in two habitats. *Science* **197**:1298–1299.

Taurog, A., C. Oliver, R. L. Eskay, J. C. Porter, and J. M. McKenzie, 1974. The role of TRH in the neoteny of the Mexican axolotl *(Ambystoma mexicanum) Gen. Comp. Endocrinol.* **24**:267–277.

Taylor, M. E., 1966. Precambrian mollusc-like fossils from Inyo County, California. *Science* **153**:198–201.

Tazima, Y., 1964. *The Genetics of the Silkworm.* Logos Press, London.

Teilhard de Chardin, P., 1965. *The Phenomenon of Man.* Colophon Books, Harper & Row Publishers, New York.

Templeton, A. R., 1979. The unit of selection in *Drosophila mercatorum.* II. Genetic revolution and the origin of coadapted genomes in parthenogenetic strains. *Genetics* **92**:1265–1282.

Templeton, A. R., 1980. The theory of speciation *via* the founder principle. *Genetics* **94**:1011–1038.

Thesleff, I., 1977. Tissue interactions in tooth development *in vitro.* In *Cell Interactions in Differentiation* (M. Karkinen-Jääskelainen, L. Saxen and L. Weiss, eds.). Academic Press, London, pp. 195–207.

Thompson, D'Arcy, 1961. *On Growth and Form,* abridged ed. (J. T. Bonner, ed.). Cambridge University Press, Cambridge.

Thompson, W. (Lord Kelvin). We have not cited individual papers because an excellent discussion of Lord Kelvin's influence on geology and evolution is available in J. D. Burchfield, *Lord Kelvin and the Age of the Earth,* Science History Publications, New York, 1975.

Tickle, C., G. Shellswell, A. Crawley, and L. Wolpert, 1976. Positional signalling by mouse limb polarizing region in the chick wing bud. *Nature* **259**:396–397.

Timberlake, W. E., D. S. Shumard, and R. B. Goldberg, 1977. Relationship between nuclear and polysomal RNA populations of *Achlya*: A simple eucaryotic system. *Cell* **10**:623–632.

Tompkins, R., 1978. Genic control of axolotl metamorphosis. *Am. Zool.* **18**:313–319.

Towe, K. M., 1970. Oxygen-collagen priority and the early metazoan fossil record. *Proc. Natl. Acad. Sci. USA* **65**:781–788.

Treadwell, A. L., 1901. The cytogeny of *Podarke obscura*. *J. Morphol.* **17**:399–486.

Tufaro, F., and B. P. Brandhorst, 1979. Similarity of proteins synthesized by isolated blastomeres of early sea urchin embryos. *Dev. Biol.* **72**:390–397.

Turing, A. M., 1952. The chemical basis of morphogenesis. *Phil. Trans. Roy. Soc. Lond. B* **237**:37–72.

Twitty, V. C., 1940. Size-controlling factors. *Growth* **4**, *Supplement:* pp. 109–120.

Ubaghs, G., 1967. General characters of Echinodermata. In *Treatise on Invertebrate Paleontology*, Part S, *Echinodermata 1*, Vol. I (R. C. Moore, ed.). Geological Society of America, New York, and University of Kansas Press, Lawrence, Kan. pp. 3–60.

Ubaghs, G., 1971. Diversité et spécialisations des plus anciens echinodermes que l'on connaisse. *Biol. Rev.* **46**:157–200.

Underhay, E. E., and E. Baldwin, 1955. Nitrogen excretion in the tadpoles of *Xenopus laevis* Daudin. *Biochem. J.* **61**:544–547.

Val, F. C., 1977. Genetic analysis of the morphological differences between two interfertile species of Hawaiian *Drosophila*. *Evolution* **31**:611–629.

Valentine, J. W., 1977. General patterns of metazoan evolution. In *Patterns of Evolution as Illustrated by the Fossil Record* (A. Hallam, ed.). Elsevier, Amsterdam, pp. 27–57.

Van Deusen, E., 1973. Experimental studies on a mutant gene (*e*) preventing the differentiation of eye and hypothalamus primordia in the axolotl. *Dev. Biol.* **34**:135–158.

Van Ness, J., I. H. Maxwell, and W. E. Hahn, 1979. Complex populations of nonpolyadenylated messenger RNA in mouse brain. *Cell* **18**:1341–1349.

Van Valen, L., 1973. A new evolutionary law. *Evol. Theory* **1**:1–30.

Vogel, O., 1977. Regionalization of segment-forming capacities during early embryogenesis in *Drosophila melanogaster*. *Roux Arch.* **182**:9–32.

Vogt, W., 1925. Gestaltungsanalyse am Amphibienkeim mit örtlicher Vitalfärbung. Vorwort über wege und Ziele. I. Methodik und Wirkungswaise der ortlichen Vitalfärbung mit Agar als Farbträger. *Roux Arch.* **106**:542–610.

Vogt, W., 1929. Gestaltungsanalyse am Amphibienkeim mit örtlicher Vitalfärbung. II. Gastrulation und Mesodermbildung bei Urodelen und Anuren. *Roux Arch.* **120**:387–706.

Von Baer, K. E., 1828. *Über Entwickelungsgeschichte der Thiere: Beobachtung und Reflexion*. Vol. I, Borntrager, Konigsberg, Translation of *Fifth Scholium* by T. H. Huxley, published as part of Fragments relating to philosophical zoology: Selected from the works of K. E. von Baer. In *Scientific Memoirs, Selected from the Transactions of Foreign Academies of Science, and from Foreign Journals: Natural History* (Arthur Henfrey and Thomas Henry Huxley, eds.), 1853. Taylor and Francis, London, pp. 176–238.

Von Ubish, L., 1938. Über Keimverschmelzungen an *Ascidiella aspersa*. *Roux Arch.* **138**:18–36.

Waddington, C. H., 1940. *Organizers and Genes*. Cambridge University Press, Cambridge.

Waddington, C. H., 1942. Canalization of development and the inheritance of acquired characters. *Nature* **150**:563–565.

Waddington, C. H., 1959. Evolutionary adaptation. In *Evolution After Darwin* (S. Tax, ed.). University of Chicago Press, pp. 381–402.

Waddington, C. H., 1966. *Principles of Development and Differentiation.* Macmillan Publishing Co., Inc., New York.

Waddington, C. H., 1966. Fields and gradients. In *Major Problems in Developmental Biology* (M. Locke, ed.). Academic Press, New York, pp. 105–124.

Wahli, W., I. B. Dawid, G. U. Ryffel, and R. Weber, 1981. Vitellogenesis and the vitellogenin gene family. *Science* **212**:298–304.

Wahli, W., I. B. Dawid, T. Wyler, R. B. Jaggi, R. Weber, and U. Ryffel, 1979. Vitellogenin in *Xenopus laevis* is encoded in a small family of genes. *Cell* **16**:535–549.

Wake, D. B., 1970. The abundance and diversity of tropical salamanders. *Amer. Natural.* **104**:211–213.

Wake, D. B., and J. F. Lynch, 1976. The distribution, ecology and evolutionary history of plethodontid salamanders in tropical America. *Nat. Hist. Mus. Los Angeles Sci. Bull.* **25**:1–65.

Wakimoto, B. T., and T. C. Kaufman, 1981. Analysis of larval segmentation in lethal genotypes associated with the Antennapedia gene complex in *Drosophila melanogaster, Dev. Biol.* **81**:51–64.

Wald, G., 1963. Phylogeny and ontogeny at the molecular level. In *Evolutionary Biochemistry: Proceedings of the 5th International Congress of Biochemistry* (A. I. Oparin, ed.). Macmillan Publishing Co., Inc., New York, pp. 12–51.

Waller, T. R., 1969. The evolution of the *Argopectin gibbus* stock (Mollusca: Bivalvia), with emphasis on the Tertiary and Quaternary species of Eastern North America. *Paleontological Society Memoir* 3, 125 pp. Menlo Park, California.

Walter, M. R., J. H. Oehler, and D. Z. Oehler, 1976. Megascopic algae 1300 million years old from the Belt Supergroup, Montana: A reinterpretation of Walcott's *Helminthoidichnites. J. Paleontol.* **50**:872–881.

Wanner, J., 1922. Neue Beiträge zur Kenntnis der Permischen Echinodermen von Timor. VII. Die Anomalien der Schizoblasten. *Mijnwezen Nederland-Oost-Indie, Jaarb., Verhandl. I, Jaarg.* **51**:163–233.

Waterman, T. H., and F. A. Chace, Jr., 1960. General crustacean biology. In *The Physiology of Crustacea*, Vol. I: *Metabolism and Growth* (T. H. Waterman, ed.). Academic Press, New York.

Weinberg, E. S., M. L. Birnstiel, I. F. Purdom, and R. Williamson, 1972. Genes coding for polysomal 9S RNA of sea urchins: Conservation and divergence. *Nature* **240**: 225–228.

Weismann, A., 1893. *The Germ-Plasm: A Theory of Heredity.* C. Scribner's Sons, New York.

Wells, D. E., R. M. Showman, W. H. Klein, and R. A. Raff, 1981. Delayed recruitment of maternal histone H3 in sea urchin embryos. *Nature* **292**:477–478.

Wessells, N. K., and J. H. Cohen, 1967. Early pancreas organogenesis: Morphogenesis, tissue interactions, and mass effects. *Dev. Biol.* **15**:237–270.

Wessells, N. K., and J. Evans, 1968. Ultrastructural studies of early morphogenesis and cytodifferentiation in the embryonic mammalian pancreas. *Dev. Biol.* **17**:413–416.

Wheeler, D. E., and H. F. Nijhout, 1981. Soldier determination in ants: New role for juvenile hormone. *Science* **213**:361–363.

White, M. J. D., 1973. *Animal Cytology and Evolution*, 3rd ed. Cambridge University Press, Cambridge.

Whiteley, A. H., and F. Baltzer, 1958. Development, respiratory rate and content of desoxyribonucleic acid in the hybrid *Paracentrotus* ♀ x *Arbacia* ♂, *Pubbl. Staz. Zool. Napoli* **30**:402–457.

Whiteley, A. H., and H. R. Whiteley, 1972. The replication and expression of maternal and paternal genomes in a blocked echinoid hybrid. *Dev. Biol.* **29**:183–198.

Whiteley, H. R., S. Mizuno, Y. R. Lee, and A. H. Whiteley, 1975. Transcripts of reiterated DNA sequences in the determination of blastomeres and early differentiation in echinoid larvae. *Am. Zool.* **15**:629–648.

Whitman, C. O., 1895. Prefatory note to the lectures for 1894. *Biological Lectures: The Marine Biological Laboratory of Woods Hole, Mass.* Ginn and Company, Boston, pp. iii–vii.

Whittaker, J. R., 1973. Segregation during ascidian embryogenesis of egg cytoplasmic information for tissue-specific enzyme development. *Proc. Natl. Acad. Sci. USA* **70**:2096–2100.

Whittaker, J. R., 1977. Segregation during cleavage of a factor determining endodermal alkaline phosphatase development in ascidian embryos. *J. Exp. Zool.* **202**:139–153.

Whittaker, J. R., 1979. Development of vestigial tail muscle acetylcholinesterase in embryos of an anural ascidian species. *Biol. Bull.* **156**:393–407.

Whittaker, J. R., G. Ortolani, and N. Farinella-Farruzza, 1977. Autonomy of acetylcholinesterase differentiation in muscle lineage cells of ascidian embryos. *Dev. Biol.* **55**:196–200.

Wiedmann, J., 1969. The heteromorphs and ammonoid extinction. *Biol. Rev. Cambridge Phil. Soc.* **44**:563–602.

Wieschaus, E., 1981. Personal communication.

Williamson, P. G., 1981. Palaeontological documentation of speciation in Cenozoic molluscs from Turkana basin. *Nature* **293**:437–443.

Williamson, V. M., E. T. Young, and M. Ciriacy, 1981. Transposable elements associated with constitutive expression of yeast alcohol dehydrogenase II. *Cell* **23**:605–614.

Wilson, A. C., S. S. Carlson, and T. J. White, 1977. Biochemical evolution. *Ann. Rev. Biochem.* **46**:573–639.

Wilson, A. C., L. R. Maxson, and V. M. Sarich, 1974. Two types of molecular evolution. Evidence from studies of interspecific hybridization. *Proc. Natl. Acad. Sci. USA* **71**:2843–2847.

Wilson, A. C., V. M. Sarich, and L. R. Maxson, 1974. The importance of gene rearrangement in evolution: Evidence from studies on rates of chromosomal, protein and anatomical evolution. *Proc. Natl. Acad. Sci. USA* **71**:3028–3030.

Wilson, A. C., G. L. Bush, S. M. Case, and M. C. King, 1975. Social structuring of mammalian populations and rate of chromosomal evolution. *Proc. Natl. Acad. Sci. USA* **72**:5061–5065.

Wilson, A. C., T. J. White, S. S. Carlson, and L. M. Cherry, 1977. Molecular evolution and cytogenetic evolution. In *Molecular Human Cytogenetics* (R. S. Sparkes, D. E. Comings, and C. F. Fox, eds.). Academic Press, New York, pp. 375–393.

Wilson, D. P., 1928. The larvae of *Polydora ciliata* Johnston and *Polydora haplura* Claparede. *J. Marine Biol. Assn. UK* **15**:567–604.

Wilson, E. B., 1898. Cell-lineage and ancestral reminiscence. *Biological Lectures: The Marine Biological Laboratory of Woods Hole, Mass.* Ginn and Company, Boston, pp. 21–42.

Wilson, E. B., 1904. Experimental studies in germinal localization. II. Experiments on the cleavage-mosaic in *Patella* and *Dentalium*. *J. Exp. Zool.* **1**:197–268.

Wilson, E. B., 1905. Studies on chromosomes II. The paired michrochromosomes, idiochromosomes and heterotropic chromosomes in Hemiptera. *J. Exp. Morph.* **2**:507–545.

Wilson, E. B., 1906. Studies on chromosomes III. The sexual difference of the chromosome groups in the Hemiptera, with some considerations on the determination and inheritance of sex. *J. Exp. Zool.* **3**:1–40.

Wilson, E. O., 1971. *The Insect Societies.* The Belknap Press of Harvard University Press, Cambridge, Mass.

Wilson, M. C., M. Melli, and M. L. Birnstiel, 1974. Reiteration frequency of histone coding sequences in man. *Biochem. Biophys. Res. Commun.* **61**:354–358.

Wold, B. J., W. H. Klein, B. R. Hough-Evans, R. J. Britten, and E. H. Davidson, 1978. Sea urchin embryo mRNA sequences expressed in the nuclear RNA of adult tissues. *Cell* **14**:941–950.

Wolff, E., 1968. Specific interactions between tissues during organogenesis. In *Current Topics in Developmental Biology*, 3rd ed. (A. A. Moscona and A. Monroy, eds.) Academic Press, New York, pp. 65–94.

Wolpert, L., 1969. Positional information and the spatial pattern of cellular differentiation. *J. Theoret. Biol.* **25**:1–47.

Wolpert, L., and J. Lewis, 1975. Towards a theory of development. *Fed. Proc.* **34**:14–20.

Wolpert, L., J. Lewis, and D. Summerbell, 1975. Morphogenesis of the vertebrate limb. In *Cell Patterning: Ciba Foundation Symposium No. 29*, London pp. 95–119.

Wood, W. G., J. B. Clegg, and D. J. Weatherall, 1977. Developmental biology of human hemoglobins. *Progr. Hematol.* **10**:43–90.

Woodland, H. R., and E. D. Adamson, 1977. The synthesis and storage of histones during the oogenesis of *Xenopus laevis*. *Dev. Biol.* **57**:118–135.

Wright, S., 1934. An analysis of variability in number of digits in an inbred strain of guinea pigs. *Genetics* **19**:506–536.

Wright, S., 1934. The results of crosses between inbred strains of guinea pigs, differing in number of digits. *Genetics* **19**:537–551.

Young, J. Z., 1959–1960. Observations on *Argonauta* and especially its method of feeding. *Zool. Soc. Lond. Proc.* **113**:471–481.

Yunis, J. J., J. R. Sawyer, and K. Dunham, 1980. The striking resemblance of high-resolution G-banded chromosomes of man and chimpanzee. *Science* **208**:1145–1148.

Zantinge, B., H. Dons, and G. H. Wessels, 1979. Comparison of poly (A)-containing RNAs in different cell types of the lower eukaryote *Schizophyllum commune*. *Eur. J. Biochem.* **101**:251–260.

Zhuravleva, I. T., 1970. Marine fauna and lower Cambrian stratigraphy. *Am J. Sci.* **269**:417–445.

Zimmerman, J. L., D. L. Fouts, and J. E. Manning, 1980. Evidence for complex class of nonadenylated mRNA in *Drosophila*. *Genetics* **95**:673–691.

Zuckerkandl, E., 1963. Perspectives in molecular anthropology. In *Classification and Human Evolution* (S. L. Washburn, ed.). Aldine Publishing Company, Chicago, pp. 243–272.

Zuckerkandl, E., 1968. Hemoglobins, Haeckel's "biogenetic law," and molecular aspects of development. In *Structural Chemistry and Molecular Biology* (A. Rich and N. Davidson, eds.), W. H. Freeman, San Francisco, pp. 256–274.

Zwilling, E., 1956. Interaction between limb bud ectoderm and mesoderm in the chick embryo. IV. Experiments with a wingless mutant. *J. Exp. Zool.* **132**:241–253.

Index

387